THE HISTORY BOOK

SIMPLY EXPLAINED

THE HISTORY BOOK

DK LONDON

PROJECT EDITORS
Alexandra Beeden, Sam Kennedy

SENIOR EDITOR
Victoria Heyworth-Dunne

US EDITOR
Christy Lusiak

EDITORIAL ASSISTANT
Kate Taylor

PROJECT ART EDITOR
Katie Cavanagh

DESIGNER
Vanessa Hamilton

DESIGN ASSISTANT
Renata Latipova

MANAGING ART EDITOR
Lee Griffiths

MANAGING EDITOR
Gareth Jones

ART DIRECTOR
Karen Self

ASSOCIATE PUBLISHING DIRECTOR
Liz Wheeler

PUBLISHING DIRECTOR
Jonathan Metcalf

JACKET DESIGNER
Natalie Godwin

JACKET EDITOR
Claire Gell

JACKET DESIGN DEVELOPMENT
MANAGER
Sophia MTT

PRODUCER, PRE-PRODUCTION
Robert Dunn

SENIOR PRODUCER
Mandy Inness

ILLUSTRATIONS
James Graham, Vanessa Hamilton

DK DELHI

PICTURE RESEARCHERS
Aditya Katyla, Deepak Negi

PICTURE RESEARCH MANAGER
Taiyaba Khatoon

JACKET DESIGNER
Dhirendra Singh

SENIOR DTP DESIGNER
Harish Aggarwal

MANAGING JACKETS EDITOR
Saloni Singh

Coproduced with
SANDS PUBLISHING SOLUTIONS

4 JENNER WAY, ECCLES, AYLESFORD,
KENT ME20 7SQ

EDITORIAL PARTNERS
David and Sylvia Tombesi-Walton

DESIGN PARTNER
Simon Murrell

original styling by
STUDIO8 DESIGN

This American Edition, 2020
First American Edition, 2016
Published in the United States by
DK Publishing
1745 Broadway, 20th Floor,
New York, NY 10019

Copyright © 2016, 2020

Dorling Kindersley Limited
DK, a Division of Penguin
Random House LLC

23 24 10 9 8 7 6 5 4 3
054—316697—Sept/2020

Published in Great Britain by Dorling
Kindersley Limited.

A catalog record for this book is available
from the Library of Congress.

ISBN 978-1-4654-9139-8

DK books are available at special discounts
when purchased in bulk for sales
promotions, premiums, fundraising, or
educational use. For details, contact:
DK Publishing Special Markets, 1745
Broadway, 20th Floor, New York, NY 10019
SpecialSales@dk.com

Printed and bound in UAE

MIX
Paper from
responsible sources
FSC™ C018179

This book was made with Forest Stewardship
Council ™ certified paper – one small step in
DK's commitment to a sustainable future.
For more information go to
www.dk.com/our-green-pledge

For the curious
www.dk.com

CONTRIBUTORS

REG GRANT, CONSULTANT EDITOR

R. G. Grant has written extensively in the fields of military history, general history, current affairs, and biography. His publications have included the DK books *Flight: 100 Years of Aviation*, *Battle at Sea*, and *World War I: The Definitive Visual Guide*.

FIONA COWARD

Dr. Fiona Coward is Senior Lecturer in Archaeology and Anthropology at Bournemouth University, UK. Her research focuses on the changes in human society, from the very small social groups of our prehistory to the global social networks that characterize people's lives today.

THOMAS CUSSANS

Thomas Cussans, writer and historian, has contributed to numerous historical works. They include DK's *Timelines of World History*, *History Year by Year*, and *History: The Ultimate Visual Guide*. He was previously the publisher of *The Times History of the World* and *The Times Atlas of European History*. His most recent published work is *The Holocaust*.

JOEL LEVY

Joel Levy is a writer specializing in history and the history of science. He is the author of more than 20 books, including *Lost Cities*, *History's Greatest Discoveries*, and *50 Weapons that Changed the World*.

PHILIP PARKER

Philip Parker is a historian specializing in the classical and medieval world. He is the author of the *DK Companion Guide to World History*, *The Empire Stops Here: A Journey Around the Frontiers of the Roman Empire*, *The Northmen's Fury: A History of the Viking World*, and general editor of *The Great Trade Routes: A History of Cargoes and Commerce Over Land and Sea*. He was a contributor to *DK History Year by Year* and *DK History of the World in 1,000 Objects*. He previously worked as a diplomat and a publisher of historical atlases.

SALLY REGAN

Sally Regan has contributed to over a dozen DK titles including *History*, *World War II*, and *Science*. She is also an award-winning documentary maker for Channel Four and the BBC in the UK.

PHILIP WILKINSON

Philip Wilkinson has written many books on historical subjects, heritage, architectural history, and the arts. As well as bestsellers such as *What The Romans Did For Us* and widely-praised titles such as *The Shock of the Old* and *Great Buildings*, he has contributed to numerous encyclopedias and popular reference books.

6

CONTENTS

8

CHANGING SOCIETIES
1776–1914

THE MODERN WORLD
1914–PRESENT

INTRODU

CTION

The ultimate aim of history is human self-knowledge. In the words of 20th-century historian R. G. Collingwood: "The value of history is that it teaches us what man has done and thus what man is." We cannot hope to understand our lives without it.

History itself has a history. From earliest times, all societies—literate or pre-literate—told stories about their origins or their past, usually imaginative tales centering around the acts of gods and heroes. The first literate civilizations also kept records of the actions of their rulers, inscribed on clay tablets or on the walls of palaces and temples. But at first these ancient societies made no attempt at a systematic inquiry

Those who cannot remember the past are condemned to repeat it.
George Santayana
The Life of Reason (1905)

into the truth of the past; they did not differentiate between what had really happened and the events manifest in myth and legend.

Ancient historical narrative

It was the Ancient Greek writers Herodotus and Thucydides in the 5th century BCE who first explored questions about the past through the collection and interpretation of evidence—the word "history," first used by Herodotus, means "inquiry" in Greek. Herodotus's work still contained a considerable mixture of myth, but Thucydides' account of the Peloponnesian War satisfies most criteria of modern historical study. It was based on interviews with eyewitnesses of the conflict and attributed events to human agency rather than the intervention and actions of the gods.

Thucydides had invented one of the most durable forms of history: the detailed narrative of war and political conflict, diplomacy, and decision-making. The subsequent rise of Rome to dominance of the Mediterranean world encouraged historians to develop another genre of broader scope: the account of "how we got to where we are today." The Hellenic historian Polybius (200–118 BCE) and the Roman historian Livy (59 BCE–17 CE) both

sought to create a narrative of the rise of Rome—a "big picture" that would help to make sense of events on a large timescale. Although restricted to the Roman world, this was the beginning of what is sometimes called "universal history," which attempts to describe progress from earliest origins to the present as a story with a goal, giving the past apparent purpose and direction.

At the same period in China, historian Sima Qian (c.145–86 BCE) was similarly tracing Chinese history over thousands of years, from the legendary Yellow Emperor (c.2697 BCE) to the Han dynasty under Emperor Wu (c.109 BCE).

Moral lessons

As well as making sense of events through narratives, historians in the ancient world established the tradition of history as a source of moral lessons and reflections. The history writing of Livy or Tacitus (56–117 CE), for instance, was in part designed to examine the behavior of heroes and villains, meditating on the strengths and weaknesses in the characters of emperors and generals, providing exemplars for the virtuous to imitate or shun. This continues to be one of the functions of history. French chronicler Jean Froissart (1337–1405) said he had

written his accounts of chivalrous knights fighting in the Hundred Years' War "so that brave men should be inspired thereby to follow such examples." Today, historical studies of Lincoln, Churchill, Gandhi, or Martin Luther King, Jr. perform the same function.

The "Dark Ages"

The rise of Christianity in the late Roman Empire fundamentally changed the concept of history in Europe. Historical events came to be viewed by Christians as divine providence, or the working out of God's will. Skeptical inquiry into what actually happened was usually neglected, and accounts of miracles and martyrdoms were generally accepted as true without question. The Muslim world, in this as in other ways, was frequently more sophisticated than Christendom in Medieval times, with the Arab historian Ibn Khaldun (1332–1406) railing against the blind, uncritical acceptance of fanciful accounts of events that could not be verified.

Neither Christian nor Muslim historians produced a work on the scale of the chronicle of Chinese history published under the Song dynasty in 1085, which recorded Chinese history spanning almost 1,400 years and filled 294 volumes.

Renaissance Humanism

Whatever the undoubted merits of other civilizations' traditions of history writing, it was in Western Europe that modern historiography evolved. The Renaissance—which began in Italy in the 15th century, then spread throughout Europe lasting until the end of the 16th century in some areas—centered upon the rediscovery of the past. Renaissance thinkers found a fertile source of inspiration in classical antiquity, in areas as diverse as architecture, philosophy, politics, and military tactics. The humanist scholars of the Renaissance period declared history one of the principal subjects in their new educational curriculum, and the antiquary

To live with men of
an earlier age is like
travelling in foreign lands.
René Descartes
Discourse on Method (1637)

became a familiar figure in elite circles, rummaging among ancient ruins and building up collections of old coins and inscriptions. At the same time, the spread of printing made history available to a much wider audience than ever before.

The Enlightenment

By the 18th century in Europe, the methodology of history—which consisted of ascertaining facts by criticizing and comparing historical sources—had reached a fair level of sophistication. European thinkers had reached general agreement on the division of the past into three main periods: Ancient, Medieval, and Modern. This periodization was at root a value judgment, with the Medieval period, dominated by the Church, viewed as a time of irrationality and barbarism and separating the dignified world of the ancient civilizations from the newly emerging, rational universe of modern Europe. Enlightenment philosophers wrote histories that ridiculed the follies of the past.

The Romantic spirit

In stark contrast, the Romantic movement that swept across Europe from the late 18th century found an intrinsic value in the difference between the past and the present. »

The Romantics drew inspiration from the Middle Ages, and instead of seeing the past as a preparation for the modern world, as had previously been the case, Romantic historians tried the imaginative exercise of entering into the spirit of past ages. Much of this was associated with nationalism. The German Romantic thinker Johann Gottfried Herder (1774–1803) burrowed into the past in search of roots of national identity and an authentic "German spirit." As nationalism triumphed in Europe in the 19th century, much of history became a celebration of national characteristics and national heroes, often veering into myth-making. Every country wanted to have its sacred heroic history, just as it had its flag and its national anthem.

The "Grand Narrative"

In the 19th century, history became increasingly important and took on the quality of destiny. Arrogantly, European civilization saw itself as the goal to which all history had been progressing and constructed narratives that made sense of the past in those terms. The German philosopher Georg Wilhelm Friedrich Hegel (1770–1831) articulated a grand scheme of history as a logical development, which culminated in the end point of the Prussian state.

Philosopher and social revolutionary Karl Marx (1818–83) later adapted Hegel's scheme into his own theory ("historical materialism"), in which he claimed that economic progress, which caused conflict between the social classes, would inevitably one day result in the proletariat seizing power from the bourgeoisie, while the capitalist world order collapsed under its own inner contradictions. Arguably, Marxism was to prove the most influential and durable of all historical "grand narratives."

Like other areas of knowledge, in the 19th century history under-went professionalization and it became an academic discipline. Academic history aspired to the status of a science, and the

History is little more than the crimes, follies, and misfortunes of mankind.
Edward Gibbon
The History of the Decline and Fall of the Roman Empire (1776)

accumulation of "facts" was its avowed purpose. A gap opened up between "serious" history—often heavy on economic statistics—and the colorful literary works of popular historians, such as Jules Michelet (1798–1874) and Thomas Macaulay (1800–59).

The rise of social history

In the 20th century, the subject matter of history—which had always focused on kings, queens, prime ministers, presidents, and generals—increasingly expanded to embrace the common people, whose role in historical events became accessible through more in-depth research. Some historians (initially those in France) chose to disregard the "history of events" altogether, preferring instead to study social structures and the patterns of everyday life, beliefs, and ways of thinking ("mentalités") of ordinary people in different historical periods.

A Eurocentric approach

Broadly speaking, until the second half of the 20th century, most world history was written as the story of the triumph of Western civilization. This approach was as implicit in Marxist versions of history as in those histories that celebrated the

progress of technology, enterprise, and liberal democracy. It did not necessarily imply optimism—there were numerous prophets of decline and doom. But it did suggest that essentially history had been made, and was still being made, by Europe and European offshoots further afield. For instance, it was deemed acceptable for respected European historians to maintain that black Africa had no significant history at all, having failed to contribute to the onward march of humanity.

Postcolonial revisionism

In the course of the second half of the 20th century, the notion of a single, purposeful, historical "grand narrative" collapsed, taking Euro-centrism with it. The postcolonial, postmodernist world was seen as requiring a multiplicity of histories told from the point of view of many different social identities. There was a surge of interest in the study of black history, women's history, and gay history, as well as histories narrated from an Asian, African, or American Indian standpoint. The marginal and oppressed in society were reassessed as "agents" of history rather than passive victims.

A riot of revisionism upturned much of the history of the world as commonly known to educated people in the West, although often without putting any satisfactory alternative version in place of the old. For example, the puzzlement that resulted can be seen in the response to the 500th anniversary in 1992 of Christopher Columbus's first voyage to the Americas. It would once have been expected to excite widespread celebration in the United States, but was in practice acknowledged with some embarrassment, if at all. People are no longer sure what to think about traditional history, its Great Men, and its epoch-making events.

A 21st-century perspective

The content of *The History Book* reflects this abandonment of "grand narratives" of human progress. It aims to present a general reader with an overview of world history through specific moments, or events, which can act as windows upon selected areas of the past. In line with contemporary concerns, this book also reflects the long-term importance of key factors such as population growth, climate, and the environment throughout human history. At the same time, it gives an account of matters of traditional popular historical interest, such as the Magna Carta, the Black Death, and the American Civil War.

The book begins with the origins of humans and "pre-history" and then progresses through different historical ages to the present day. In reality of course there were no such clear breaks between epochs, and where there is an overlap on dates, entries are included in the most appropriate ideological era.

As this book illustrates, history is a process rather than a series of unconnected events. We can only speculate on how the events we experience today will shape the history of tomorrow. No one in the early 21st century can possibly claim to make sense of history, but it remains the fundamental discipline for anyone who believes, as the poet Alexander Pope did, that "the proper study of Mankind is Man." ∎

We are not makers of history. We are made by history.
Martin Luther King, Jr.
Strength to Love **(1963)**

HUMAN O
200,000 YEARS

RIGINS
AGO—3500 BCE

The **first humans** (*Homo sapiens*) emerge in **East Africa**; Neanderthals (*Homo neanderthalensis*) are living in **Europe and West Asia**.

Paleolithic people start to **create art** (sculptures of animals and cave paintings) and **artifacts** (jewelry and decorative tools and weapons).

A period of intense cold, known as the "**Big Freeze**," occurs. People and animals in northern regions die out or migrate southward.

Jericho (in the modern-day West Bank) is settled; to this day it remains one of the **oldest towns in the world** still to be inhabited today.

c.200,000 YEARS AGO   **c.40,000 YEARS AGO**   **c.23,000 YEARS AGO**   **c.9000 BCE**

c.45,000 YEARS AGO   **c.35,000 YEARS AGO**   **c.15,000 YEARS AGO**   **c.7500 BCE**

Humans have **spread** across the globe and inhabit most of **Eurasia and Australia**, which they have reached by boat from Southeast Asia.

The first examples of **human figurines** emerge, usually representing **women** and carved or sculpted from bone, ivory, terracotta, or stone.

Humans start to arrive in **North America**, either across the **land bridge** connecting Asia and North America (now the Bering Strait) or by **sea**.

A **settlement** at Çatalhöyük, central Turkey, is established; evidence of complex **rituals** indicates **social cohesion**.

It is widely believed that the origins of the human race lie in Africa. By the usual processes of biological evolution and natural selection, the genus *Homo* evolved in East Africa over millions of years alongside the chimpanzees, its near relatives. By the same biological processes, *Homo sapiens*—modern humans—evolved alongside other hominins (the relatives of humans, including Neanderthals, who died out 40,000 years ago).

About 100,000 years ago or so, the scattered bands of hunting and foraging humans would have been almost indistinguishable from the other great apes. But at some point (precisely when is hard to define) humans began to change in a new way, not by the process of biological evolution but by cultural evolution. They developed the ability to alter their way of life through the creation of tools, languages, beliefs, social customs, and art. By the time they were painting exquisite pictures of animals on the walls of caves and carving or sculpting figurines out of stone or bone, they had marked themselves out uniquely from other animals. Their transformation was slow in the early years, but it was set to gather incredible momentum over millennia. Humans had become the only animals with a history.

Discovering history
The early development of human cultures and societies presents a particular problem to historians. The first writing was not invented until quite late in the human story— about 5,000 years ago. Traditionally, the period before writing tended to be dismissed as "pre-history," since it left no documents for historians to study. However, in recent years a wide range of new scientific methods—including the study of genetic material and radiocarbon dating of organic remains—have been added to the long-established techniques of archaeology, enabling scholars to shine at least a flickering light upon the pre-literate era.

The narrative of the distant human past is under constant revision as new discoveries and research—its findings frequently disputed—create radical shifts in perspective. The fresh investigation of a single cave, a burial site, or a human skull can still throw large areas of accepted knowledge into question. However, in the 21st century much of the history of early humans can be described with a reasonable degree of confidence.

There is evidence of **copper smelting** in Serbia and the **wheel** is invented in the Near East, probably for the production of pottery rather than for transport.

The **Bronze Age** begins in the **Near East**, and the **Indus Valley Civilization** emerges on the Indian subcontinent.

Cuneiform script, one of the world's **oldest** writing system, is invented in **Sumer**, in southern Mesopotamia (modern-day Iraq).

Stones are raised at Britain's **Stonehenge**, at the center of an earthwork enclosure constructed 500 years previously; the stones are later rearranged.

c.5000 BCE **c.3300 BCE** **c.3000 BCE** **c.2500 BCE**

c.4000 BCE **c.3100 BCE** **c.2700 BCE** **c.1800 BCE**

Civilizations develop in **Mesopotamia**, in the Tigris–Euphrates valley (modern-day Iraq, Syria, and Kuwait), where **irrigated agriculture** is established.

Narmer **unifies** Upper and Lower **Egypt**, becoming king of the **First Dynasty**; Egyptian **hieroglyphs** are prevalent.

The first stone **pyramids** are constructed as monumental **tombs** in **Egypt**; the Great Pyramid of Giza is built two centuries later.

Alphabetic writing (Proto-Sinaitic script, based on hieroglyphs) **emerges in Egypt**; it is the ancestor of most modern alphabets.

Nomadic hunter-gatherers

All historians agree that until about 12,000 years ago humans were hunter-gatherers, using stone tools and living in small, mobile groups. This period is referred to as the Paleolithic Era (or Old Stone Age). Humans were a successful species, expanding their numbers to perhaps 10 million and spreading to most parts of the Earth. Generally, they adapted well to the major natural climate changes that occurred over tens of thousands of years, although they were temporarily driven out of northerly areas, such as Britain and Scandinavia, during the coldest phase of what is popularly known as the Ice Age.

Humans existed in an intimate relationship with their natural environment, but their effect on that environment even at this early stage was not necessarily benign. There is a disturbing coincidence between the spread of human hunters across the planet and the extinction of megafauna such as woolly mammoths and mastodons. Although human hunting is far from being identified as the sole cause of these extinctions—natural climate change may well have been a contributing factor—from our modern perspective they can seem to set a troubling precedent.

The farming revolution

The hunter-gatherer lifestyle, which can reasonably be described as "natural" to human beings, appears to have had much to recommend it. Examination of human remains from early hunter-gatherer societies has suggested that our ancestors usually enjoyed abundant food, obtainable without excessive effort, and suffered very few diseases. If this is true, it is not clear what then motivated so many human beings all over the world to settle in permanent villages and develop agriculture, growing crops and domesticating animals: cultivating fields was grindingly hard work, and it was in farming villages that epidemic diseases first took root.

Whatever its immediate effect on the quality of life for humans, the development of settlements and agriculture indisputably led to a high increase in population density. Sometimes known as the Neolithic Revolution (or New Stone Age), this period was a major turning point in human development, opening the way to the growth of the first towns and cities, and eventually leading to settled "civilizations." ∎

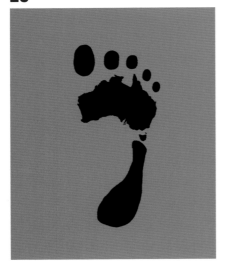

AT LEAST AS IMPORTANT AS COLUMBUS'S JOURNEY TO AMERICA OR THE *APOLLO 11* EXPEDITION
THE FIRST HUMANS ARRIVE IN AUSTRALIA (c.60,000–45,000 YEARS AGO)

IN CONTEXT

FOCUS
Migration

BEFORE
c.200,000 years ago *Homo sapiens* (modern human) evolves in Africa.

c.125,000–45,000 years ago Groups of *Homo sapiens* expand out of Africa.

AFTER
c.50,000–30,000 years ago Denisovan hominins are present in south-central Russia.

45,000 years ago *Homo sapiens* arrives in Europe.

c.40,000 years ago The Neanderthals die out. Their last known sites are on the Iberian peninsula.

c.18,000 years ago *Homo floresiensis* fossils date from this time.

c.13,000 years ago Humans are present near Clovis, New Mexico, but may not be the continent's first humans.

Homo sapiens evolves in Africa.

Homo sapiens spreads into the **Near East** but retreats to Africa, only later reaching **Europe and western Asia**.

After moving into **southern Asia**, *Homo sapiens* groups follow the coastline to **Southeast Asia**.

In western Eurasia, *Homo sapiens* encounters **other hominin species**, the **Neanderthals and Denisovans**.

Homo sapiens **arrives in Australia.**

All hominin species except *Homo sapiens* **die out**.

Modern humans are the only truly global mammal species. Since evolving in Africa around 200,000 years ago, *Homo sapiens* has rapidly expanded across the world—testament to our species' curiosity in exploring its surroundings and creativity in adapting to different habitats. In particular, many researchers think that humans' ability to exploit coastal environments was key to their rapid spread along the southern coasts of Asia.

Even the radically different flora and fauna of Australia proved no barrier; humans may have arrived on the continent as early as 60,000

See also: Cave paintings at Altamira 22–27 ▪ The Big Freeze 28–29 ▪ The settlement at Çatalhöyük 30–31

Remains of *Homo floresiensis* were found on the Indonesian island of Flores in 2003. Some studies suggest that its small size was due to disease rather than indicating a new species.

years ago, although the earliest dates are controversial. Small groups may have visited much earlier, but the bulk of the evidence suggests widespread colonization of Australia only around 45,000 years ago, at much the same time as *Homo sapiens* arrived in Europe.

Other hominin species

Homo sapiens was the first hominin to arrive in Australia. However, in parts of Eurasia, humans did face competition. By the time humans reached Europe, Neanderthals had already been there for around 250,000 years, having evolved from an ancestor they shared with modern humans, *Homo heidelbergensis*, and they were well adapted to life in the region.

Further east, at Denisova Cave in Russia's Altai Mountains, there is evidence of a mysterious species—the Denisovans—known only from their DNA. And on the island of Flores in Southeast Asia, fossils of another possible species—the short, small-brained *Homo floresiensis*—date from just 18,000 years ago, although some researchers believe these were simply modern humans afflicted with some form of disease.

Of all these species, *Homo sapiens* is the only one to have survived and gone on to colonize the New World. Beringia, a land-bridge between Russia and Alaska, exposed when sea levels dropped as a result of the Ice Age, allowed humans to reach the Americas from northeast Asia. The exact date remains controversial: stone tools

The human blitzkrieg across America testifies to the incomparable ingenuity and the unsurpassed adaptability of *Homo sapiens*.
Yuval Noah Harari
Sapiens **(2011)**

from the c.13,000-year-old "Clovis culture" were once thought to have belonged to the earliest humans in the New World. Older sites are now known, but many of the earlier dates, particularly in South America, remain highly contentious.

The social network

Until more evidence is found, the fates of the Denisovans and *Homo floresiensis* remain unknown, while the most recent research suggests Neanderthals died out around 40,000 years ago. Many researchers believe the resourcefulness of *Homo sapiens* was crucial to its success in other species' home territories in the face of climate change around the time of the Last Glacial Maximum. In particular, it is thought that they could also rely on more extensive social networks than those other species—an asset that would have proved crucial both to survival in lean times and to helping them colonize the unfamiliar environments they encountered as they expanded across the globe, perhaps following animal herds. ▪

Homo sapiens: the only remaining hominin

There is no evidence of violence between humans and other species. Indeed, modern human DNA shows small traces of Neanderthal and Denisovan genes, suggesting that a few individuals from each species interbred, albeit rarely.

Although Neanderthals were skilled manufacturers of stone tools and excellent hunters, modern humans may have been quicker to adapt, and therefore better able to cope with the rapid climatic changes occurring as the Ice Age progressed. They developed new stone tools, as well as techniques that made use of resources such as bone and antler. They also established extensive networks of support, enabling various groups to pool resources across large distances, enhancing their chances of survival. This cultural adaptability may have been what allowed humans to outcompete their cousins for access to increasingly unpredictable resources.

EVERYTHING WAS SO BEAUTIFUL, SO FRESH

CAVE PAINTINGS AT ALTAMIRA (c.40,000 YEARS AGO)

IN CONTEXT

FOCUS
Paleolithic culture

BEFORE
c.45,000 years ago Modern humans arrive in Europe.

c.40,000 years ago The earliest currently known examples of art in Europe are made, such as the sculpture of the Lion Man of Hohlenstein-Stadel, Germany.

AFTER
c.26,000 years ago
A triple burial is carried out at Dolní Věstonice, in the Czech Republic.

c.23,500 years ago The Arene Candide "prince" is buried in Italy, richly adorned with dentalium shell jewelry.

c.18,000 years ago The last Ice Age reaches its height.

The Altamira cave complex, near Santander on the northern coast of Spain, comprises a series of passages and chambers extending for nearly 984ft (300m) that boast some of the best examples of Stone Age, or Paleolithic, cave art yet discovered. So impressive are the paintings that when the cave was discovered in 1880, they were widely considered fakes and took nearly 20 years to be accepted as the genuine creations of prehistoric hunter-gatherers. Some of the early artistic activity here may date from more than 35,000 years ago, although most of the famous paintings were probably created much later, around 22,000 years ago. These include the images in the famous Bison Chamber: here the low ceiling is covered in representations of animals including multicolored, lifelike images of bison, expertly painted across the natural undulations of the rock in such a way as to make them appear almost three-dimensional.

The artistic impetus

Other stunning displays of cave art are also known, concentrated in southwest France and northern Spain. They include not only finely detailed images of animals, but also engraved and painted signs, symbols, and handprints. Archaeologists remain divided over the meaning and function of Stone Age art. One explanation is simply that these people appreciated the aesthetic qualities of art—just as their descendants do today. Others suggest that the incredible detail of some of the images—the sex of the animal or the season in which it was observed can still be determined, for example—may mean the paintings were a means of conveying vital survival information, such as which animals to hunt, and when and how they could be found and targeted.

Hunting rituals

Alternatively, cave art might be linked to the world views or religions of Paleolithic people. Even today, many societies still living mainly by hunting and gathering share animistic beliefs, meaning they believe entities such as animals, plants, and parts of the landscape have spirits with which humans interact during their daily life. Many such societies' religious specialists, or shamans, believe

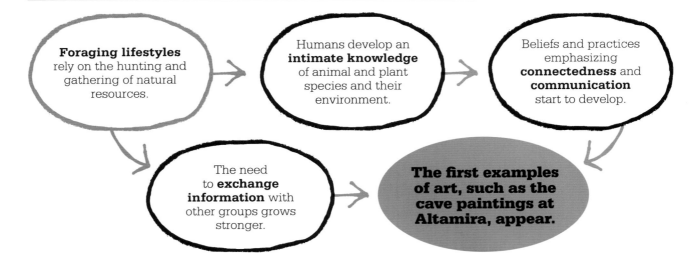

Foraging lifestyles rely on the hunting and gathering of natural resources.

Humans develop an **intimate knowledge** of animal and plant species and their environment.

Beliefs and practices emphasizing **connectedness** and **communication** start to develop.

The need to **exchange information** with other groups grows stronger.

The first examples of art, such as the cave paintings at Altamira, appear.

See also: The first humans arrive in Australia 20–21 ▪ The Big Freeze 28–29 ▪ The settlement at Çatalhöyük 30–31

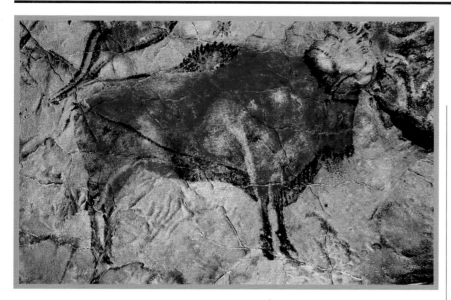

Handprints in the cave of Fuente del Salín, in Spain's Cantabria region, were probably left by youngsters, suggesting that venturing underground might have been a coming-of-age ritual.

The undulating structure of the rock cave at Altamira enhances, rather than detracts from, the art, with the animals in the Bison Chamber acquiring an almost three-dimensional quality.

they are able to communicate with these spirits to help sick or injured people, and historically, rock art has been created by shamans during states of altered consciousness, or trances, as part of this communication, leading some researchers to suggest that Paleolithic societies may have had similar beliefs. Shamans are also often thought to be able to transform themselves into animals to encourage them to give themselves up to the hunter, which could also explain depictions combining human and animal characteristics, such as the Lion Man of Hohlenstein-Stadel, in Germany, or the Sorcerer of Les Trois Frères Cave in France, a human-like figure with antlers.

Creating images of animals may have also been part of "magic" rituals designed to improve the chances of success during hunting. For societies dependent on animal resources for a significant part of their diet, the importance of such rituals cannot be overstated.

Initiation ceremonies

Other researchers have noted that many of the handprints and footprints found beside the art in the caves seem to belong to quite young individuals. Traveling down into dark, damp, and potentially dangerous caves with only a lamp filled with animal fat might have been a form of initiation test for young people— one that would have required a great deal of courage to endure.

Burials and the afterlife

More evidence of human beings engaging in religious or ritual practices at this time comes from burials. At the site of Dolní Věstonice, in the Czech Republic, for example, three bodies were buried together in a sexually suggestive pose, with one of the male individuals flanking a female skeleton reaching toward her pelvis, and the male on the other side buried face down. A red pigment known as ochre had been sprinkled across their heads and across the female's pelvis. Interestingly, all three individuals »

People everywhere and throughout time have shared the basic instinct to represent themselves and their world through images and symbols.
Jill Cook
Ice Age Art **(2013)**

share the same rare skeletal deformities and may therefore have been related. Although the reasons why these bodies were arranged this way will probably always be a mystery, it is clear that there was more to this burial than just the functional disposal of remains.

At other sites, some individuals were buried with many "grave goods"—for example, the complex jewelry made from dentalium shells at Arene Candide, in Italy, and the striking spears fashioned from mammoth ivory at the burial site of two young children in Sunghir, in Russia. Some researchers have suggested that these richly adorned individuals—especially the young ones, who would not have had time in their short life to establish a reputation that might account for special treatment in death—imply that hierarchies and status distinctions were beginning to develop in some groups. However, they do not appear to have become widespread until much later. It is clear, however, that for the first time, people were now increasingly concerned with what happened after death, and about how the dead should enter into the afterlife.

Marking territory

Other researchers note that most "classic" Paleolithic cave art is concentrated in southwest France and northern Spain. This region would have been a relatively favorable place to live: even at the height of the Last Glacial Maximum, more southerly, warmer climates and hence more productive habitats attracted dense herds of animals. As a result, people may have lived here in fairly large numbers, packed closely together, leading to greater social tensions among groups vying for territory and resources.

Just as human groups today—whether it be football supporters or nation states—use symbols such as flags, costumes, and markings of

> People thought of themselves as part of a living world, where animals, plants, and even landmarks and inanimate objects had lives of their own.
> **Brian Fagan**
> *Cro-Magnon* (2010)

borders, territories, and group identities, so European Paleolithic groups may have decorated caves for similar reasons at a time when there was the potential for intense competition for resources.

Cooperation to survive

Such complex social interactions may help explain how *Homo sapiens* was able to survive in the harsh environments of Ice Age Europe. Hunter-gatherers probably lived in small groups scattered at relatively low densities across the landscape. Most archaeological sites from this time do not demonstrate any evidence of complex buildings or structures, suggesting that people moved around a lot, according to the weather and the local environment, often following large herds of animals like reindeer as they migrated with the seasons.

Homo sapiens' ability to forge new relationships readily allowed groups of hunters to combine as and when necessary. When resources were plentiful, they would hunt together—for example, intercepting migrating herds of reindeer at places in the landscape where they were most vulnerable, such as in

Historians are still unsure whether or not there are precise meanings behind the majority of cave art. Their best guesses are that they may relate to any one or more of several possibilities: art for art's sake; spirituality; initiation rites; the marking of territory; and a method of imparting valuable information about hunting.

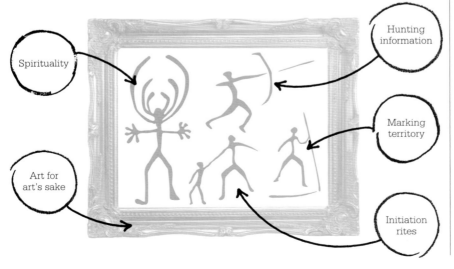

Hunting information

Spirituality

Marking territory

Art for art's sake

Initiation rites

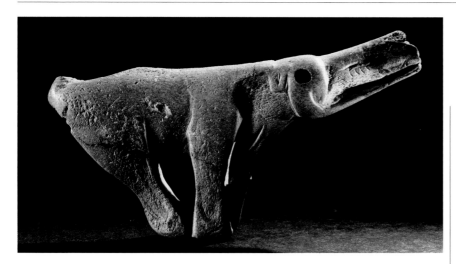

Hunting tools, such as this spear-thrower, were often carved in the shape of the animals they were used to kill, probably as a sort of "magic ritual" to improve chances of success in the hunt.

narrow valleys or at river crossings. In leaner times, these groups would split up again and range far across the landscape to find enough wild resources to sustain themselves.

Early technologies

These hunter-gatherers expended considerable effort on hunting technology, since it could spell the difference between life and death. They hafted elaborately worked stone tips on to spears that were then launched at the target using atlatls, or spear-throwers, designed to increase the distance over which a spear could travel and the force with which it hit its target. These tools were crucial to hunting success, so it is no surprise that some of these atlatls were beautifully carved and decorated, often with representations of the animals being hunted. Similarly, they also painstakingly carved complex barbed harpoons from bone and antler for fishing.

First seeds of a society

Delicately worked bone awls and needles suggest Stone Age humans also made warm clothes out of animal skin and fur with much more care than their predecessors, and they made many other items—from jewelry finely crafted from animal teeth and shell, to figurines carved from stone or sculpted from clay. Many of these may also have been traded, gifted, or exchanged with individuals from other groups as part of large-scale social networks.

The unpredictable environments of Europe during the Last Glacial Maximum meant sharing resources with other groups in times of plenty could pay off significantly at a later date: if a group struggled to find resources in one area, others elsewhere who had previously benefited from their generosity would be more inclined to return the favor. These kinds of exchange relationships probably linked even very far-flung groups together into complex networks of individual and group relationships that were fundamental to survival in such a tough environment. ∎

Venus figurines

Figurines of women carved or sculpted from stone, ivory, or clay are a type of Paleolithic art found widely across Europe. These figurines share many striking similarities. While details such as facial features and feet are largely ignored, feminine sexual characteristics (breasts, belly, hips, thighs, and vulva) are often exaggerated. The focus on features related to sexuality and fertility, and the round body shapes depicted (during the Ice Age fat would have been a precious commodity) suggest that the figurines may have played a symbolic role as a charm relating to childbirth or, more generally, fertility.

Some researchers believe that the figures represent a "mother goddess," but there is no real evidence for such an interpretation. Others have focused instead on the fact that the figurines demonstrate widely shared cultural ideas and symbols. These would have been crucial to social interactions and exchanges of resources, information, and potential marriage partners in the Ice Age world.

THE FOUNDATIONS OF TODAY'S EUROPE WERE FORGED IN THE EVENTS OF THE LATE ICE AGE
THE BIG FREEZE (c.21000 BCE)

IN CONTEXT

FOCUS
Climate change

BEFORE
c.2.58 million years ago
The Pleistocene, or Ice Age, begins.

c.200,000 years ago *Homo sapiens* emerges as a species.

AFTER
c.9700 BCE The Pleistocene ends, marking the beginning of today's relatively warm and stable climates—the Holocene.

c.9000–8000 BCE Agriculture becomes established in the Near East.

c.5000 BCE Sea level reaches near-modern levels; low-lying land is submerged.

c.2000 BCE The last mammoths are thought to have died out, on Wrangel Island, Russia.

Climate change results from shifts in the earth's position and orientation relative to the sun.

The Big Freeze expands ice caps, lowering sea levels.

Habitats change, and plant and animal species' **ranges alter** for survival.

Animals and humans colonize **newly exposed low-lying land**, only to be isolated when sea levels rise again.

Human groups are faced with **new opportunities** and constraints.

Scientists have only recently begun to appreciate how the two-way relationship that exists between humans and our environments has affected the development of our societies. Humans evolved during the last Ice Age, living through periodic shifts between very cold climatic conditions (glacials) and warmer periods more like those of today (interglacials). However, toward the end of the Ice Age, these shifts became more pronounced and occurred at shorter intervals, culminating around 21000 BCE in a "Big Freeze," a period of intense cold known as the Last Glacial Maximum. People and animals living in northern regions died out or retreated south as ice caps expanded to reach southern

See also: The first humans arrive in Australia 20–21 ▪ Cave paintings at Altamira 22–27 ▪
The settlement at Çatalhöyük 30–31 ▪ The Law Code of Hammurabi 36–37

An entire mammoth was unearthed in Siberia, Russia, in 1900—the first complete example ever found. A cast of it is on display in St. Petersburg's Natural History Museum.

England. Such huge amounts of sea water froze that sea levels dropped, exposing low-lying land such as Beringia, the continental shelf that connects North America and Asia—and the route by which humans first reached the Americas.

Rising temperatures
Temperatures eventually rose again, and today's relatively warm and stable climate had become established by around 7000 BCE. The ice caps melted, and rising sea levels separated Eurasia from the Americas, turned Southeast Asia into an archipelago, and made islands out of peninsulas such as Japan and Britain, thereby isolating many human groups. The impact on ecosystems was particularly severe for the large animals known as megafauna—mammoths, for example. The open glacial steppe grasslands in which megafauna thrived were replaced by expanding forests, and across the globe the combination of environmental change and human hunting drove many species to extinction.

The forests and wetlands of the new post-glacial world offered humans many new opportunities. They hunted large forest animals such as red deer and wild boar, as well as smaller mammals like rabbits, and they foraged for a range of aquatic and coastal food sources. Migratory fish like salmon, sea mammals such as seals, and shellfish, seasonal wildfowl, and a range of fruits, tubers, nuts, and seeds all became important dietary staples.

Changing lifestyles
In areas that were particularly rich in natural resources, human groups may not have settled in one place, sending small bands on forays further afield to target specific resources. The Natufian communities of the Eastern Mediterranean, for example, were able to exploit abundant stands of wild cereals in the Near East. Some groups began to manipulate their environments, burning vegetation and cutting down trees to encourage their preferred

Few humans have ever lived in a world of such extreme climatic and environmental change.
Brian Fagan
Expert in human prehistory

plant and animal species to thrive. They started to select and care for productive plant species and sowed the seeds of favored strains, while managing and controlling certain animals. This manipulation led to these species becoming ever more reliant on human input—and to the development of agriculture, a radical change in the human way of life that has since resulted in even more dramatic human impact on the environment. ▪

Ice cores and past environments

Paleoclimatologists study the elemental composition of the sediments laid down over time on ocean floors to understand how climates have changed in the past. Tiny sea creatures known as foraminifera absorb two different forms of oxygen, ^{16}O and ^{18}O, from sea water. Because ^{16}O is the lighter of the two, it evaporates into the air more easily, but during warmer periods it falls as rain and drains back to the sea. So ^{16}O and ^{18}O exist in sea water and appear in the shells of foraminifera, in roughly equal ratios. However, in cold conditions most of the evaporated ^{16}O does not return to the ocean but freezes as ice, so sea water contains more ^{18}O than ^{16}O. When foraminifera die, their shells sink to the ocean floor, building up over time. Paleoclimatologists drill into the ocean floor to extract cores of sediment and study the changing proportions of ^{16}O and ^{18}O in different layers to see how climates have changed over time.

A GREAT CIVILIZATION AROSE ON THE ANATOLIAN PLAIN
THE SETTLEMENT AT ÇATALHÖYÜK (10,000 YEARS AGO)

Hunter-gatherers interact closely with **animal and plant species**.

The **climate and environment** stabilize after the **Ice Age**.

Humans start to **manage and control** some **animals and plants**, domesticating them.

↔

Human populations begin to grow.

The **cultivation of land and crops** and the stockpiling of harvests **reduce mobility**.

People build larger settlements, such as the one at Çatalhöyük.

T he Neolithic town of Çatalhöyük on the Konya Plain in Turkey was discovered by James Mellaart in the 1960s. It has become one of the most famous archaeological sites in the world due to its size, density of settlement, spectacular wall paintings, and evidence of complex religious and ritual behavior.

Since its discovery, several other large settlements across West Asia have been found that attest to the growing scale of human communities during the shift from foraging to agricultural lifestyles, or "Neolithic revolution," that occurred between around 10000 BCE and 7000 BCE. Whether rising populations forced people to find

See also: The first humans arrive in Australia 20–21 ▪ Cave paintings at Altamira 22–27 ▪ The Big Freeze 28–29 ▪ The Law Code of Hammurabi 36–37

This illustration shows the way in which humans lived and worked close to each other at the Çatalhöyük site, with their domesticated animals also kept nearby.

more stable means of subsistence or farming allowed people to have more children, the sizes of many settlements increased substantially and became more permanent. New ways had to be found to resolve social stresses such as disputes between neighbors.

Early villagers invested time and effort in planting and cultivating crops, then in storing the harvest to last the year, so they could no longer simply move as foragers had.

Community cohesion

It is thought that the development of more formal religious organization and group ritual practices may have helped community cohesion. At many sites, buildings were set aside for such purposes; these were larger than domestic structures, with unusual features such as lime plaster benches and more evidence of symbolic and representational art: Çatalhöyük boasts murals and figurines of a range of subjects including wild animals such as bulls, leopards, and vultures. At many sites, some inhabitants

remained in the community even when they died; they were buried under the floors of the houses. Sometimes they were later dug up and their skulls removed; facial features were molded on some in plaster and painted with ochre for display. At sites like Ain Ghazal in Jordan, large statues made of lime plaster have been found, and there are many examples of clay figurines of animals and (mainly female)

humans. It is not clear whether these decorated skulls, statues, and figurines represent specific individuals or heads of households or lineages, or perhaps mythical ancestors or gods, but they may have been part of the communal ideologies, rituals, and social practices that helped smooth over tensions between individuals and broader regional groups, who were establishing more formal links with one another for long-distance trade and exchange of goods. Some of the success of Çatalhöyük may have been due to its role as a center for the large-scale trade of items made from the obsidian, or volcanic glass, of Hasan Dağ.

The many dramatic social and economic changes that came with the Neolithic revolution have helped shape both human history and the world's ecosystems ever since. ▪

Farming and health

The adoption of farming established a plentiful and stable long-term source of food, allowing for population growth. However, there were negative consequences, too. Farmers may have had to work harder at times than hunter-gatherers did, and their more limited diets—focused on just a few crops and animal species—led to nutritional deficiencies.

The health of early farmers also suffered in other ways. Living at close quarters with animals meant that some animal diseases spread to humans—for example, smallpox, anthrax, tuberculosis, and the flu. Larger communities living at higher densities allowed for such diseases to be more easily passed around. It also caused problems in disposing of human and animal waste and thus a rise in intestinal complaints and waterborne diseases such as cholera and typhoid, while irrigation created breeding grounds for mosquitoes and parasites, infecting humans with diseases like malaria.

ANCIENT CIVILIZAT

6000 BCE—500

ONS
E

Hammurabi, one of the great kings of **Mesopotamia**, writes a **law code**—the **earliest** known written legal system in history.

Egyptian pharaoh **Ramesses II** builds two vast temples at **Abu Simbel** to **glorify** the pharaohs and assert **dominance** in Nubia.

Democracy is introduced in **Athens** by Cleisthenes. All Athenian citizens are **allowed to vote** directly on Athenian policy.

The start of the **Persian Wars** between **Greece** and the **Persian Empire**; military successes influence the development of classical Greek identity.

1780 BCE   **1264 BCE**   **507 BCE**   **490 BCE**

1700 BCE   **650 BCE**   **c.500 BCE**   **c.334 BCE**

Knossos palace is built on Crete by the **Minoans**—the first civilization in Europe to produce a **system of writing** (known as the **Linear A** syllabary).

The high point of a **Celtic culture**, which developed around Halstatt, **Austria**, and spread to France, Romania, Bohemia, and Slovakia.

Siddhartha Gautama (known as **Buddha**) rejects material life to seek **enlightenment** and preach **Buddhism** in India.

Macedonian king **Alexander the Great** invades Asia Minor and creates a vast empire; **Greek culture** spreads **eastward**.

About 5,000 years ago, humans began to form societies of unprecedented complexity. These "civilizations" typically had state structures and social hierarchies, they built cities and monuments such as temples, palaces, and pyramids, and used some form of writing. The basis for the development of civilizations was progress in agriculture. When only part of the population was required to work in the fields to produce food, the rest could inhabit towns and palaces, performing a range of specialty functions as bureaucrats, traders, scribes, and priests. The invention of civilization undoubtedly raised human life to a new level in many ways—in technology, the arts, astronomy, the measurement of time, literature, and philosophy—but also established inequality and exploitation as the basis of society, leading to larger-scale warfare as states expanded into empires.

Emerging civilizations

The earliest civilizations developed in areas where it was possible to practice intensive agriculture, usually involving use of irrigation systems—for instance, along the rivers of the Tigris and Euphrates in Mesopotamia (modern Iraq), the Nile in Egypt, the Indus in northern India and Pakistan, and the Yangtze and Yellow rivers in China. Although these civilizations of Eurasia and North Africa seem to have been founded independently of one another, they developed multiple contacts over time, sharing ideas, technology, and even diseases. All followed a pattern in which stone tools (the Stone Age) were replaced by bronze (the Bronze Age) and then predominantly iron (the Iron Age). In the Americas, where the Olmec and Maya developed the civilizations of Mesoamerica, the use of stone tools persisted and most of the epidemic diseases that plagued Eurasia were unknown.

Writing and philosophy

From around 1000 BCE, Eurasian civilizations found an innovative momentum. The use of writing evolved from practical record-keeping to the creation of sacred books and classic literary texts that embodied the founding myths and beliefs of different societies, from the Homeric tales in Greece to the Five Classics of Confucianism in China and the Hindu Vedas in India. Forms of writing using an alphabet developed in the eastern

Qin Shi Huangdi **unites China**, previously a region of warring states, and begins major projects, including building the **Terracotta Army**.

Julius Caesar is **assassinated** in Rome by senators who believe he is becoming increasingly power-hungry.

The **Maya Classical Period** begins; many cities, temples, and monuments are built throughout **Mexico** and **Guatemala**.

Rome falls to the **Visigoths**; the Roman Empire shrinks and much of Europe is invaded by **Barbarian tribes**.

221 BCE   **44 BCE**   **250 CE**   **410 CE**

218 BCE   **43 CE**   **312 CE**   **486 CE**

Military commander **Hannibal**, from **Carthage** (north Africa), crosses the Alps to **invade Italy**. Unable to capture Rome, he returns to Africa.

A **Roman army** led by General Aulus Plautius **invades** southern **England**; later, Roman rule extends to **Wales** and the **Scottish** border.

Roman **emperor Constantine** adopts **Christianity** after victory at the Battle of Milvian Bridge; Christianity rapidly gains **popularity**.

Clovis, leader of the Salian Franks, **defeats the Romans** in Gaul and **unites France** north of the Loire under his dynasty.

Mediterranean region and were spread by the Phoenicians—a race of traders and sailors.

The Greek city-states became a test-bed for new forms of political organization, including democracy, and the source of new ideas in the arts and philosophy. The influence of Greek culture spread as far as northern India, while India itself was the birthplace of Buddhism— the first "world religion," winning converts beyond its society of origin.

Growing populations

The ancient world reached the peak of its classical period around 2,000 years ago. The world's population had grown from around 20 million at the time of the first civilizations to an estimated 200 million. About 50 million of these lived in a united Han China, while about the same number were under the governance of the Roman Empire, which had extended its rule to the shores of the Atlantic and the borders of Persia. In large part, the empires were successful because of efficient communications by land and water, and the ruthless deployment of military power. Long-distance trade routes linked Europe to India and China, and cities had expanded to a great degree—Rome's population was estimated at over 1 million.

Civilizations in decline

The causes of the decline of these powerful classical empires from the 3rd century CE have long been disputed among historians. Bred in overcrowded cities and transmitted along trade routes, epidemic diseases certainly played a part. Internal power struggles were also a major factor, leading to political fragmentation and a decline in the quality of government. But perhaps most crucial was the geographical limitation of the civilized areas of Eurasia. Both the Roman and Han empires built walls to mark and defend the borders of their empires, beyond which lived mostly nomadic or semi-nomadic "barbarian" tribes. The civilized societies had little or no military advantage over these peoples, who increasingly raided or settled within their territories. The eastern part of the Christianized Roman Empire survived until 1453, and Chinese civilization revived to full vigor under the Tang dynasty from 618, but Western Europe would take centuries to recover the levels of population and organization that it had known under the rule of Rome. ■

TO BRING ABOUT THE RULE OF RIGHTEOUSNESS IN THE LAND

THE LAW CODE OF HAMMURABI (c.1780 BCE)

IN CONTEXT

FOCUS
Origins of civilizations

BEFORE
c.5000 BCE Copper and gold smelting is common in Mesopotamia and beyond.

c.4500 BCE Uruk in Mesopotamia is the first settlement large enough to be called a city.

c.3800 BCE Upper and Lower Kingdoms of Egypt established along the Nile Valley.

c.3500 BCE Development of the Indus Valley civilizations.

c.3350 BCE Stone circles erected in west and north Europe.

c.2000 BCE Shang dynasty builds the first cities in China.

AFTER
c.1500 BCE Rise of Olmec culture in Mesoamerica.

c.600 CE Emergence of the Mayan civilization.

Agriculture, population, and **urbanization** increase.

Local **networks break down** and mechanisms for dispute resolution weaken.

Hammurabi writes a new code of law to cement his control over the region.

Need grows for **tools of governance:** laws, permanent records, and judiciary.

Cylinder seals (to control transactions), writing, judicial institutions, and **written laws develop**.

In 1901, a six-foot-tall slab of black stone was found in the ruins of the city of Susa. Carved onto its face were 280 "judgments," or laws, constituting the earliest known written legal code in history. The slab had originally been erected in Babylon, in around 1750 BCE, by Hammurabi, one of the greatest kings of ancient Mesopotamia.

Bronze Age Revolution

Mesopotamia, which means "between two rivers," lies between the Euphrates and the Tigris, and it is considered to be the first human civilization ever. Its writing, math, and astronomy were also the first known, and its cities arguably the world's first true examples. Growth of its population and wealth led to

See also: The settlement at Çatalhöyük 30–31 ▪ The temples of Abu Simbel 38–39 ▪ The palace at Knossos 42–43 ▪ The conquests of Alexander the Great 52–53 ▪ The founding of Baghdad 86–93 ▪ The foundation of Tenochtitlan 112–17

Hammurabi the Law-Giver

In around 2000 BCE, the Amorites (Westerners), a semi-nomadic people from Syria, swept across Mesopotamia, replacing local rulers with Amorite sheikh dynasties in many of the city-states. By the early 18th century BCE, the three most powerful Amorite kings were pre-eminent Shamshi-Adad in the north, Rim-Sin in Larsa in the south, and Hammurabi in Babylon in the center. Over the course of his long reign, Hammurabi consolidated all of southern Mesopotamia into his kingdom and eventually extended his power as far up the Tigris as Nineveh, and as far up the Euphrates as Tuttul, on the junction with the river Balikh. He personally supervised the construction of many temples and other buildings.

The prelude to his code, a tribute to Hammurabi, and a long historical record of his conquests, boasts that his leadership was divinely sanctioned by the gods who passed control of humanity to Marduk (deity of Babylon), and so to its king. It also reveals he saw his role as the guarantor of a just and orderly society.

the emergence of a hierarchy in society, led by rulers, courtiers, and priests at the top, through merchants and artisans, to servants and laborers at the bottom. This is often referred to as "specialization": members of society having different tasks, rather than all producing food as had been the case in previous subsistence societies.

Mesopotamian communities coordinated manpower to build large structures such as defensive walls and huge temples, and to mobilize armies. They utilized hydrological engineering to divert river water and irrigate the alluvial floodplains. Administrative needs such as bookkeeping led to the development of cuneiform writing, the first known script, and of complex mathematical concepts such as fractions, equations, and geometry. Sophisticated astronomy developed for calendric purposes. Sometimes called the Bronze Age Revolution, this great step forward can be seen as the most important change in the human world before the Industrial Revolution.

Mesopotamian unification

For much of the 4th to the 2nd millennia BCE, Mesopotamia was a mosaic of competing kingdoms and city-states such as Uruk, Isin, Lagash, Ur, Nippur, and Larsa. Hammurabi, the Amorite king of Babylon, unified the region through a combination of guile, diplomacy, opportunism, military might, and longevity. As was traditional with conquering kings, Hammurabi used previous edicts as the basis for his laws, but these laws were

When Marduk sent me to rule over men... [I] brought about the well-being of the oppressed.
Hammurabi

distinguished by the reach of his empire, and by the fact that they were inscribed on stelae (stone slabs), and so recorded in perpetuity.

Hammurabi's laws and their detailed prelude reveal much about life in what is known as the Old Babylonian Period. They contain judgments on matters ranging from property disputes and violence against the person, to runaway slaves and witchcraft.

Hammurabi's legacy

Although Hammurabi's laws seem to have carried little weight and were rarely followed at the time, and despite the fact that his empire disintegrated soon after his death, his reign was a turning point for southern Mesopotamia. He firmly established the ideal of a unified state, centered in Babylon, and his laws were copied by Mesopotamian scribes until at least the 6th century BCE. They show many points of similarity with, and may have influenced, laws of the Hebrew Bible, which in turn influence laws in many societies today. ▪

ALL THE LANDS HAVE FALLEN PROSTRATE BENEATH HIS SANDALS FOR ETERNITY

THE TEMPLES OF ABU SIMBEL (c.1264 BCE)

IN CONTEXT

FOCUS
Pharaonic Egypt

BEFORE
c.3050 BCE Narmer unifies the kingdoms of Upper and Lower Egypt.

c.2680 BCE Khufu begins construction of the Great Pyramid in Giza— it is the largest pyramid in history.

c.1480 BCE Thutmose III conquers Syria, extending his empire as far as the Euphrates.

AFTER
c.1160 BCE Ramesses III fights off invasions of Egypt by Libyans and raiding tribes known as the Sea People.

c.1085 BCE Collapse of the New Kingdom; Egypt is divided with Libyan rulers in the north and Theban priest-kings ruling in the south.

7th century BCE Egypt is invaded by Assyrians and then Persians.

A round 1264 BCE, the Egyptian pharaoh Ramesses II (c.1278–1237 BCE) had two mighty temples hewn out of the cliffs on the west bank of the Nile in southern Egypt. The entrance was guarded by four vast statues of the pharaoh, seated in glory and wearing the symbols of divine kingship, including the double crown that signified his authority over Upper and Lower Egypt. The temples were designed to signify and embody the unique status, ambition, and power of the ancient Egyptian pharaohs.

The pharaonic tradition
Ramesses II inherited a tradition that was already very ancient: about 1,800 years earlier, King Narmer (called Menes by the ancient Greek historian Herodotus) first unified the kingdoms of the Upper (southern) and Lower (northern) Nile. Narmer's deeds were recorded on a stone palette, which was recovered from a temple at Hierakonpolis in the 19th century and provides one of the earliest known depictions of an Egyptian king. The palette is inscribed with many of the symbols and traditions that would come to typify the

pharaohs for the next three millennia. For instance, Narmer is shown holding an enemy by the hair, about to smite him, and Ramesses II was often depicted in the same way—military might and supernatural strength were hallmarks of Egyptian kingship. The pharaoh, like the gods, was frequently shown much larger than ordinary mortals.

The geographical situation of Egypt—with its stark contrasts between the fertile Nile Valley and its delta, which empties in the

The magnificent temple complex at Abu Simbel was, remarkably, moved 656 ft (200 m) inland and 213 ft (65 m) higher up in 1964–68 to rescue it from the rising waters of the Nile during the construction of the High Aswan Dam.

See also: The Law Code of Hammurabi 36–37 ▪ The palace at Knossos 42–43 ▪ The conquests of Alexander the Great 52–53 ▪ The assassination of Julius Caesar 58–65

I, [the creator], give you Ramesses II, constant harvests... [your] sheaves are as plentiful as the sand, your granaries approach heaven and your grain heaps are like mountains.
Inscription in temple at Abu Simbel, c.1264 BCE

north into the Mediterranean Sea, and the surrounding expanses of uninhabitable desert—gave rise to the kingdom's unique culture and civilization. The pharaoh was viewed as a living god who could control the order of the cosmos, including the annual flooding of the Nile, which brought fertilizing silt to replenish the soil. Pharaohs were also often depicted as farmers in agricultural scenes, representing their role as guardians of the land.

The Old Kingdom
The Old Kingdom that followed Narmer was ruled by a succession of dynasties that were led by powerful pharaohs, who channeled the bureaucratic and economic might of the unified kingdom into monumental building projects, such as the construction of the pyramids. These, in turn, stimulated scientific, technological, and economic development, increasing trade with other kingdoms in the Near East

and the Mediterranean. In the Old Kingdom the predominant gods were Ra, the sun god; Osiris, the god of the dead; and Ptah, the creator. In the Middle and New Kingdoms that followed, which were ruled by families from Thebes, Amon became the main deity. As supreme ruler, the pharaoh was closely associated with the gods, and was believed to be the living incarnation of certain deities.

The New Kingdom
In the 23rd century BCE, the Old Kingdom collapsed. After what is known as the Intermediate Period, the Middle Kingdom dynasties restored unified control of Egypt from 2134 BCE until around 1750 BCE, when they were invaded by the Hyksos (probably Semites from Syria). The Hyksos, in turn, were expelled from Egypt in about

1550 BCE, with the XVIII dynasty— arguably the greatest and most important—coming to power and establishing the New Kingdom. By this time, immortality was believed to be available not just to the pharaoh, but to priests, scribes, and others who could afford offerings, spells, and mummification, and many tombs were dug into the Valley of the Kings to be filled with extraordinarily rich grave goods.

Under expansionist pharaohs, such as Thutmose III and Ramesses II, Egyptian control was extended into Asia as far as the Euphrates River, and up the Nile into Nubia. It was no coincidence that Ramesses built Abu Simbel in Nubia: as well as representing the divine glory of Egypt's pharaohs generally, the temple was a symbol of Ramesses' control over the recently conquered territory. ▪

The **Nile Valley** is bordered by **inhospitable desert**, but is **highly fertile** because the longest river in the world flows through it and **irrigates it**.

A **sophisticated, coherent, and unified** civilization develops over a **vast stretch** of terrain.

Trade and conquest boost the economy and **population levels**. A large, prosperous kingdom emerges.

Vast monuments, such as the Abu Simbel temple complex, are constructed, reflecting Egypt's power, wealth, and belief systems.

ATTACHMENT IS THE ROOT OF SUFFERING
SIDDHARTHA GAUTAMA
PREACHES BUDDHISM (c.500 BCE)

IN CONTEXT

FOCUS
The spread of Buddhism

BEFORE
1200 BCE Vedic (aka Aryan) culture extends across northern and central India.

1200–800 BCE Oral Vedic traditions are written down in Sanskrit as the Vedas.

c.600 BCE The Mahajanapadas, the 16 competing kingdoms of Vedic India, emerge.

AFTER
322 BCE Chandragupta Maurya founds the Mauryan Empire.

3rd century BCE Sri Lanka converts to Buddhism.

185 BCE The Mauryan Empire collapses.

1st century CE Buddhism arrives in China and Japan.

7th century Buddhist missionaries are invited to establish a monastery in Tibet.

Siddhartha rejects material life and preaches Buddhist philosophy.

Ashoka the Great conquers India and **unifies the empire**.

Ashoka **makes Buddhism the state religion** and spreads it across South and East Asia.

After the collapse of the Mauryan Empire, **Buddhism declines in India**.

Buddhism flourishes in Sri Lanka, Southeast Asia, China, Japan, Tibet, and Central Asia.

Siddhartha Gautama, better known as the Buddha, was born at the end of the Vedic Age (1800–600 BCE) into a South Asia in transition. In the country's caste system, the priestly Brahmins and the warrior-elite Kshatriyas ranked highest, and it was into this latter group that Siddhartha Gautama was born.

India was then a ferment of sects and new ideologies, some of which espoused a philosophy renouncing the material world. Siddhartha developed a similar philosophy based on mystical Hinduism, but he also rejected the increasingly rigid strictures of Vedic ritual and the inherited piety of the Brahmins. Renouncing material possessions, he sought and eventually found enlightenment, and became the Buddha. He preached in northeast India and founded the Sangha—the monastic order of Buddhism—to continue his ministry.

See also: The conquests of Alexander the Great 52–53 ▪ The Indus Valley Civilization collapses ▪ The construction of Angkor Wat 108–09 ▪ The conquests of Akbar the Great 170–71

Given that separation is certain in this world, is it not better to separate oneself voluntarily for the sake of religion?
Siddhartha Gautama

For the next two to three centuries, Buddhism remained one among several minor sects but, under the Mauryan emperor Ashoka the Great (304–232 BCE), it became India's state religion. Ashoka's reign had proceeded initially through bloody conquest, but in around 261 BCE he had a change of heart. From then he embraced a new model of kingship and religious philosophy based on a creed of tolerance and non-violence.

He extended Mauryan control and, his Buddhism proving a powerful unifying force, succeeded in joining all of India, except the southern tip, into an empire of 30 million people.

A world religion

Having established Buddhism as the state religion, Ashoka founded monasteries, and sponsored scholarship. He sent Buddhist missionaries to every corner of the subcontinent and abroad as far as Greece, Syria, and Egypt. His missions established Buddhism initially as an elite pursuit, but the religion went on to take root at all levels of society in Sri Lanka, Southeast Asia, along the Silk Road in the Indo–Greek kingdoms (in modern-day Pakistan and Afghanistan), and later in China, Japan, and Tibet. In India—its birthplace—Buddhism started to decline after Ashoka's death in 232 BCE, affected by a resurgence of Hinduism and then the arrival of Islam. Outside India, however, its tradition and scholarship flourished, evolving into multiple strands

Stone reliefs depicting the life of Buddha decorate gateways of The Great Stupa at Sanchi, commissioned by the emperor Ashoka in the 3rd century BCE.

including Zen Buddhism, Theravada or Hinayana Buddhism, Mahayana Buddhism, and Varayana Buddhism.

The first religion to have spread widely beyond the society in which it originated—so the first "world religion"—Buddhism is also one of the oldest, having been practiced since the 6th century BCE. ▪

The Buddha

The life history of Siddhartha Gautama is obscured by the myth and legend that has grown up around him. Different traditions give different chronologies for his birth and death, but many agree on 563–483 BCE. Said to have been born miraculously through the side of his mother, Siddhartha was raised in luxury in the palace of his father, King Suddhodana Tharu, leader of the Shakya clan.

Aged 29, Siddhartha rejected this luxurious life and left his wife and child, renouncing material things to seek enlightenment through asceticism. Having spent six years wandering and meditating, he achieved enlightenment and became the Buddha, but instead of ascending to nirvana, the transcendent state that is the goal of Buddhism, he chose to remain and preach his new message, the *dharma*.

Gathering followers who formed the Sangha, a monastic order, the Buddha pursued his ministry until he died, at age 80. He urged his disciples to follow the *dharma*, instructing them: "All individual things pass away. Strive on, untiringly."

A CLUE TO THE EXISTENCE OF A SYSTEM OF PICTURE-WRITING IN THE GREEK LANDS
THE PALACE AT KNOSSOS (c.1700 BCE)

IN CONTEXT

FOCUS
Minoan Crete

BEFORE
c.7000 BCE Initial colonization of Crete.

c.3500 BCE Beginning of the Bronze Age in Crete.

AFTER
c.1640 BCE Massive eruption of volcano Thera devastates Minoan colonies and coastline.

c.1500 BCE Deeper stratification of Minoan culture; local administration is devolved to large villas.

c.1450 BCE The Mycenaean invasion of Crete.

c.1100 BCE The Sea Peoples terrorize the Mediterranean world, leading to the final decline of Minoan civilization.

1900 CE Arthur Evans begins the excavation of Knossos.

1908 Italian archaeologist Luigi Pernier discovers the Phaistos disc.

Minoan society becomes highly prosperous through **agriculture and trade**.

Social stratification develops, with a wealthy elite controlling trade.

Elaborate **palace complexes** are built to store commodities for redistribution.

The need for **record-keeping** gives rise to "writing" in the form of **hieroglyphs**.

Hieroglyphs evolve into Linear A syllabary at Knossos.

I n the 1890s, British historian Arthur Evans came across some ancient clay seals for sale in Athens. They originated from the relatively unexplored Mediterranean island of Crete, and for Evans they offered a tantalizing hint at the existence of the first writing system in Europe.

Following the seals to their Cretan source, Evans decided to excavate a promising parcel of land at Knossos, in the north of the island, where he uncovered a vast palace complex. The iconography of the palace centered on a bull-cult, including frescoes that depicted the sport of bull-leaping. Evans named the civilization "Minoan" after the mythical Cretan King Minos, who—according to Greek legend—built a labyrinth to contain the Minotaur: a fearsome half-man, half-bull creature. In the process, Evans discovered that the Minoans had indeed invented an early type of alphabet, which he called Linear A.

The Palatial Period
The Minoans were a people of unknown origin (possibly from Anatolia), who settled on Crete in the Neolithic era, in about 7000 BCE. They farmed crops, herded sheep,

See also: The settlement at Çatalhöyük 30–31 ▪ The Law Code of Hammurabi 36–37 ▪ The Persian Wars 44–45 ▪ Athenian democracy 46–51 ▪ King Sejong introduces a new script 130–31 ▪ The fall of Constantinople 138–41

and worshipped in caves, on top of mountains, and at springs, but by 2400 BCE they had begun to build large palace complexes. By 1900 BCE, in what is known as the Palatial Period of the Minoan civilization, palaces at Knossos, Phaistos, Malia, and Chania had been constructed in broadly similar forms, with the one at Knossos being the largest. It was destroyed, possibly by fire or perhaps a tsunami, around 1700 BCE, but it was rebuilt soon after on the same site. At its peak, in about 1500 BCE, Knossos palace and the city that grew up around it covered 185 acres (75 hectares) and had a population of up to 12,000.

The Minoan palaces all had large central courts, flanked by many-chambered buildings, and were highly decorated with frescoes of flora and fauna. In the extensive magazines (storehouses), the rulers—who may have served dual roles as priest-kings or priest-queens—gathered many commodities for redistribution. Minoan rulers also

This bull-leaping fresco in the palace at Knossos in Crete is the most completely restored of several taureador stucco panels. Bull-handling was a common theme in art at this time.

controlled trade with other Bronze Age civilizations around the Mediterranean, such as Byblos in Phoenicia (now Lebanon), Ugarit in Syria, pharaonic Egypt, and Mycenaean Greek settlements in the Cyclades and further afield.

Linear A script

The Minoans developed their own script, probably initially for record-keeping and administration purposes. It began as hieroglyphic picture-writing, but later evolved into the Linear A syllabary, in which symbols denote syllables (rather than letters, as is the case with the alphabet). The Minoan language as recorded in Linear A script remains undeciphered to this day, but in around 1450 BCE the Minoans were invaded by the Mycenaeans from mainland Greece, who adapted the Minoan script into Linear B, which was used to write archaic Greek.

Not long after the Mycenaeans invaded Crete, Minoan civilization collapsed completely. However, the legacy of Minoan writing persisted through its connection with the Mycanean alphabet, which, in turn, would come to influence the characters used by the Greeks of the mainland. ▪

The Phaistos disc

Found in 1908 in the ruins of the Minoan palace at Phaistos, southern Crete, the Phaistos disc (shown above), made from fired clay and about 6in (15cm) across, is printed with symbols in an unknown script. Although dated to 1700 BCE, it was made using the technique of woodblock printing, which was not thought to have been invented for another 2,000 years or so (in China), making the disc one of the great archaeological mysteries. The symbols, many of which are recognizable as everyday objects, are arranged in a spiral and divided into words by vertical lines. Some scholars have drawn parallels between certain symbols in Cretan hieroglyphics and Linear A, suggesting that the writing on the disc may be an elaborated form of an existing Minoan script. There are many theories about the disc's significance—some consider the inscription is a hymn to a goddess, others that it tells a story, or that the disc is a calendar or a game. Some experts even believe the disc to be a clever fake.

IN TIMES OF PEACE, SONS BURY THEIR FATHERS, BUT IN WAR IT IS THE FATHERS WHO BURY THEIR SONS

THE PERSIAN WARS (490–449 BCE)

IN CONTEXT

FOCUS
The Persian Empire

BEFORE
7th century BCE The Medes establish a powerful kingdom in modern-day Iran.

c.550 BCE Cyrus the Great rebels against Median rule and founds the Achaemenid Persian Empire.

c.499 BCE Greek city-states rebel against Persian control, but their revolt fails.

AFTER
431 BCE Athens and Sparta clash for supremacy in Greece in the Peloponnesian War.

404 BCE Artaxerxes II becomes ruler of the Achaemenid Empire.

331 BCE Alexander the Great defeats Darius III and conquers the Persian Empire.

312 BCE Persia becomes part of the Seleucid Empire, founded by one of Alexander's generals.

Leonidas of Sparta stood before his band of 300 warriors facing the mightiest army the world had ever seen. The envoy of his enemy demanded that he lay down his arms at the feet of the Persian god-king. "Come and take them" was Leonidas's laconic reply.

The Persian Wars (490–449 BCE), also known as the Greco–Persian Wars, pitted a vast and cosmopolitan empire against a small band of city-states in the south of Greece. The conflict profoundly influenced the development of Classical Greek identity and culture, leaving a vivid trail in Western literature and myth. By contrast, the story of the Persian Achaemenid Empire remains comparatively neglected, belying the significance of that great Middle Eastern civilization.

The Achaemenids

The first Persian Empire, ruled by the dynasty known as the Achaemenids, grew rapidly. At its height it may have ruled over half the world's population. It began in around 550 BCE, when the Persian king Cyrus the Great overthrew the ruling Medes, going on to conquer Babylonia, and Lydia (now in

A hoplite—or Greek citizen-solider—vanquishes his Persian adversary in this decoration inside a 460 BCE wine cup. The winged horse Pegasus adorns the victor's shield.

Turkey), which brought the Ionian Greeks under Persian rule. Cyrus's successors Cambyses II and Darius extended the empire into Egypt and the Balkans, where Thrace and Macedon gave the Persians a foothold in Europe.

The Achaemenids established Persian rule as a model for later empires. Despite its vast size, the state embraced a degree of multiculturalism, allowing conquered peoples to keep liberty of religion, language, and culture. There was investment in infrastructure—like

See also: The Law Code of Hammurabi 36–37 ▪ Athenian democracy 46–51 ▪ The conquests of Alexander the Great 52–53 ▪ The Peloponnesian Wars 70 ▪ Muhammad receives the divine revelation 78–81

the Romans, the Persians built a network of roads to hold their empire together—and the military, and devolution of administration to local provinces. Under the Achaemenids, the Middle East was united under a single umbrella culture for the first time.

Conflict with the independent Greeks arose after the city-states of Athens and Eretrea supported an unsuccessful revolt by the Ionians against Persian rule in 499 BCE. Darius responded by invading mainland Greece, but was defeated by the Athenians and their allies at Marathon in 490 BCE. He planned an even larger invasion, but it was only after his death that his son Xerxes began mustering a huge army to execute the plan.

Father of Lies

The main source for the Greco–Persian Wars is the ancient Greek historian Herodotus of Halicarnassus, known as both the Father of History and the Father of Lies. Herodotus estimated that Xerxes' land army was made up

> All other expeditions... are as nothing compared with this. For was there a nation in all Asia which Xerxes did not bring with him against Greece?
> **Herodotus**

of 1,700,000 men—but modern historians believe the maximum figure to be closer to 200,000.

The second Persian invasion, in 480 BCE, was held up by the heroic defense of Leonidas and his 300 Spartans at Thermopylae, and by Greek naval resistance at Artemisium. Later the Athenian navy lured the Persian fleet into a trap at Salamis. Xerxes returned to

Persia, leaving a large force to carry on the fight, but at the Battle of Plataea in 479 BCE the Greeks, led by the Spartans, crushed the Persians, who also lost to the Spartans at Mycale. Greek success can probably be ascribed to Xerxes' difficulties in keeping his vast army supplied and supported after naval defeat, although Herodotus ascribed it to the moral superiority of their cause.

The Delian League

The Greeks now began to go on the offensive, forming the Delian League to oppose Persia. In 449 BCE, the Persians finally concluded peace, conceding the independence of the Ionian states.

The Persian War had reinforced Greek identity and bolstered cultural and military confidence, most significantly in Athens. The country's rising power sparked conflict with Sparta, leading to the Peloponnesian War of 431–404 BCE. The Persian Empire had reached the limits of its expansion, but remained strong until defeated by Alexander the Great in 331 BCE. ▪

Cyrus the Great

The founder of the Achaemenid Empire was Cyrus II, later known as "the Great." In around 557 BCE, he became king of Anshan, a vassal of the Median king.

According to legend, he won the Persian army's support by making them spend one day clearing thorn bushes, and the next banqueting, then asking why they remained slaves to the Medes when, by backing his revolt, they could live in luxury.

Some ten years later he had conquered Media, and Sardis and Lydia in Asia Minor. He conquered Babylon seven years after that by

diverting the Euphrates and marching his army along the dry riverbed into the great city. This victory brought him the lands of the neo-Babylonian Empire, including Assyria, Syria, and Palestine. He liberated the Jews from their Babylonian bondage and allowed them to rebuild the Temple in Jerusalem. The Greek writer Xenophon saw him as an example of the ideal ruler.

Cyrus died in 530 BCE while on campaign in Central Asia. He was buried in a great tomb inside the royal palace he had built at Pasargadae in Persia.

ADMINISTRATION IS IN THE HANDS OF THE MANY AND NOT OF THE FEW

ATHENIAN DEMOCRACY (c.507 BCE)

IN CONTEXT

FOCUS
**Greek politics
and philosophy**

BEFORE
14th–13th centuries BCE
Mycenaean settlement at
Athens, with fortification
of the Acropolis.

c.900 BCE Political union of
small towns in Attica into a
city-state centered on Athens.

c.590 BCE Reforms of Solon
open the political machinery
of Athens to all citizens,
regardless of class.

AFTER
86 BCE Athens sacked by
Romans under General Sulla.

c.50 BCE Beginning of the
Roman philhellenic movement;
Athens becomes the focus of
imperial benefactors.

529 CE Christian Emperor
Justinian I closes Plato's school
and drives out pagan scholars.

The term "democracy"
comes from the Greek
words *demos* (people) and
kratos (rule). The democracy that
developed in ancient Athens
around 507 BCE and flourished in
its purest form from 462 to 322 BCE,
albeit with some interruptions,
provided the model for what has
become the dominant form of
government in the world: by 2015,
125 of the world's 195 countries
were electoral democracies. The
democracy of ancient Athens,
however, differed from its modern
form, reflecting the history of
Athens and the warring Greek
states of the age.

Oligarchs and hoplites

After the chaos of the ancient
Greek Dark Ages—a period that
followed the breakdown of
Mycenaean civilization around
1100 BCE and lasted until about
the 9th century BCE—most of the
emergent city-states evolved into
oligarchies, with powerful nobles
monopolizing government and
serving their own interests. In
Athens, the Areopagus—a council
and law court consisting of men of
aristocratic birth—controlled the

To the Athenian the
fruits of other countries
are as familiar a luxury
as those of his own.
Pericles

machinery of state, appointing
officials and serving as a civil
court, while the lower classes
(*thetes*) were excluded from office.

However, the development of the
"hoplite" model of citizen-soldiery in
the 8th to 7th centuries BCE proved
disruptive to those who were in
power, as it led to a certain level of
egalitarianism. Hoplites were men
in the heavy infantry, mainly free
citizens, whose primary tactic was
the phalanx—a military formation
in which soldiers stood in tightly
packed ranks, with each man's
shield protecting the hoplite to his
left. Any man who could afford the
arms and armor would be putting
his life on the line to defend the
state. As a result, a kind of middle
class emerged, which declared that
service should bring full citizenship
and political representation. At the
same time, the lower classes were
also making demands, and tensions
between them and the higher
orders over key issues, such as land
reform and debt slavery, threatened
to lead to civil breakdown.

Solon and Cleisthenes

In Athens, some of these tensions
were eased around 594 BCE by the
reforms of the statesman Solon. He

Pericles

Pericles (c.495–429 BCE) became
Athens' most famous democrat
and the leading man of the
city-state for about 30 years.
He came to prominence around
462 BCE, when he helped the
politician Ephialtes dismantle
the Areopagus—the last bastion
of oligarchic control. After
Ephialtes' death, Pericles
undertook further reforms,
including the introduction of pay
for those serving in the courts,
making it possible for even the
poorest citizen to have his say.

He is also believed to have
helped drive Athens' assertive
foreign policy as the city sought
to exploit its dominance of the
Delian League. During the
440s and 430s BCE, Pericles
was involved in an ambitious
public building program that
provoked controversy at home,
where he fought off revolt, and
abroad, where he was
condemned for requisitioning
money from the Delian League
to pay for the Parthenon.
Nonetheless, he was popular
and was elected as general
every year from 443 BCE.

See also: The Law Code of Hammurabi 36–37 ▪ The palace at Knossos 42–43 ▪ The Persian Wars 44–45 ▪
The conquests of Alexander the Great 52–53 ▪ The Peloponnesian Wars 70 ▪ The fall of Constantinople 138–41

The Parthenon, built in 447–438 BCE as a temple dedicated to the goddess Athena, is often seen as a symbol of democracy and Western civilization.

established a law that declared all citizens could vote in matters of state, and that a law court should admit all citizens. At the same time, however, he mollified the upper classes by introducing a graded oligarchy in which power corresponded to wealth—the aristocracy was to control the highest offices, the middle class the lesser offices, and the poor could be selected by lot to serve on juries.

In the late 6th century BCE, Athens fell under the sway of the tyrant Pisistratus and his sons. In response, a faction of aristocrats led by Cleisthenes allied with lower-ranking members of society to take power. The institution of true democracy in Athens is traditionally dated to this point— around 507 BCE. Cleisthenes introduced true popular government, or direct democracy, enabling all citizens of Athens to vote directly on Athenian policy (unlike in a contemporary representative democracy, in which the people elect representatives to act as the legislature). He also reorganized the citizenry into units by geography rather than kinship, breaking the

traditional ties that underpinned Athenian aristocratic society, and established sortition—the random selection of citizens for government positions rather than basing the choice on heredity. In addition, he restructured the Boule—a council of 500, which drew up legislation and proposed laws to the assembly of voters (*Ecclesia*). In 501 BCE, command of the military was transferred to popularly elected generals (*strategoi*).

In 462 BCE, Ephialtes became leader of the democratic movement in Athens and, together with his deputy Pericles, he dismantled the

Areopagus council, transferring the majority of its powers to the Boule, the Ecclesia, and the citizen courts. Ephialtes was assassinated in 461 BCE and Pericles took over the political leadership, becoming one of the most influential rulers in the history of ancient Greece.

A perfect democracy?
Athens now had a genuine direct democracy, but many people were not allowed to participate in the system as they were not considered true citizens. Political rights were restricted to adult male Athenians; women, foreigners, and slaves were **»**

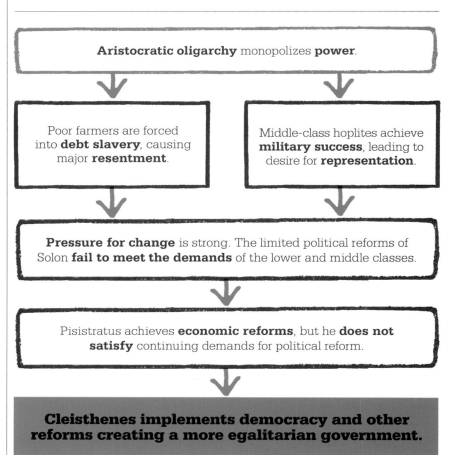

Aristocratic oligarchy monopolizes **power**.

Poor farmers are forced into **debt slavery**, causing major **resentment**.

Middle-class hoplites achieve **military success**, leading to desire for **representation**.

Pressure for change is strong. The limited political reforms of Solon **fail to meet the demands** of the lower and middle classes.

Pisistratus achieves **economic reforms**, but he **does not satisfy** continuing demands for political reform.

Cleisthenes implements democracy and other reforms creating a more egalitarian government.

The Athenian constitution relied on a careful separation of powers. This was essential to make the practical operation of direct democracy possible. It also ensured that all citizens (men aged 20 and above) could serve and that power could not be abused.

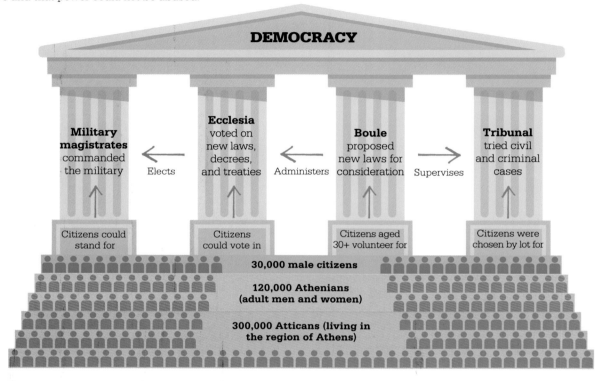

DEMOCRACY

Military magistrates commanded the military

Elects

Ecclesia voted on new laws, decrees, and treaties

Administers

Boule proposed new laws for consideration

Supervises

Tribunal tried civil and criminal cases

Citizens could stand for

Citizens could vote in

Citizens aged 30+ volunteer for

Citizens were chosen by lot for

30,000 male citizens

120,000 Athenians (adult men and women)

300,000 Atticans (living in the region of Athens)

excluded. In the 4th century BCE, out of the 300,000-strong total population of Attica—the region of Greece controlled by Athens—just 30,000 men comprised the voting population. In theory, men became voting citizens at age 18, but as they were generally liable for two years of military service they were not enrolled on the rosters of the council until they turned 20, and did not come into their full political rights until the age of 30.

During the "Pentekontaetia"—the decades between Greek victory in the Persian War (479 BCE) and the start of the Peloponnesian War (431 BCE)—Athens reached the height of its glory. In 447 BCE, Pericles appropriated the treasury

of the Delian League (the anti-Persian confederation that had become a vessel for Athenian hegemony) to build a magnificent

> Our ordinary citizens, though occupied with the pursuits of industry, are still fair judges of public matters.
> **Pericles**

temple (the Parthenon) on the rocky hill known as the Acropolis. Citizenship of Athens was highly coveted, and in 451 BCE Pericles passed a law restricting it to men whose parents were both Athenians.

A center of philosophy
As well as being the most powerful city-state in ancient Greece, Athens was also the crucible of a revolutionary new direction in philosophy, in large part due to Socrates (c.469–399 BCE). Earlier Greek philosophers, collectively known as the pre-Socratics, had introduced a revolution of their own in human thought in the 5th and 6th centuries BCE. They rejected supernatural explanations for the

world, the explanatory power of mythology, and the authority of tradition, and set out to discover the origins and workings of the natural world through reason and observation. The pre-Socratic natural philosophers developed theories about the elements, classifications of nature, and mathematical and geometric proofs.

Socrates turned his enquiries inward to more human matters—as Cicero said of him, "he brought philosophy down from heaven." Socrates' method was simply to ask questions—What is friendship? What is justice? What is knowledge? The Socratic method tended to lay bare the limits of existing thinking, often making people look foolish or pompous. Accordingly, Socrates was unpopular and eventually he was accused of two crimes by his enemies—corrupting youth by encouraging them to go against the government, and impiety, or lack of respect for the gods. Consequently, he was sentenced to death.

Socrates' successors

The fate of Socrates was taken as an indictment of democracy by his successors, particularly Plato (c.428–348 BCE), who saw him as a martyr for truth. Plato ran a school (the Academy) and developed ideas about universal truths and metaphysics that have shaped all subsequent religion and philosophy in the Western world. His student Aristotle (384–322 BCE) became equally influential, setting up the Lyceum school and writing on such diverse topics as politics, ethics, law, and natural sciences.

Plato opposed democracy, as he believed that the people were not sufficiently equipped with philosophical grace to legislate and if governance were left in the hands of the ordinary, citizen tyranny

would emerge. In his ideal republic, enlightened philosophers would rule as kings. He also challenged the basic principle of democracy—that of liberty (*eleutheria*)—which he believed could divert people from the proper pursuit of ethics and cause social disunity.

The fall of democracy

During the Peloponnesian War (431–404 BCE), in which Athens was ultimately defeated by the Spartans, Athenian democracy was twice suspended, in 411 and 404 BCE. Athenian oligarchs claimed that

Dictatorship naturally arises out of democracy, and the most aggravated form of tyranny and slavery out of the most extreme liberty.
Plato

An Audience in Athens (1884), by Sir William Blake, captures the atmosphere at the Greek tragedy Agamemnon by Aeschylus c.450 BCE. This period is regarded as the Golden Age of drama in ancient Greece.

Athens' weak position was due to democracy and led a counter-revolution to replace democratic rule with an extreme oligarchy. In both cases, democratic rule was restored within one year.

Democracy flourished for the next eight decades. However, after the Macedonian conquest of Athens under Philip II and his son Alexander (later Alexander the Great) in 322 BCE, Athenian democracy was abolished. It was intermittently restored in the Hellenistic age in the 1st and 2nd centuries BCE, but the Roman conquest of Greece in 146 BCE effectively killed it off.

Although democratic rule had been quashed, Athenian science and philosophy lived on. The renown and influence of Plato and Aristotle endured through the ages that followed, and much of their work continues to influence Western thought to this day. ∎

THERE IS NOTHING IMPOSSIBLE TO HE WHO WILL TRY

THE CONQUESTS OF ALEXANDER THE GREAT (4TH CENTURY BCE)

IN CONTEXT

FOCUS
Hellenistic world

BEFORE
359 BCE Philip II of Macedon begins his rise to power and develops innovative military technology and tactics.

338 BCE Philip II defeats the Greek states and becomes undisputed leader of Greece.

336 BCE On Philip's death, his son Alexander is proclaimed king of Macedon.

AFTER
321 BCE After Alexander's death, squabbling between his generals breaks out into widespread civil war.

278 BCE Alexander's generals establish three Hellenistic kingdoms in Greece, the Middle East, and Europe.

30 BCE The Roman emperor Octavian annexes Egypt, the last Hellenistic kingdom.

I n one of the fastest and most daring military expansions in history, Alexander the Great, the young king of Macedon in the Balkans, blazed a trail of conquest across most of the known world of his day, and set in motion a process of Hellenization—the spread of Greek culture and its fusion with non-Greek, Eastern traditions—which endured for centuries.

Alexander's father, Philip II, had transformed this peripheral state into a formidable military power, and had waged campaigns against his neighbors that culminated in Macedon's domination over all of

In this late Roman mosiac, Darius III is shown fighting at Issus in 333 BCE. Alexander conquered the Persian king's empire and destroyed its capital in Persepolis without suffering a defeat.

Greece. When he was assassinated in 336 BCE, Philip had been planning an expedition to West Asia, to free the former Greek city-states that had been reconquered by the Persian Empire. After securing the Macedonian throne by destroying his rivals, Alexander set about pursuing his father's quest, while satisfying his own thirst for glory.

King of the world
After forcing the other Greek city-states to accept his authority, in 334 BCE Alexander marched into Asia Minor (modern-day Turkey) at the head of an army of 43,000 foot soldiers and 5,500 cavalry. At its heart lay the Macedonian phalanx, a well-drilled, tight-knit corps of 15,000 men armed with the *sarissa*, a pike that was up to 23ft (7m) long. When combined with the shocking cavalry charge provided by the king's personal bodyguard, the Companions, the formation proved irresistible.

After an initial victory over the Persians at the River Granicus in the northwest, Alexander pressed on across Asia Minor. He stopped at Gordium in the central kingdom of Phrygia, where tradition held that he who could untie a complex

See also: The Persian Wars 44–45 ▪ Athenian democracy 46–51 ▪ The assassination of Julius Caesar 58–65 ▪ Belisarius retakes Rome 76–77 ▪ The founding of Baghdad 86–93 ▪ The fall of Constantinople 138–41

East–West **cultural exchange** begins from an era of **Persian Wars**, with western provinces of Persian Empire becoming **Hellenized** and Macedonians adopting aspects of **Persian culture**.

Alexander's conquests force the rapid synthesis of Greek and Asian cultures, laying seeds of the Hellenistic age.

Hellenistic learning survives the fall of Rome in the **Byzantine Empire** and in the **Translation Movement** of the **Islamic Caliphate**.

Hellenized societies in Egypt and West Asia **assimilated** into Roman Empire.

---

knot made by the city's founder, would conquer the entire continent. Alexander, in a typically forthright move, cut the knot with his sword. He went on to twice defeat the far superior forces gathered by Darius III, the Persian emperor—at Issus (on the southern coast of Asia Minor) in 333 BCE and Gaugamela (in modern Iraq) in 331 BCE, subduing Egypt in the interval.

Having forced the Persians into submission, Alexander drove his troops eastward, across mountains, deserts, and rivers into Afghanistan and Central Asia, and on to the Indian Punjab, ruthlessly crushing all resistance. He would have pushed further into India, but in 325 BCE his exhausted men refused to go on.

The Hellenistic legacy

Alexander was now the king of a vast and ethnically diverse empire that included 70 newly founded cities, united by a common Greek culture, customs, and language, and linked by trade routes; although the process of Hellenization was already underway in the western half of Persia before his expedition, Alexander had accelerated its spread throughout the Middle East.

In 323 BCE, Alexander died—most likely from disease but perhaps by poisoning—without naming a successor. His empire was carved up by his leading generals, but some of the Hellenistic dynasties they founded, notably Selucid Syria and Babylon and Ptolomeic Egypt, survived until Roman times. ▪

Alexander the Great

Throughout antiquity, Alexander was widely regarded as the most remarkable man who ever lived, and in terms of the breadth and duration of his renown, which saw him become a key figure in national literatures from Central Asia to Western Europe, he is one of the most famous men in history.

Born in 356 BCE, to parents who claimed descent from demigods and heroes, Alexander's education under the philosopher Aristotle ensured he was steeped in Greek legend, and he came to believe he was invincible, even divine. As a general he was decisive, bold to the point of recklessness—with his own life and those of his men—and a brilliant tactician. He maintained the loyalty of his forces throughout his long and arduous campaign, but his quick and violent temper, fueled by his heavy drinking, occasionally spurred him to eliminate those closest to him, including his friends. Alexander died at age 32, at the height of his power. His funeral cortège was hijacked by Ptolemy, one of his generals, and diverted to Alexandria in Egypt, where his tomb was later visited by Julius Caesar, but is now lost.

IF THE QIN SHOULD EVER GET HIS WAY WITH THE WORLD, THEN THE WHOLE WORLD WILL END UP HIS PRISONER

THE FIRST EMPEROR UNIFIES CHINA (221 BCE)

IN CONTEXT

FOCUS
Han China

BEFORE
1600–1046 BCE Shang dynasty rules.

c.1046–771 BCE Western Zhou dynasty.

771–476 BCE Spring and Autumn Period (the first half of the Eastern Zhou dynasty).

551–479 BCE Life of Kong Fuzi (known as Confucius).

476–221 BCE Warring States Period (the second half of the Eastern Zhou dynasty).

AFTER
140–87 BCE Reign of Han Emperor Wudi (Liu Che)—a time of imperial expansion.

220–581 CE Three Kingdoms and Six Dynasties Period.

581–618 Sui dynasty.

618–907 Tang dynasty.

China is probably the most enduring coherent state in world history, and to an extraordinary extent this is due to the will of one man: Qin Shi Huangdi, the self-styled First Emperor. Before he unified ancient China in 221 BCE, it was a region of diverse states, differing in culture, ethnicity, and language. During the era known to Chinese historians as the Spring and Autumn Period (771–476 BCE), the region was nominally under the control of Zhou dynasty kings, but in reality their feudal system of government meant that only a token authority rested with the royal throne, while feudal lords held genuine power over what

See also: Emperor Wu claims the Mandate of Heaven 70 ▪ China is divided into Three Kingdoms 71 ▪ The An Lushan revolt 84–85 ▪ Kublai Khan conquers the Song 102–03 ▪ Hongwu founds the Ming dynasty 120–27

> When [Qin Shi Huangdi]
> is in difficulty he readily
> humbles himself
> before others, but when
> he has got his way,
> then he thinks nothing
> of eating others alive.
> **Sima Qian**
> **Han historian**

were effectively autonomous states. Up to 140 small states competed for power and territory.

The Spring and Autumn Period gave way to the Warring States Period (476–221 BCE), in which power was consolidated into the hands of seven kingdoms: Qi, Chu, Yan, Han, Zhao, Wei, and Qin. At this point in China's history it was by no means certain that an overarching Chinese identity or state would emerge. If anything, it was more likely that the considerable geographical, climatic, cultural, and ethnic differences between the various kingdoms would see the region develop in a similar fashion to Europe many centuries later, with multiple distinct and divergent national entities.

The rise of Qin

In 247 BCE, a 13-year-old prince named Ying Zheng succeeded to the throne of Qin. He inherited a militarized state, in which effective bureaucracy, powerful armies, and competent generals combined to produce a formidable and ruthless war machine. Zheng had rivals executed or exiled, appointed very effective generals and counselors, and conquered the six other states in the region, so that by 221 BCE all seven states were unified under his rule. Disdaining the old title of king (Wang), he styled himself as emperor (Huangdi). Since he was the first (Shi) emperor of the Qin dynasty, he was known as Qin Shi Huangdi.

The governing philosophy of the Qin state had been legalism: strict centralization of power and severity in enforcing adherence to the law. The emperor now set about applying this philosophy throughout the whole of China, ruthlessly imposing cultural, linguistic, economic, and technological unity. All scripts except Xiaozhuan (small seal script) were banned. In addition, according to legend, the emperor gave orders for 400 Confucian scholars to be buried alive and all existing books to be burned; his reign was to mark a new "Year One" in the history and culture of China. He also introduced a host of economic reforms—there was to be a single, unified system of weights and measures, a uniform coinage, and even the gauge of cart tracks was standardized so that axle-widths could be the same across the empire.

The new order

The new social and political order of the empire reflected changes that had been underway since the Spring and Autumn Period. The feudal system was abolished, so that the mass of peasantry now owed their allegiance to the state rather than feudal or clan lords. Over 100,000 noble families were »

Qin Shi Huangdi

As First Emperor of China, Ying Zheng (later known as Qin Shi Huangdi, 260–210 BCE) was a truly pivotal figure in Chinese history, uniting the country and ushering in a period of imperial rule that lasted nearly 2,000 years. He was a brutal despot but was also innovative, dynamic, and energetic—reports claim that he needed just one hour's sleep per night and he set himself a daily work quota, measured by the weight of papers that he needed to go through. He regularly walked the city streets in disguise to keep tabs on the populace, and he made five great tours of inspection of the empire. Highly paranoid and fearful of possible attempts on his life (he survived at least one assassination attempt), the emperor became obsessed with the quest for immortality, sponsoring expeditions to look for magic ingredients and mystics who could brew an elixir of life, to enable him to live forever. Ironically, his death at the age of 50 might well have been linked to his consumption of toxic mercury-based potions that he had taken to extend his life.

relocated to the emperor's capital city Xianyang (near Xi'an, in the Shaanxi province), and their arms were confiscated, melted down, and then cast into giant statues. During the Warring States Period, the pressure of incessant military competition had generally favored the emergence of more meritocratic avenues for advancement, thereby facilitating social mobility while undermining the importance of noble lineage. In the Qin dynasty, aristocratic rule was replaced with a centralized bureaucratic administration and the country was divided into 36 commanderies, which were administrative divisions controlled by appointed (not hereditary) governors. Censors, or inspectors, traveled the country to enforce adherence to Qin law.

The Qin dynasty also saw the emergence of a new scheme of social stratification, with society divided into four classes: gentlemen (*Shi*), peasants (*Nong*), and two new classes that had emerged during the Zhou dynasty—artisans (*Gong*) and merchants (*Shang*). The educated gentry would replace the nobility as the main source of

> With his puffed-out chest like a hawk and voice of a jackal, Qin is a man of scant mercy who has the heart of a wolf.
> **Sima Qian**
> **Han historian**

state officials. The merchant class was officially the lowest and most despised of the orders, and was subject to legal discrimination; however, wealthy merchants were able to use their financial muscle to become important political players.

Great works
Among Qin Shi Huangdi's greatest achievements were his ambitious civil engineering projects, although there was a great human cost as many lost their lives in the process.

He is traditionally credited with building the first part of the Great Wall of China, to keep out nomadic tribes in the north, by connecting parts of old walls erected by the Warring States and then adding thousands of miles of new wall. Other projects included constructing the Lingqu canal, which linked the Xiang and Li rivers so military supplies could be transported from northern to southern China, and building military roads including "the Straight Road," which was 497 miles (800 km) long and ran from Xianyang to the Great Wall.

Most famous of all the emperor's ventures was the construction of his own elaborate mausoleum complex, which took 38 years and over 700,000 workers to construct. It consisted of a giant pyramid covered in earth to create an immense mound, 328 ft (100 m) high and 1,640 ft (500 m) across. Within the pyramid was a tomb in which his beloved empire was recreated in miniature, complete with liquid mercury rivers and seas. Buried around the tomb were large pits filled with thousands of life-sized terracotta warriors, bureaucrats, and entertainers, all intended to serve the emperor in the afterlife. Workers on the tomb were killed after completing their tasks so the secrets of the mausoleum's location and contents died with them, and the tomb remained undiscovered for over 2,000 years.

Despite the megalomaniacal exertions of the First Emperor, the Qin dynasty was to prove short-lived. Peasant unrest caused

Guarding the tomb of Emperor Qin Shi Huangdi, these life-sized terracotta soldiers were discovered in 1974 by workers digging a well. The figures were originally brightly painted and each has a unique facial expression.

```
┌─────────────────┐          ╭──────────────╮          ╭──────────────────╮
│ Large region    │          │  Qin state   │          │ Chinese unity is │
│ comprises many  │          │  conquers the│          │ strengthened     │
│ small, culturally│         │  other six   │          │ further.         │
│ diverse states. │          │  states.     │          ╰──────────────────╯
└─────────────────┘          ╰──────────────╯
        ╭──────────────────────╮     ╭──────────────────────╮
        │   Seven major states │     │  Qin Shi Huangdi     │
        │ emerge and engage in │     │ imposes unification, │
        │ frequent warfare to  │     │ standardization, and │
        │ gain power and       │     │   homogeneity.       │
        │ territory.           │     ╰──────────────────────╯
        ╰──────────────────────╯
```

Large region comprises many small, **culturally diverse** states.

Seven major states emerge and engage in frequent **warfare** to gain **power** and **territory**.

Qin state conquers the other six states.

Qin Shi Huangdi imposes unification, standardization, and homogeneity.

Chinese unity is strengthened further.

by deep-seated resentment over the brutal extortions of money and the many years of forced labor, plus bankruptcy as a result of over-ambitious civil works, combined to undermine the carefully ordered administration of the emperor and his leading counselors, chief among them the chancellor Li Si.

When the First Emperor died in 210 BCE his youngest son, Hu Hai, under the influence of advisor and former tutor Zhao Gao, seized the throne and exiled—and later executed—Li Si. Hu Hai was subsequently murdered after just three years of being in power and his successor, Zi Ying, found his authority so reduced that he adopted the title of king, rather than emperor.

The Han Dynasty

China collapsed into rebellion and civil unrest, and a few days after Zi Ying's accession, the Han general Liu Bang marched into Xianyang. The following year, in 206 BCE, he declared himself emperor of the Han dynasty, which would go on to rule China for 400 years, shaping its subsequent history to such an extent that the main ethnic group in China is now known as the Han.

The Han expanded Chinese territory in all directions—west to Xinjiang and Central Asia, northeast to Manchuria and Korea, and south to Yunnan, Hainan, and Vietnam. Most importantly, they consumed the powerful Xiongnu Empire in the north. They also reintroduced Confucianism as the official state philosophy: Confucian education and ethics soon became the cornerstones of the scholar-bureaucracy, eventually forming the basis for the all-important civil service examination system, which would give a meritocratic basis to imperial institutions and combat the power of the aristocracy for millennia to come.

Han success in building and maintaining a unified, centralized China was based on the foundations that had been laid down by the First Emperor. The Han dynasty finally collapsed in 220 CE, amid a foment of civil unrest and natural disasters that convinced the Chinese that their dynasty had lost "the mandate of heaven," giving way to the violent and chaotic era known as the Three Kingdoms and Six Dynasties Period. Despite the devastating cost of this breakdown, which saw the Chinese population plummet from 54 million in 156 CE to 16 million in 280 CE, the concept of a unified China survived 360 years of division, enabling the Sui dynasty to reunify China in 581.

The influence of the First Emperor is still felt in modern China, and Chairman Mao Zedong (1893–1976) explicitly drew on the emperor for inspiration. "You accuse us of acting like Qin Shi Huangdi," Mao thundered in a 1958 tirade against intellectual critics. "You are wrong. We surpass him a hundred times. When you berate us for imitating his despotism, we are happy to agree! Your mistake was that you did not say so enough." ■

Confucius is generally considered to be the most influential philosopher in Chinese history. His teachings emphasized the importance of morality, integrity, humility, and self-discipline.

THUS PERISH ALL TYRANTS

THE ASSASSINATION OF JULIUS CAESAR (44 BCE)

IN CONTEXT

FOCUS
Fall of the Roman Republic

BEFORE
509 BCE Rome becomes a republic in which a small number of wealthy families share power.

202 BCE Rome defeats Carthage in North Africa and the empire expands rapidly.

88–82 BCE Civil war between rival generals Sulla and Marius tips the republic into crisis.

AFTER
31 BCE Octavian's victory at the Battle of Actium leads to his accession as Rome's first emperor and the end of the republic.

79 CE Vesuvius erupts, destroying Pompeii.

2nd century CE The Roman Empire reaches its greatest extent, with a population of around 60 million people.

The **oligarchic** political system in the **Roman Republic** is corrupt and decaying.

Rome's **nobility** dominate the **Senate**, protecting their privileges at the expense of political change, leading to a **crisis of the republic**.

After successful military campaigns, **Julius Caesar becomes dictator** and forces **political and social reforms** on the nobility.

Fearing Caesar's popularity and power, a group of senators assassinate him.

Octavian wins the **civil war** to determine Caesar's heir. Calling himself **Augustus**, he becomes the **first emperor of Rome**.

Augustus ensures that **the office of emperor** survives by making **Tiberius** his heir, transforming Rome into a **hereditary monarchy**.

On March 15, 44 BCE, the life of Julius Caesar, dictator of Rome, came to a bloody end at the hands of a faction of aristocratic senators who were determined to rescue the Roman Republic from what they saw as Caesar's tyranny. In reality, the dictator's death did not save the republic: it merely unleashed the latest in a series of civil wars, which exhausted the Roman state. It was left powerless to resist the rise to absolute power of Caesar's great-nephew Octavian. Taking the title Augustus, Octavian created a new political arrangement that enabled him to rule as emperor, bringing the 500-year-old Roman Republic to an end in all but name.

Republican origins

From its ancient beginnings as a cluster of small villages on seven hills by the River Tiber, Rome grew into a city-state that was just one of many on the Italian peninsula. According to legend, Rome was first ruled by kings, but in 509 BCE, the monarchy was overthrown and it became a republic. A new constitution allowed the election of two top officials, known as consuls, to run the state, but in order to prevent abuse of power, their term was limited to one year. The office of king was also prohibited, and special provision was made for the appointment of a dictator to replace the consuls in times of crisis—his term being limited to six months.

The fledgling Roman Republic proved remarkably successful: between 500 and 300 BCE, it increased its extent and power

See also: Athenian democracy 46–51 ▪ The conquests of Alexander the Great 52–53 ▪ The Battle of Milvian Bridge 66–67 ▪ The Sack of Rome 68–69 ▪ Belisarius retakes Rome 76–77 ▪ The crowning of Charlemagne 82–83 ▪ The fall of Constantinople 138–41

Trajan's column in Rome is one of the most valuable sources of information about the Roman army—it is decorated with a spiraling relief depicting the well-drilled legions on campaign.

through a combination of conquest and diplomacy until it incorporated the whole of Italy. Between 202 and 120 BCE, Rome came to dominate parts of North Africa, the Iberian Peninsula, Greece, and what is now southern France. Its conquered territories were organized into provinces, ruled by short-term governors who maintained order and oversaw the collection of taxes.

By the 1st century BCE, Rome was a Mediterranean superpower, yet its long tradition of collective government, in which no individual could gain too much control, was being challenged by the personal ambitions of a few immensely powerful military men. A series of bloody civil wars, internal political struggles, and civil unrest culminated with the dictatorship of Julius Caesar, a brilliant general and statesman, whose murder at the hands of his political enemies led to the demise of the republic and the birth of the Roman Empire.

The republic crumbles
In the period in which Julius Caesar came to prominence on the Roman political scene (around 70 BCE), Rome was in turmoil: beset with ever worsening social and economic problems and torn by political conflict. Early in Rome's history, the non-slave population had been officially split into two classes: the patricians (members of the ancient hereditary nobility and wealthy landowners) and the plebeians, or plebs for short (the

common people). On the formation of the republic, only patricians had been entitled to hold office in the Senate—Rome's governing and advisory council—but in 368–367 BCE, a constitutional amendment allowed the election of wealthier plebs too, and the result was a power-sharing arrangement.

However, in reality, a small group of patrician families known as the Optimates (the "Best Men") had long dominated the Senate and jealously guarded their privileges. In the late Roman Republic, those who championed the rights of the plebs—the Populares (the "People's Men")—sought popular support against the Optimates, either in the interests of the people themselves, or more often, in pursuit of their own careers. The self-interested Optimates resisted making the social and economic reforms that were urgently required to meet the

changing needs of the Roman people. In Italy and the provinces an unequal system of taxation and corrupt governance were causing social unrest, while in the city of Rome itself, the infrastructure was barely able to cope with a growing population. The empire's rapid expansion had brought a flood of »

In Caesar were combined genius, method, memory, literature, prudence, deliberation, and industry.
Cicero
2nd Philippic, section 116

slave labor from the provinces, driving many Roman farm workers and smallholders off the land and into the city in search of work.

The rise of Julius Caesar

Meanwhile, a handful of military leaders in Rome's provinces had begun to use their armies to jockey for political prominence. Among them was Julius Caesar, a highly intelligent and ambitious general from a patrician family who had aligned himself with the Populares and risen swiftly through the political ranks. Caesar was intent on making the reforms necessary to meet the challenges of the republic, and so he maneuvered himself into a position that would allow him to achieve his goal.

In 60 BCE, Caesar became consul, and two years later he was appointed governor of the province of Gaul, a role which enabled him to remain abreast of developments in the Senate while also offering a springboard to military glory. In a series of masterful campaigns over the next eight years, he conquered Gaul, bringing the whole of what is now France, along with parts of Germany and Belgium, under his

rule. He also led two expeditions to Britain, in 55 and 54 BCE. Caesar's heroic military exploits left him immensely rich and increased his personal prestige; he enjoyed the loyalty of his armies and the love of the Roman mob, upon whom he could now afford to lavish feasts, games, and money.

Buoyed by his achievements, Caesar attempted to dictate the terms on which he would return to Roman politics, demanding to be allowed to stand for a second consulship while remaining in command at Gaul. This put him on a collision course with the Optimates in the Senate, since Roman law required military leaders to relinquish control of their armies before entering Rome, a prerequisite for running for office. Caesar knew that if he agreed to enter the city as a private citizen, without his armies, his political opponents would most likely attempt to try him for abuse of power during his first consulship.

Back in Rome, the Optimates, alarmed by the implications of Caesar's meteoric rise, allied themselves to one of his main political rivals, the renowned

Even yet we may draw back, but once across that little bridge, and the whole issue is with the sword.
Julius Caesar
Speaking to his army before crossing the Rubicon

general Pompey. The Senate passed laws intending to strip Caesar of his command when he returned from Gaul, and in 49 BCE they declared him *hostis*, or public enemy. In response to this direct threat, Caesar did the unthinkable: he marched his army on Rome. En route, he paused at the border between the Gallic provinces and Italy proper: a small river called the Rubicon. Caesar was acutely aware that crossing the river would constitute a declaration of war against the Senate but, quoting the

Julius Caesar

Gaius Julius Caesar was born in Rome in 100 BCE, to a patrician family of distinguished ancestry. From an early age, he grasped that money was the key to power in a political system that had become hopelessly corrupt. He also quickly learned that forging a network of alliance and patronage would be crucial to his success.

After serving in the war to crush the slave revolt led by Spartacus in 72 BCE, Caesar was briefly taken hostage by pirates. Once he returned to Rome in 60 BCE, Caesar spent vast sums on buying influence and positions,

eventually teaming up with the two other leading men in Rome, Crassus and Pompey, to form the so-called First Triumvirate. Between 58 and 50 BCE, he formed a provincial power base in Gaul where, without the sanction of the Senate, he launched a series of campaigns that made him master of Western Europe, with fabulous wealth and powerful armies. However, these campaigns also earned him many opponents among the governing classes, who would eventually cut short both his career and his life.

Athenian poet Menander, he announced *alea iacta est* ("let the dice roll") and led his men onward.

Caesar's new order

In the ensuing civil war, Caesar finally triumphed over Pompey's forces at the Battle of Pharsalus in northern Greece in 48 BCE. The defeated Pompey fled to Egypt for sanctuary, where he was assassinated. After crushing the remaining pockets of resistance, Caesar finally returned to Rome in 45 BCE, to consolidate his political position. In 46 BCE he accepted the dictatorship for 10 years; two years later, he was granted the office for life. Now in a position to begin the monumental task of reconstructing the Roman state and restoring stability to the empire, Caesar initiated far-reaching social and political reforms. He extended Roman citizenship; he enlarged the Senate, bringing in allies from among the provincial aristocracy; he established colonies outside Italy, to help spread Roman culture and knit the empire together; he spent lavishly on grandiose public works and buildings; he cut taxes; and he even reformed the Roman calendar, introducing the system of leap years that is still in use today.

A murder plot

Caesar's pragmatic solutions for re-establishing unity in the empire after years of chaos found favor with many parts of society, yet at the same time, his increasingly autocratic attitude to power was alienating fellow members of the ruling class. They felt that Caesar was trying to destroy the cherished traditions of the Roman state, and to undermine the prestige of the nobility, and spread the rumor that he was planning to make himself king. Unfortunately, Caesar failed to

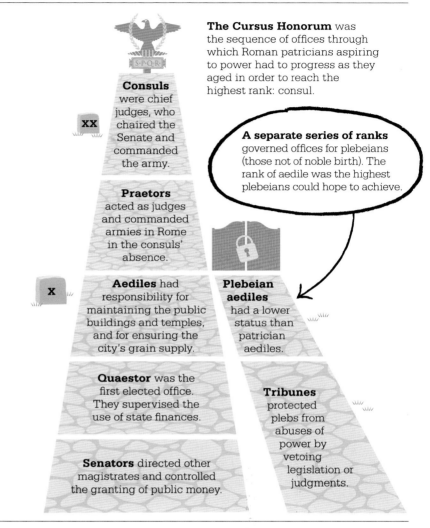

The Cursus Honorum was the sequence of offices through which Roman patricians aspiring to power had to progress as they aged in order to reach the highest rank: consul.

Consuls were chief judges, who chaired the Senate and commanded the army.

A separate series of ranks governed offices for plebeians (those not of noble birth). The rank of aedile was the highest plebeians could hope to achieve.

Praetors acted as judges and commanded armies in Rome in the consuls' absence.

Aediles had responsibility for maintaining the public buildings and temples, and for ensuring the city's grain supply.

Plebeian aediles had a lower status than patrician aediles.

Quaestor was the first elected office. They supervised the use of state finances.

Tribunes protected plebs from abuses of power by vetoing legislation or judgments.

Senators directed other magistrates and controlled the granting of public money.

quell the suspicions. He accepted unprecedented honors, such as assuming the title "Imperator" ("Victorious General") as a family name; he also allowed temples and statues to be erected in his honor, and had coins minted bearing his image. And when he adopted his grand-nephew, Octavian, there were fears that he was trying to establish a dynastic succession. Some members of the Senate concluded that the only solution to the problem was to assassinate Caesar, and so they hatched a conspiracy to carry it out.

Representing those opposed to the dictator's reforms—and the leading agent in the plot to murder him—Gaius Cassius Longinus was a general who had risen to political prominence during a largely disastrous campaign in Persia. Ancient Roman historians argued that Cassius's involvement was prompted by a combination of jealousy and greed. He is also said to have recruited the most important conspirator, Marcus Junius Brutus, a trusted colleague and confidante of Caesar, opposed to the dictator's presumed monarchic ambitions. »

Death of a dictator

The assassination plot grew rapidly, eventually including 60 senators, among them many of Caesar's close colleagues. The plotters decided to strike at a meeting of the Senate that had been called for March 15, (the Ides of March). On the day, they gathered at Cassius's home, each senator concealing a dagger beneath his robes, before moving on to Pompey's Theatre—part of a great civic complex that Caesar's old rival had constructed—where the Senate was meeting. A group of gladiators had been stationed in the theatre itself, to help control any crowd problems. However, many of the conspirators were nervous and ready to flee, convinced that the plot had been uncovered.

Caesar had indeed been warned: a list of the plotters had been thrust into his hands, but he ignored it. His wife pleaded with him not to attend the Senate meeting, but one of the conspirators, stationed at Caesar's house, helped calm her fears. When Caesar arrived at the meeting, a conspirator distracted his deputy, Mark Anthony, delaying him outside the theatre. As Caesar took his seat, the conspirators drew their daggers and struck, stabbing him 23 times. In an ironic twist,

I found Rome a city of bricks, and left it a city of marble.
Augustus
According to Suetonius, Augustus's biographer

Caesar, like a most gentle physician, had been assigned to the Romans by Heaven itself.
Plutarch
Parallel Lives

Caesar breathed his last slumped against the base of a statue of his old rival Pompey.

The Second Triumvirate

Seized with manic fervor, the conspirators dipped their hands in Caesar's blood and rushed out into the Forum to proclaim their tyrannicide. In the power vacuum that followed, Mark Anthony, and Caesar's heir, Octavian, promptly assumed control of the state, forming in 43 BCE a triumvirate (a group of three men holding power) with Lepidus, one of Caesar's former allies.

Needing to gather enough funds to stabilize their authority, and to remove political opposition, the triumvirate drew up a list of those who had supported Caesar's murderers, and declared them outlaws. Around 200 senators and more than 2,000 equites ("knights" or minor nobility) were either killed or had their estates confiscated. The treasury's coffers now filled, the triumvirate hunted down and destroyed Brutus and Cassius. In 40 BCE, the triumvirs met again, this time to carve up the Roman world. Africa was given to Lepidus, the East to Mark Anthony, and the West to Octavian. However, it was

not long before Octavian went to war against Anthony in north Africa, and, after defeating his forces at Actium in western Greece in 31 BCE, Octavian became the master of the Roman world.

Rome's first emperor

Octavian returned to Rome in 28 BCE and, instead of following Caesar's example, he renounced the dictatorial powers granted to him in order to wage his war against Antony. In 27 BCE, in gratitude for his service to Rome, the Senate bestowed on him the name Augustus ("revered personage") and granted him wide-ranging legal powers. Eventually, through political sleight of hand, he became Rome's sole ruler, controlling all aspects of the Roman state and command of the army.

An emperor in all but name (he was careful to spurn such titles, styling himself instead as princeps, or "first citizen"), over the next four decades, Augustus set about transforming the ruins of the republican system into an imperial autocracy, all the while maintaining the illusion that his authority was dependent on the will of the people. He loosely established the boundaries of the empire, pushed through reforms to clean up both private and public life, and crushed dissent. After the long periods of exhausting civil war, many in the empire were grateful for peace.

The Pax Romana

Indeed the might of the Roman military and the consequent improvements in security and stability across a vast stretch of territory, in what became known as the Pax Romana ("Roman Peace"), led to a growth in trade, economic activity, population, and general prosperity. The arts and culture flourished, public and private building works proliferated, and the provinces outside Italy underwent a process of Romanization, in which the Roman language, culture, laws, and institutions were embedded into diverse societies and across ethnic boundaries. Provincials were even granted full Roman citizenship after a period of military service.

However, for the regions beyond the bounds of empire, Augustus's Pax Romana often meant just the opposite. Even after reducing the army from 80 legions to a permanent force of just 28, Augustus had to find employment for 150,000 soldiers. He launched a series of campaigns to extend borders, suppress and harry rebels and "barbarians," and seize slaves from conquered areas.

The Ara Pacis Augustae altar in Rome is dedicated to Pax, the Roman goddess of peace. The processional frieze shows members of the Roman Senate with a priest.

Bear with me the hope that when I die, that the foundations which have I laid for [Rome's] future government, will stand firm and stable.
Augustus

An imperial legacy

By the end of his life in 14 CE, Augustus had established a new imperial system that would endure for centuries. For some years before his death, Augustus had prepared the way for an heir to succeed him, and retain control of the state. His step-son Tiberius was gradually granted powers until he could effectively be considered to be a co-emperor. This smoothed the transition of authority on Augustus' death, preventing a vacuum of power and ensuring continuity.

Augustus thus established the principle of direct succession and ensured the survival of the office of emperor. The system continued through multiple dynasties, with the empire reaching its height under the Nerva-Antonine dynasty when the emperor Hadrian ordered the building of a wall in northern Britain to mark the empire's outer limit.

The transition from republic to monarchy, while drastic, gave Rome a new stability. Masquerading as a democrat, Augustus created a new autocratic system of government, which, despite restricting political participation, was much better able to resist the compulsive upheavals that had plagued the Roman Republic a generation before. ∎

BY THIS SIGN CONQUER

THE BATTLE OF MILVIAN BRIDGE (312 CE)

IN CONTEXT

FOCUS
The spread of Christianity

BEFORE
33 CE Crucifixion of Jesus.

46–57 Missionary journeys of St. Paul the Apostle.

64–68 CE When a fire breaks out in Rome, Emperor Nero kills hundreds of Christians as scapegoats; martyrdom of saints Peter and Paul.

284–305 Diocletian and Galerius suppress Christianity throughout the empire.

AFTER
325 The first Council of Nicaea defines the nature of orthodox Christian belief.

c.340 Ulfilas, the "Moses of the Goths," begins to spread Arian Christianity to the Germanic tribes.

380 Christianity becomes the Roman Empire's official faith.

391 Pagan worship is banned in the Roman Empire.

In October 312 CE, Emperor Constantine I was stationed at the Milvian Bridge near Rome, waiting to join battle with Maxentius, his rival for control of the Western Roman Empire (in 285, the empire had been split into two halves, eastern and western, each ruled by an emperor and a deputy). Tradition says that in the days before the encounter, Constantine had a vision of a flaming cross in the heavens bearing the inscription *in hoc signo vinces* ("by this sign conquer"). This convinced him that he had the support of the god of the Christians, and this belief was upheld when his army went on to

Constantine I's adoption of Christianity after his victory at the Milvian Bridge gave the faith a huge boost: it rapidly gained more followers and began edging out the pagan cults.

defeat Maxentius's men. In fact, the Christian god was not the first deity Constantine had auditioned; an earlier version of his vision had involved the Greek and Roman god Apollo. He appears to have been looking for theological "back-up" to legitimize his ambition to become sole emperor, and a monotheistic supreme being may have seemed to him a good fit: a heavenly mirror image of his own position on Earth.

See also: The Sack of Rome 68–69 ▪ Belisarius retakes Rome 76–77 ▪ The crowning of Charlemagne 82–83 ▪ The Investiture Controversy 96–97 ▪ The fall of Jerusalem 106–07 ▪ The fall of Constantinople 138–41 ▪ Martin Luther's 95 theses 160–63

Despite the legend of his divine vision, Constantine's conversion to Christianity seems to have been gradual rather than immediate—he was not baptized until many years later, on his deathbed. However, soon after his victory at the Milvian Bridge, he began the process of rehabilitating, and then exalting, Christianity; in 313 CE, he issued the Edict of Milan, a proclamation that established religious toleration for Christianity within the empire.

A multi-faith empire

For almost 300 years after the life of Jesus Christ, the religion based on his teachings remained a minor sect within the Roman Empire, practiced alongside many other faiths, both mono- and polytheistic. Some aspects of Christianity, such as its egalitarian nature, made it suspicious to the imperial authorities however, and Christians were periodically persecuted.

All across the ancient world at this time, changing social, political, and economic conditions were reflected in cultural and religious changes; Christianity was just one of a number of monotheisms gaining popularity in the Roman Empire, including the Persian cult of Mithraism, with which it had much in common.

The rise of Christianity

In 324, after disposing of the emperor in the East, Constantine became sole ruler of the Roman Empire, and then sought to use Christianity as a unifying force across his diverse and fractious realm. To make the increasingly dominant eastern half easier to govern, he founded a new city called Constantinople (now Istanbul),

consecrating it with both Christian and pagan rites, but allowing only Christian churches to be built. Although it would take time for all Roman citizens to convert to Christianity, in Constantine's reign, the higher ranks of society, seeking political advancement and personal favor with the emperor, flocked to the Church, and the emperor built basilicas across the empire.

Christianity, however, was not a single, uniform religion at this time, and splits, or schisms, formed. In 325, Constantine convened the Council of Nicaea—the first universal council of the Christian Church—mainly to settle the Arian schism, a theological dispute over whether Jesus was of the same substance as God.

Rome is Christianized

In the mid-300s, Emperor Julian, an adherent of the old religion, tried to revive paganism, but it was too late: Christians had become a majority, at least in the East. The faith was increasingly bound up with empire, as the Roman state adopted and molded the Church into an instrument of social and political control, unity, and stability.

Under Emperor Theodosius I (reigned 379–395), pagan temples and cults were suppressed, heresy was outlawed, and Christianity became the official religion of the Roman Empire. Eventually, it also became the faith of the barbarian successor states in the Roman Western Empire, as well as of the Byzantine Empire in the East. Over the course of many centuries, the western (Catholic) and eastern (Orthodox) churches grew apart in doctrine and organization, but Christianity endured. ▪

Roman emperors derive authority and legitimacy from **pagan religions**.

Christianity's egalitarianism threatens to **disrupt the strict social order** of the Roman Empire.

Constantine sees Christianity, with its one supreme deity, as a **tool for unity**, and a **validation of imperial authority**.

After the Battle of Milvian Bridge, Constantine adopts Christianity. It later becomes the official religion of the Roman Empire.

The Church is refashioned in the image of the **Roman state**, with a **strict hierarchy** and **centralization of dogma**.

THE CITY WHICH HAD TAKEN THE WHOLE WORLD WAS ITSELF TAKEN

THE SACK OF ROME (410 CE)

IN CONTEXT

FOCUS
Nomad invasion

BEFORE
9 CE Germanic tribes secure their independence with victory at Teutoburg Forest.

285 The Roman Empire is divided into East and West.

372 The Huns defeat the Ostrogoths in Eastern Europe.

378 Visigoths destroy a Roman army and kill the emperor at the Battle of Adrianople.

402 The Western Roman capital moves to Ravenna.

AFTER
451 A Roman–German coalition defeats the Huns at the Battle of Chalons.

455 Vandal pirates loot Rome.

476 The last Western Roman emperor is deposed.

489 Theodoric of the Ostrogoths conquers Italy, with Byzantine consent.

Western Roman Empire declines in economic and military strength.

Steppe nomads are forced to migrate.

Imperial authority weakens and borders are **breached**.

Germanic **tribes are displaced by** migrating nomads.

Barbarian invasions begin, culminating in the Sack of Rome.

Germanic tribes create **new kingdoms in Western Europe**.

In 410 CE, Rome fell to an army of nomadic Germanic peoples—Visigoths—who pillaged the city over the course of three days. Although Rome had already ceased to be the capital of the Western Roman Empire and the destruction was relatively restrained, the sack sent shock waves across the world.

Changes known as the Migration Period, or the Barbarian Invasions, were then taking place, with great movements of peoples across all of Eurasia, from China to Britain. Barbarian peoples began to invade settled empires such as those of Rome and China from around 300 to 650. They carved out new

See also: The assassination of Julius Caesar 58–65 ▪ Belisarius retakes Rome 76–77 ▪ Clovis unites Gaul 71 ▪
The crowning of Charlemagne 82–83 ▪ Kublai Khan conquers the Song 102–03

The barbarian "other"

Barbarian was a Greek word signifying the unintelligible babble of those who did not speak Greek, and therefore could not be considered civilized. Romans adopted this "them-and-us" construction. However, by the 4th century, the boundaries between Rome and its barbarian neighbors were blurred, both culturally and geopolitically: the barbarians had become more like the Romans, and vice versa. The Roman army comprised mostly barbarians—either Germanic auxiliaries and mercenaries, or Roman citizens who were actually Gauls, Britons, or one of the hundreds of other groups and ethnicities. Nevertheless, much Roman culture survived the invasions. For instance, although much of Italy, Gaul, and Spain fell under the sway of the "Germanic" Goths, Suevi, and Vandals, their languages resisted Germanic influence and remain Romance languages—that is, languages that have evolved from the Latin spoken by Romans in Rome.

kingdoms, which in many cases gave rise to the nations of the modern era. Climatic changes in Central Asia drove the nomadic horse tribes of the steppes to seek better pastures, which in turn forced neighboring nomads to invade the so-called civilized empires. China was ravaged by the Xiongnu, Persia by the Hepthalites, and India by the White Huns.

Barbarians at the gates

In Europe, the arrival of the Huns in the lands east of the Rhine and north of the Danube displaced Germanic tribes who had long lived in delicate balance with the Roman Empire. The Visigoths moved into Roman lands, eventually storming Rome in 410, while other tribes including the Vandals, Suevi, Alans, Franks, Burgundians, and Alemanni invaded and settled territory from Gaul to Spain to North Africa. In the 440s the Huns, under Attila, ravaged Eastern Europe before being defeated by a coalition of Romans and Germans. The Western Roman Empire shrank to encompass little more than Italy itself, its puppet emperors controlled by barbarian generals. In 476, the last nominal emperor was deposed by one such general, Odoacer, marking the end of the Roman Empire in the west.

The Western Empire had however been in decline since at least the 3rd century. Its population and economy had diminished, making it increasingly financially dependent on the Eastern Empire; weakening central authority had given more autonomy to the provinces. The military, obliged to recruit from barbarian tribes, was losing its core strength. In reality, the Barbarian Invasions were probably part of a process: a transition, rather than a fall. Roman customs, culture, language, and particularly its religion in the form of Christianity, endured across the provinces, and many of the new ruling elite saw themselves as continuing in the tradition of Rome. The city itself survived sack by Alaric and his Visigoths, and by the Vandals in 455, and flourished under Theodoric the Ostrogoth (489–526).

In their turn, the successor states formed by Germanic tribes over the following centuries eventually found themselves under attack by further waves of invaders such as the Magyars and Vikings. ▪

In *Destruction* (c.1935) by Thomas Cole, invaders overrun a once-great city often likened to Rome. Citizens' bodies litter monuments that were built to celebrate the now fallen civilization.

FURTHER EVENTS

THE INDUS VALLEY CIVILIZATION COLLAPSES
c.1900–1700 BCE

The Indus Valley Civilization (c.3300–c.1700 BCE) was based around large cities with planned streets and impressive drainage and water-supply systems in what are now Pakistan and northwestern India. By 1900 BCE, this civilization was in decline and no longer producing the elaborate jewelry and fine seals for which it was famous. By c.1700 BCE, the great Indus cities of Harappa and Mohenjo-Daro were virtually empty. The reason is unclear, but the most likely explanation is a combination of crop failure and a decline in trade with Egypt and Mesopotamia. There is also some evidence of flooding due to a change in the course of the Indus River.

EMPEROR WU CLAIMS THE MANDATE OF HEAVEN
(1046 BCE)

The idea that the emperor of China rules with the approval of heaven dates to the Zhou dynasty, which was founded when Wu and his ally Jiang Ziya defeated the long-ruling Shang at the Battle of Muye in 1046 BCE. The Shang had presided over a long period of peace and prosperity but by the 1040s had become corrupt. The Zhou concept of the Mandate of Heaven aimed to prevent this from happening, placing good government above noble birth, and sanctioning others

to overthrow the ruler if he did not display these qualities. It influenced the way the Chinese regarded their rulers for thousands of years.

JUDAH DEFIES THE ASSYRIANS
(c.700 BCE)

In the 9th century BCE, the Hebrew state of Judah (west of the Dead Sea) was part of the large Assyrian empire. In the 8th century, the Judaean ruler Hezekiah refused to pay tribute to the Assyrians. The Assyrian king, Sennacherib, laid siege to Jerusalem (an event described in the Bible), but the Judaeans resisted their mighty enemies, who failed to take the city. Although this was a relatively small setback for Assyria, it was a triumph for the Judaeans, who attributed their victory to Yahweh. This was a major factor in the Hebrew peoples' adoption of monotheistic religion soon after.

CELTIC CULTURE FLOURISHES AT HALLSTATT
(c.650 BCE)

In the 8th century BCE, a distinctive culture developed around Hallstatt, southeast of modern Salzburg, in Austria. Hallstatt's people were Celts, probably originally from Russia, and by the time their civilization reached its highest point, around 650 BCE, it had spread west to eastern France, east into Romania, and north to Bohemia and Slovakia. Its people produced fine

tools and ornamental objects in bronze, but they were also among the first in Europe to use iron for items such as swords. Their striking bronze jewelry bore intricate patterns featuring spirals, knotwork, and animal designs, which had a lasting influence on later Celtic art.

THE PELOPONNESIAN WARS
(431–404 BCE)

The Peloponnesian Wars were fought between Athens (initially the most powerful Greek city-state and the center of classical civilization) and the more militaristic Sparta. Sparta first launched land-based attacks on Athens, while Athens used its superior sea power to suppress revolts along the coast. In 413 BCE, an attack on Syracuse, Sicily, went wrong, leaving most of the Athenian force destroyed. Then the Spartans, allied with Persia, supported rebellions in a number of Athenian subject states and finally wiped out the Athenian fleet at Aegospotam (405 BCE). The war deeply damaged Athens, ending the golden age of Greek culture and leaving Sparta dominant.

HANNIBAL INVADES ITALY
(218 BCE)

By the 3rd century BCE, Carthage, in Tunisia, had established itself as a major regional power, extending along the coast of North Africa before invading Spain in the 230s BCE. In 218 BCE, Hannibal, Carthage's commander in Spain, took his army across the Alps to attack Italy. Despite a series of victories in what became known as the Second Punic War, Hannibal could not take Rome itself and in 202 BCE, he returned to Africa. The Romans had proved their strength, put an end to the idea that Carthage was invincible in the Mediterranean, and paved the way for their own rise to power.

VERCINGETORIX IS DEFEATED AT ALESIA
(52 BCE)

In 52 BCE, the Gaulish chieftain Vercingetorix led a revolt of local tribes against the Roman conquest of Gaul (modern France). At the Battle of Alesia, in Burgundy (eastern France), Roman forces under Julius Caesar built an ingenious donut-shaped fortification around the town, blocking Vercingetorix inside while also creating a stronghold against Gaulish reinforcements. The chieftain was forced to surrender, and after five years in captivity he was strangled on Caesar's orders. The battle resulted in an all-embracing Roman Empire stretching right across Europe.

ROMANS OCCUPY BRITAIN
(43 CE)

In 43 CE, on the command of the emperor Claudius, a Roman invasion force landed in Britain. In spite of opposition from local chieftains such as Caratacus, and a later revolt by the Iceni tribe under their leader Boudica, Roman rule eventually extended across England to the Scottish border and into Wales. The Romans governed Britain until c.410, founding towns, developing a system of roads, and introducing such innovations as underfloor heating and the use of concrete for building. Many Britons benefited from Roman rule and from strong trading links with the empire in products such as metals and grain.

CHINA IS DIVIDED INTO THREE KINGDOMS
(220 CE)

The last years of the Han dynasty in China were marked by bitter divisions and fighting that led in 220 CE to the country being divided between three rival emperors, all claiming to be the rightful successors of the Han. These Three Kingdoms—the Wei in the north, the Wu in the south, and the Shu in the west—reached a fairly stable agreement over territory until fighting broke out from 263, when the rival Jin dynasty challenged and then conquered them. The wars had a devastating impact on the population.

THE MAYA CLASSICAL PERIOD BEGINS
(250 CE)

The Maya civilization reached its Classic phase in the 3rd century CE, with a large number of cities across Mexico and Guatemala that featured distinctive temples shaped as stepped pyramids, carved monuments inscribed with dates from the complex Maya calendar, and a large and extensive trade network. The largest city was Teotihaucan in central Mexico, although lowland cities such as Tikal were also powerful. Maya civilization left a lasting mark in North and Central America, its culture influencing later peoples, such as the Aztecs.

OBELISKS ARE ERECTED IN THE KINGDOM OF AXUM
(4th century CE)

In the 4th century CE, the people of the Ethiopian city of Axum erected tall stone obelisks that would be a feature of their civilization. Axum dominated the maritime trade routes around the Horn of Arabia and into the Indian Ocean, offering traders a vital link between Asia and the Mediterranean making the kingdom an impressive income. The obelisks are up to 108 ft (33 m) tall and are thought to be memorials to prominent people. They testify to the power of this early African kingdom and its development of a distinctive civilization. The obelisks have become symbols of enduring African culture.

CLOVIS UNITES GAUL
(late 5th century CE)

The end of Roman rule in Gaul (modern France) came about when Clovis, leader of the Salian Franks, defeated the Roman leader Syagrius in 486 CE. This victory, which added to those of Clovis' father Childeric, brought virtually all of France north of the Loire under the rule of his dynasty, called the Merovingians, after his grandfather Merovech. The Merovingians ruled France for some 300 years, making real the idea of a united France independent of outside rulers.

THE MED
WORLD
500–1492

EVAL

The army of the **Eastern Roman Empire**, led by Belisarius, **retakes Rome**, driving out the **Ostrogoths**.

The Abassid caliph al-Mansur's **founding of Baghdad** marks the start of the **Islamic golden age**. The city is a center of **Muslim scholarship**.

Frankish king **Charlemagne** is crowned **emperor** in Rome. As secular leader of **Christendom**, he **unites** much of Western Europe.

In **Cambodia**, work begins on the vast Hindu temple **Angkor Wat**, which becomes the world's **largest religious structure**.

536   **762**   **800**   **1120**

c.610   **793**   **1099**   **1192**

Muhammad announces that he has received a divine revelation and **founds Islam**. Within 20 years, the religion will come to **dominate** the Arabian peninsula.

Viking warriors mount a brutal **raid** on a monastery on the holy island of **Lindisfarne**, northern England—the first of many Viking raids.

Christian knights seize **Jerusalem** from the **Muslims**, and go on to found **crusader states** in Palestine and Syria.

Minamoto Yoritomo becomes **shogun**, establishing a line of **military rulers** who would govern Japan for **650 years**.

Historians call the period from 500 to 1500 "the Middle Ages," seeing it as a separate era sandwiched between the ancient world and modern times. In reality, there was never a clear break with the ancient world. In the eastern Mediterranean, the Roman Empire continued for almost 1,000 years after the fall of Rome, although it was rebranded by historians as the Byzantine Empire. The ancient tradition of a united China ruled by an emperor was revived in the 6th century and continued to the Ming dynasty, albeit with interruptions. Even in Western Europe, where the breakdown after the Roman Empire's collapse was most evident, Christian religion survived in Rome as the key marker for the distinction between what were considered "civilized" and "barbarian" societies.

The rise of Islam

The dominance of two mutually hostile monotheistic religions—Christianity and Islam—was the most distinctive characteristic of this period across much of Eurasia. The founding of Islam in the 7th century was a transformative event, and Arab armies inspired by the faith altered the political landscape: Muslim rule spread from Spain in the west to central Asia in the east.

Although a united Islamic caliphate could not be sustained, the religion ensured a continuity of civilization even when power shifted from the Arabs to other peoples such as the Turks. The great cities of the Muslim world surpassed any in Christendom in size and sophistication, and Muslim scholars preserved the science of the ancient Greeks and built upon it. Islamic civilization remained dynamic and expansive throughout the entire medieval period.

Western European fortunes

In Western Europe, civilization fell drastically from the level achieved under the Roman Empire. Warrior kings ruled over a thinly spread population sustained by subsistence agriculture, and the area remained prey to non-Christian raiders and invaders, such as the Vikings and the Magyars, into the 10th century.

A nostalgia for ancient Rome led to King Charlemagne being crowned emperor in 800, but the Holy Roman Empire, based on the tradition Charlemagne founded, failed to unify Western Europe politically. In the absence of strong centralized state systems, feudal relationships held societies together.

King John of England signs the **Magna Carta**, which asserts that all individuals, including the king, are **subject to the law** of the land.

Mansa Musa, the wealthy ruler of **Mali**, makes a high-profile *hajj* **to Mecca**, resulting in the spread of **Islam** in **West Africa**.

The **bubonic plague** arrives in Europe, probably originating in Asia. Within two years, it **kills over one-third** of Europe's population.

Korean king Sejong declares the creation of a new, simpler **alphabet** for the **Korean language**, to encourage literacy.

1215

1324

1347

1443

1275

1325

1368

1492

The Venetian merchant **Marco Polo** arrives at the court of **Kublai Khan**; the Mongol ruler will go on to **conquer southern China** four years later.

The **Aztecs** found their capital city Tenochtitlan in central **Mexico**. Meanwhile, the **Incas** establish a civilization in **Peru**.

Hongwu is proclaimed the first emperor of the **Ming dynasty**, having ousted the Yuan dynasty. Almost 300 years of **prosperity and stability** follow.

King Ferdinand and Queen Isabella of Spain **seize Granada**, ending 800 years of **Muslim rule** on the Iberian peninsula.

From the 11th century, a revival of Western European culture, trade, and urban life gathered pace. The "Medieval Warm Period" (950–1250), when Europe experienced above-average temperatures, improved yields from agriculture; it was also a time when great cathedrals and castles were constructed. But even when the Christian crusaders fought their way to Jerusalem at the heart of the Muslim world, the flow of civilization was the other way, with Islamic scholars far advanced in medicine, philosophy, astronomy, and geography.

Expansion and contraction

By the 13th century, the world's population is believed to have risen to around 400 million—double its total at the high point of the ancient empires. A wide-ranging network linked Europe to China and the thriving trading kingdoms of Asia, by land along the Silk Road and by sea across the Indian Ocean. Cairo and Venice both became wealthy cities as focal points at the western end of this trade.

However, civilized life remained precarious. The Mongols—nomadic warriors from the Asian steppes—seized major cities from the Middle East to southern China, carrying out large-scale massacres. Lethal diseases were also highly prevalent. Carried along the trade routes in the mid-14th century, the Black Death epidemic may have killed a quarter of the world's population.

Inventions and progress

Technological progress was slow but cumulatively substantial. As the world's most advanced country, China was the ultimate source of most inventions, from paper and printing to the magnetic compass and gunpowder. Even relatively backward Europe benefited from improvements in shipbuilding and metalworking, and the invention and spread of the plow and the windmill transformed agriculture.

By the end of the Middle Ages, Western European kingdoms had developed from "feudal" states, based on oaths of loyalty, to more stable and centralized states, able to channel their key resources into the large projects of colonization and exploration. In the Americas, meanwhile, civilizations such as the Aztecs and Incas continued to evolve independently, untouched by developments in Eurasia and Africa, until the Spanish conquistadors arrived in the 16th century. ■

SEEK TO ENLARGE THE EMPIRE AND MAKE IT MORE GLORIOUS
BELISARIUS RETAKES ROME (536 CE)

IN CONTEXT

FOCUS
The Byzantine Empire

BEFORE
476 CE Barbarian general Odoacer deposes the last emperor of the Western Roman Empire and rules as independent king in Italy.

493 CE Ostrogothic ruler Theoderic overthrows Odoacer and becomes king, notionally subject to Byzantine rule.

534 CE Byzantines end Vandal rule in North Africa.

AFTER
549 CE Byzantines recapture Rome from the Goths for the third and final time.

568 CE Lombards (a barbarian tribe) invade Italy and seize land that Justinian had recaptured for the Byzantines.

751 CE Lombards capture Ravenna—the last remaining major Byzantine holding in northern Italy.

On December 9, 536 CE, the army of the Eastern Roman (or Byzantine) Empire, led by general Belisarius, entered the city of Rome through the ancient Porta Asinaria gate. The Byzantine's arrival forced the rapid departure of the city's current defenders, the barbarian Ostrogoths, who were fleeing northward through the Porta Flaminia. Almost precisely 60 years after Italy had fallen out of imperial hands, it appeared that the empire's ancient birthplace might be restored to Roman rule.

The survival of Byzantium

While the Western Roman Empire finally fell in 476 after a century of barbarian invasions, the eastern portion—the Byzantine Empire, with its capital at Constantinople (modern Istanbul)—weathered the storm, its retention of rich provinces, such as Egypt, enabling it to mount a successful defense of its territory. However, the loss of the empire's birthplace was a blow to the prestige of the Byzantine emperors, who refused to accept it. In 488, the Emperor Zeno despatched one tribe of Germanic barbarian mercenaries, the Ostrogoths, to remove another, led by Odoacer, who had been responsible for deposing the last Western Roman emperor. In return the Ostrogoths would be allowed to rule Italy as subjects of the Byzantine Emperor. Furthermore, the Goths had been encroaching on imperial lands, and so Zeno hoped their removal to Italy would neutralize both problems.

The Gothic War

For the following 40 years, the Goths' rule of Italy was relatively untroubled. However, the accession of Justinian (c.482–565) as Byzantine emperor in 527 changed things.

To find money in Italy for the war is impossible, since the country has been largely reconquered by the enemy.
Belisarius, 545

See also: The Battle of Milvian Bridge 66–67 ▪ The Sack of Rome 68–69 ▪ The fall of Jerusalem 106–07 ▪ The Great Schism 132 ▪ The fall of Constantinople 138–41

Emperor Justinian, a man of great energy, set about an ambitious, wide-ranging program of expansion and reform in order to restore the Roman Empire to its former glory.

He was determined to restore Roman dignity, and this meant reconquering the lost Roman provinces. He began in 533 by despatching an army to North Africa under general Belisarius, who swiftly defeated the Vandals (a Germanic people who had ruled there since the 430s).

Emboldened by his success, Justinian ordered an invasion of Italy in 535. Belisarius's army made rapid progress and in 536 they successfully recaptured Rome. Byzantine euphoria at the recovery of their ancient capital was rudely shattered, however, when the Gothic King Witigis counterattacked and subjected Rome to a grinding, year-long siege.

Stalemate in Italy

Belisarius launched a fresh assault, but he was recalled after Justinian began to fear that he would set himself up as an independent king in Italy. The country passed back and forth between the two sides as the war in Italy dragged on for almost 20 years.

Twice the Goths retook Rome but, lacking the resources to hold it, lost it again both times to the Romans. Finally, the last major Gothic army was defeated in 552.

The impact of the war

Although the Byzantines had won the war, the victory was hollow. Italy was devastated—the cities had lost much of their population and the rural economy was in tatters. The traditional Latin-speaking ruling classes found that Greek-speakers from Constantinople were given all the key positions. Rome was treated as a provincial outpost of the Byzantine Empire, and hopes that the city might be restored as the center of imperial power were dashed.

The effects of the war, together with a plague that killed one-third of the empire's inhabitants in 542, made it hard to find troops that could garrison Italy. The new province provided little tax revenue and it became a major financial drain The optimism that greeted the capture of Rome was replaced by a profound gloom—a mood confirmed when in 568 the Lombards, another barbarian group, invaded Italy and took most of the Byzantine land in north and central Italy.

Although the Byzantine Empire survived a further nine centuries, it was never again able to make another serious attempt to restore the Roman Empire in the west. Instead, it focused on defending its Greek-speaking core in the east, leaving the Germanic kingdoms in Italy, France, and Spain free to develop unhindered. ▪

Tensions between Byzantine Empire and **unstable** Gothic Kingdom of Italy **grow**.

Byzantines invade Italy and capture Rome.

War **devastates Italy**, making it harder to raise **tax revenues** to fund its defense.

New **barbarian invasions** are able to penetrate borders of an empire **crippled by debt** and effects of the **plague**.

Byzantine **expansion** in the west **stops** and the empire **turns inward**.

TRUTH HAS COME AND FALSEHOOD HAS VANISHED

MUHAMMAD RECEIVES THE DIVINE REVELATION (c.610)

IN CONTEXT

FOCUS
The rise of Islam

BEFORE
c.550 CE Fall of the Himyarite Kingdom in southern Arabia.

570 Birth of Muhammad.

611 Persian Shah Khusrau conquers the Byzantines in Egypt, Palestine, and Syria.

AFTER
622 Muhammad and his followers flee Mecca and take up residence in Medina.

637 Muslim army captures Jerusalem after a siege.

640 Muslim general Amr ibn al-As conquers Egypt.

661 Umayyad caliphate established by Muawiya at Damascus, Syria.

711 Muslim armies cross into Spain and conquer the Christian Visigothic kingdom.

A round 610 CE, in a cave in the hills above the town of Mecca, central Arabia, Muhammad—a 40-year-old man from a merchant family—declared that he had received a divine message from the angel Gabriel. This was followed by similar revelations over the coming months and years and led to the founding of a new monotheistic religion: Islam. Within 20 years, this creed had come to dominate the Arabian peninsula, and a century later its followers had shattered the ancient Byzantine and Persian Empires, creating a state that stretched from Spain in the far west to Central Asia in the east.

See also: The founding of Baghdad 86–93 ▪ The fall of Jerusalem 106–07 ▪ Mansa Musa's *hajj* to Mecca 110–11 ▪ The Arab advance is halted at Tours 132 ▪ The fall of Constantinople 138–41 ▪ The conquests of Akbar the Great 170–71

In this 16th-century miniature The Kaaba, considered the house of God and the holiest shrine in Islam, is decorated by angels on the occasion of the Prophet Muhammad's birth.

Arabia before Islam

From the first millennium BCE there were sophisticated kingdoms in southern Arabia, which derived their wealth from the spice trade. In the early days, the trade routes ran along the northwest coast, but by the 7th century these had diminished as merchants increasingly used a maritime route up the Red Sea, leaving many places that had been relatively prosperous in decline. There were a few scattered towns, such as Medina (then known as Yathrib) and Mecca, which were dependent on more local trade in wool and leather, along with a few key imports such as grain and olive oil. The central desert regions of the Arabian peninsula were very poor: Bedouin tribes followed a nomadic lifestyle, and competition for scarce resources shaped a society in which primary loyalty was to a kinship group, or tribe.

At the time of Muhammad, Arabia was in a state of religious and political ferment. Strong Jewish communities had become established in Yemen in the south and in northwestern oasis towns, such as Medina, while Christianity had gained footholds in Yemen and eastern Arabia. Although monotheistic faiths were making inroads against the traditional polytheistic paganism of the Bedouin Arabs, paganism still remained strong. Conflict between tribes was also common, and in Mecca, in the sacred enclosure known as the *haram*, a truce was enforced so men of different tribes could trade freely without violence.

Muhammad in Mecca

The Meccan *haram* was controlled by the powerful Quraysh clan, of which Muhammad was a member. Muhammad's rejection of paganism, and his bold proclamation that there was but a single God, and that believers needed to follow a prescribed set of religious observances—including praying five times a day and fasting during Ramadan—set his followers apart. His preaching of a single religious community that cut across social boundaries was perceived as threatening by the traditional leaders, who felt it undermined the source of their authority.

The flight to Medina

By 622, the atmosphere in Mecca had become so tense that Muhammad and his handful of followers fled north to Medina— an event called the *hijra* (meaning emigration), which marked the real foundation of the Islamic community. The Medinans, who resented the power of the Mecca-based Qurayshi, were sympathetic to Muhammad's cause and allowed him to preach freely, giving him the opportunity to attract further converts.

The Qurayshi were not content to see Muhammad's powerbase grow in Mecca and within two »

Muhammad

The Prophet Muhammad was born in Mecca around 570 CE to a branch of the influential Quraysh tribe. Tradition relates that he was an orphan, whose first marriage to a wealthy widow named Khadijah secured his economic future. The religious revelations that were imparted to Muhammad over a period of over 20 years from about 610—and which would later be written down as the Qur'an—caused a rupture with the traditional Meccan elites when he began to preach against pagan polytheism and practices such as female infanticide. Muhammad's flight to Medina in 622 marked a key moment in the spread of Islam, as its acceptance outside Mecca showed that its appeal might transcend traditional kinship structures. Muhammad proved an inspirational leader, and his adept handling of the challenges facing the new religion meant that by the time of his death in 632, two years after imposing his rule on Mecca, its adherents had spread throughout Arabia.

The Battle of Uhud (in 625) was one of several bloody conflicts fought between the Muslims of Medina, led by Muhammad, and the larger Qurayshi army from Mecca.

Conquests beyond Arabia

Having secured their position, Muhammad's successors, in particular Umar (634–44), initiated campaigns of conquest further afield. They were fortunate in that profound changes had occurred on the northern fringes of Arabia. Between 602 and 628, the two long-established empires in the area—the Byzantines to the northwest and the Persian Sassanids to the northeast—had been engaged in a long, vicious war that ended in catastrophe for both parties. Their coffers had been drained by the costs of the conflict and some regions within their territories had been utterly devastated. Both sides had also become reliant on Arabs to defend their borders and small, semi-independent Arab states had emerged on the peripheries of the two empires.

Rapid defeat

The Arab armies that swept northward in the 630s faced far less resistance than they would have half a century earlier. Provinces fell easily as weakened garrisons and the doubtful loyalty of citizens undermined resistance. Although relatively small in number and lightly armed, the Arab armies were very mobile and did not need to defend fixed positions, giving them a huge advantage over their opponents. When they defeated the Byzantines at Yarmuk in 636, the whole edifice of imperial control in Palestine and Syria came crashing down. In the case of Persia, it took Arab generals just nine years to dismember the Sassanid Empire.

years violence had broken out between the established powers there and Muhammad's supporters. Muhammad outmaneuvered the Qurayshi, first by raiding their caravans, then defeating them in a pitched battle in 627, and finally negotiating the right to return to Mecca on a pilgrimage in 629. By the time he died in 632, Muhammad was re-established in Mecca, and his diplomatic and military successes in attracting other tribes to his cause had made his position unassailable. As his authority spread, so too did the reach of his religious message and the numbers of new Muslim converts.

After Muhammad's death, Islam entered a crisis and the fledgling religion might easily have been crushed. Tribes in the east broke away from the Muslim religious community (the *umma*) and declared allegiance to one of their commanders, while the Medinans were unhappy about the dominance of Meccans in the movement. The choice of Abu Bakr, Muhammad's father-in-law, as caliph (successor) overcame the division within the Islamic community and this decision, together with a series of successful military campaigns against the malcontents, enabled the *umma* to survive.

Islamic society

The newly conquered lands became part of an Islamic caliphate. Many of its inhabitants converted, while those who did not were tolerated if they were Christians, Jews, or Zoroastrians, provided they paid a special tax. Islam transformed the lands it absorbed in many ways. As well as sweeping away the old imperial structures, it imparted a new sense of religious community, often uniting the conquerors and the conquered. Islamic scholars resurrected the works of Greek philosophers and scientists that had languished forgotten for centuries, translating them into Arabic, and beautiful mosques began to adorn the towns. Areas that had been marginalized under the Byzantine or Sassanid Empires now found themselves at the heart of a new, vibrant civilization.

Success, however, brought its own problems for Islam. Acquiring lands far more urbanized than Arabia meant that the caliphs had to adapt from being warrior chiefs commanding a tight-knit group of followers, to monarchs ruling over a huge area with complex economies and societies. In addition to this, Muslims were initially in the minority, and not wholly united.

Recite in the name of your Lord who created, Created man from a blood-clot.
Qur'an (Surah 96)
The first words revealed to Muhammad (c.610 ce)

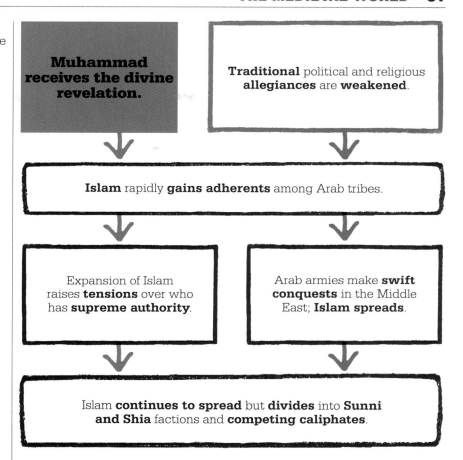

Muhammad receives the divine revelation.

Traditional political and religious **allegiances** are **weakened**.

Islam rapidly **gains adherents** among Arab tribes.

Expansion of Islam raises **tensions** over who has **supreme authority**.

Arab armies make **swift conquests** in the Middle East; **Islam spreads**.

Islam **continues to spread** but **divides** into **Sunni and Shia** factions and **competing caliphates**.

Growing divisions

Tensions over the succession to the caliphate resulted in a major schism in Islam. A struggle between Ali, Muhammad's son-in-law, and Muawiya, the Governor of Syria, led to a civil war that ended in Ali's murder and Muawiya taking control of the caliphate in 661. While Muawiya's descendants (the Umayyads) ruled from the Syrian city of Damascus, Ali's followers opposed their authority, claiming the caliph should be chosen from among Ali's offspring. After the murder of Ali's son Husayn at Karbala in 680, the split between the Shia (those who supported the right of Ali's descendents to rule the caliphate) and the more mainstream Sunni (who rejected this) became definitive—a division that continues to this day.

Islamic unity was fractured in other ways too; ruling over such a vast empire was almost impossible when messages from the eastern and western extremities might take months to reach the caliph's court. Independent Muslim dynasties emerged in peripheral areas and rival caliphs appeared in the 10th century in Spain, and Egypt. Yet even though its political unity had been shattered, and its religious unity compromised, Muhammad's creed was so popular and successful that by the 21st century there were about 1.5 billion Muslims worldwide. ■

A LEADER IN WHOSE SHADOW THE CHRISTIAN NATION IS AT PEACE

THE CROWNING OF CHARLEMAGNE (800)

On Christmas Day, 800 an extraordinary event took place at St. Peter's Basilica, Rome. Pope Leo III crowned the Frankish King Charlemagne with an imperial diadem, coronating the first emperor in the west for three centuries. The imperial crown established Charlemagne and his successors as secular rivals to the Papacy's claim (as spiritual head of the church) to a position of authority over western rulers. In due course, Charlemagne's empire (which later became known as the Holy Roman Empire) expanded to cover a vast area and laid the foundations for some of the future nation-states of Western Europe.

New rulers
In the half century before the Western Roman Empire finally collapsed in 476 CE, most of its

The **Roman Empire** in the West **collapses**.

Charlemagne **expands** the **Frankish state**.

Weak **pope seeks allies** outside Italy.

The pope crowns Charlemagne as emperor in Rome—the first in 300 years.

The notion of the **emperor** as the **secular leader of Christendom** allows the office to **survive divisions** of the **Frankish kingdom**.

See also: The Battle of Milvian Bridge 66–67 ▪ The Sack of Rome 68–69 ▪ Belisarius retakes Rome 76–77 ▪ The Investiture Controversy 96–97 ▪ The fall of Jerusalem 106–07 ▪ Martin Luther's 95 theses 160–63

He cultivated the liberal arts most studiously and, greatly respecting those who taught them, he granted them great honours.
Einhard
Frankish scholar and courtier
(c.770–840)

provinces were invaded by barbarian tribes who established smaller kingdoms on its former territory. At first, the Eastern Roman emperors did not recognize the legitimacy of these new kings' right to rule in nominally Roman territory. But as the new kingdoms, particularly that of the Franks, became stronger and more unified, Eastern Roman recognition ceased to matter.

From kingdom to empire

Charlemagne, who came to the Frankish throne in 768, expanded his dominions extensively over time, conquering northern Italy and Saxony, gaining some areas from the Arabs in northern Spain, and taking Avar territories in the Danube. He strengthened Frankish administration, establishing a network of *missi domenici*—royal agents who would enforce his will in the provinces. For the first time in centuries, a powerful ruler controlled most of the former Western Roman Empire's lands, turning them into a single political entity.

By contrast, the Papacy had experienced difficult times in the 8th century, snared in petty power politics as various Roman noble families sought to secure positions in the ecclesiastical hierarchy. After Leo was assaulted in Rome in 799, he fled across the Alps to seek help from Charlemagne, inviting him to bring order to Italy and restore the status of the church. A year later, Leo crowned Charlemagne, creating a Western emperor alongside the Eastern one.

Carolingian Renaissance

Charlemagne pushed forward his program of reforms, issuing an edict in 802 that required an oath of loyalty to be sworn and laying out the duties of his vassals. He also invited distinguished scholars to court, and encouraged academic disciplines that had languished since the collapse of the Roman Empire, including grammar, rhetoric, and astronomy. Music, literature, art, and architecture also flourished during his reign.

After Charlemagne's death, divisions were rife. The Frankish custom of dividing the kingdom between several heirs weakened central authority and led to civil wars; it also allowed the emergence of powerful landowners, who often challenged royal authority. Ultimately, the empire split into two main portions, which roughly equate to France and Germany today. The title of emperor was passed down to the immediate descendants of Charlemagne and then, from the 10th century, to more distantly related German princes. In this form, as the Holy Roman Empire, it was to survive to the early 19th century. ▪

Charlemagne

Charlemagne (c.747–814) was the eldest son of Pippin III, who in 751 deposed the last Merovingian king of the Franks and assumed the royal office himself. Energetic and visionary, Charlemagne greatly expanded the Frankish kingdom. He was also a very strong ruler, implementing reforms that enhanced the authority of the monarchy and the church. In addition, he reformed the kingdom's economy by introducing a new monetary system, standardizing weights and measures, and unifying an array of different currencies to encourage commerce and trade. His acquisition of the imperial title in 800 further consolidated his power, but at first he made no plans to pass it on. His first decision on the succession, in 806, divided the realm between three of his sons but made no mention of the office of emperor. However, the deaths of two of his sons led Charlemagne to bequeath his lands and title to a single heir—Louis the Pious.

THE RULER IS WEALTHY BUT THE STATE IS DESTROYED
THE AN LUSHAN REVOLT (756 CE)

IN CONTEXT

FOCUS
Tang China

BEFORE
618 Li Yuan becomes first emperor of the Tang dynasty.

632–35 Chinese armies capture Kashgar, Kokand, and Yarkand in Central Asia.

751 Tang armies are defeated by Arab forces at the Battle of Talas River (Kyrgyzstan).

AFTER
762 Luoyang is retaken by the Tang, and in 763 the last Yan emperor commits suicide, ending the An Lushan revolt.

874 The faction-torn Tang court is unable to resist the first of a series of revolts by over-taxed peasants.

907 The last Tang emperor is overthrown by rebel leader Zhu Wen, who founds the Later Liang dynasty.

960 China is reunified under the Song dynasty.

Protecting China's borders requires a **larger military**, leading to the rise of powerful army commanders and **increased taxation**.

Civil service reforms **reduce** the political power traditionally wielded by **noble families**.

Tensions and competition for power in the Tang court between aristocrats, bureaucrats, and military commanders leads to An Lushan revolt.

Tang authority is restored, but **central control is weakened**, eventually leading to the break up of China.

In 618, the Tang dynasty succeeded the Sui as rulers of China, ushering in one of the most glittering eras in the country's history. The early Tang emperors directed military campaigns that pushed China's frontiers deep into Central Asia, and established a centralized government with a highly competent bureaucracy to administer the empire. Later rulers presided over long periods of peace, relative political stability, and economic growth that triggered a cultural and artistic renaissance and technological innovation.

But in 755, this golden age was violently interrupted by An Lushan, a discontented army general who led an internal rebellion against the Tang that plunged northern China into a devastating war, after which the dynasty was never again fully in control of the country.

See also: The First Emperor unifies China 54–57 ▪ Kublai Khan conquers the Song 102–03 ▪
Marco Polo reaches Shangdu 104–05 ▪ Hongwu founds the Ming dynasty 120–27

The seeds of rebellion

Under Xuanzong (712–56), the Tang dynasty reached the zenith of its power and prestige, yet several key economic, social, and political issues threatened to destabilize it.

Firstly, the state was struggling to raise sufficient taxes to fund a sharp rise in military expenditure. The *fu-bing*, the cost-effective and self-supporting national militia system in which soldiers worked the land when not required for active military duty, was proving inadequate in the face of repeated invasions by neighboring groups. Xuanzong was forced to establish military provinces along China's northern frontiers, headed by local governors who commanded huge armies, and who came to acquire considerable power and autonomy.

The Tang's coffers were drained further by the failure of the "equal field" system, a program of land distribution and tax collection that protected small farmers from the depredations of wealthy landowners by periodically reallocating land to them. Its gradual demise enabled the nobility to grab land to increase their regional power bases, and led to unrest among the peasantry.

Lastly, earlier reforms made by Emperor Taizong (reigned 626–649) to the examination system used to recruit civil servants, which opened it up to able men from humbler backgrounds without connections, had created a bureaucracy based on merit that eroded the power and influence of the aristocracy. Xuanzong now had to manage rival factions in his court—potentially

Ten thousand houses with stabbed hearts emit the smoke of desolation.
Wang Wei
Tang poet (756)

An Lushan's rebels conquered and occupied Chang'an, but the general himself remained in Luoyang, where he was later assassinated by one of his sons in a dispute over the succession.

rebellious nobles, ambitious professional bureaucrats, and military governors, some of whom had begun to intervene in politics.

However, it was a series of military debacles that provided the spark for revolt against the Tang, including the defeat by Abbasid Arabs in 751 that halted China's expansion into Central Asia.

Turning on the Tang

Discontent exploded among the military, which saw its position threatened now that the era of conquest was over. An Lushan, a prominent military governor who had become a court favorite, rose up against his masters. Claiming that the emperor had asked him to remove Yang Guozhong (the court's chief minister, with whom An Lushan was engaged in an intense power struggle), he mobilized a rebel army and marched south.

At first the revolt looked set for success: it captured the eastern capital, Luoyang, early in 756— where An Lushan declared a rival dynasty, the Yan—before storming Chang'an, the primary capital. Xuanzong fled from his court, only just escaping An Lushan's clutches.

After eight years of war, the Tang finally crushed the revolt, but the effort had fatally weakened it. Over the next century it lost more political power to the military, and further rebellions broke out. By 907, the empire had fragmented into local dynasties and kingdoms that vied for power for 50 years. ▪

A SURGE IN SPIRIT

AND AN AWAKENING IN INTELLIGENCE

THE FOUNDING OF BAGHDAD (762)

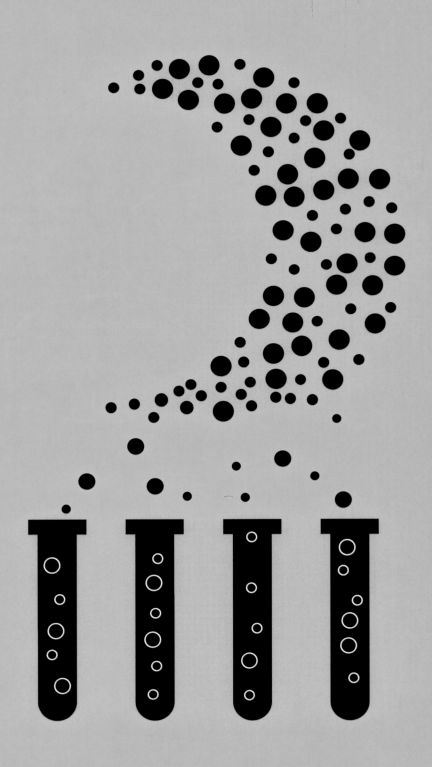

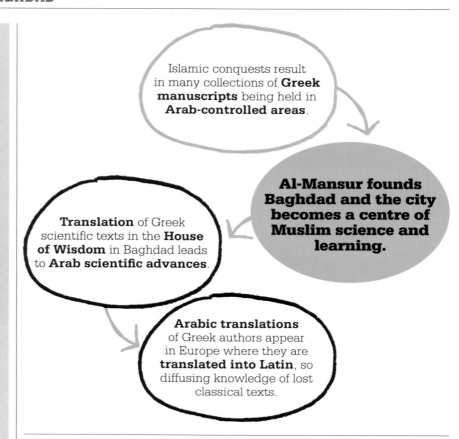

Islamic conquests result in many collections of **Greek manuscripts** being held in **Arab-controlled areas**.

Al-Mansur founds Baghdad and the city becomes a centre of Muslim science and learning.

Translation of Greek scientific texts in the **House of Wisdom** in Baghdad leads to **Arab scientific advances**.

Arabic translations of Greek authors appear in Europe where they are **translated into Latin**, so diffusing knowledge of lost classical texts.

I n 762 the second ruler of the newly ascendant Abbasid dynasty moved the capital of the powerful Islamic Caliphate from Damascus to the newly-founded city of Baghdad. The move is often seen as marking the beginning of an Islamic golden age in which science, art, and culture flourished. The extent of Muslim technological development was demonstrated in 802 when the Abbasid caliph Harun al-Rashid dispatched an embassy to the Frankish ruler, Charlemagne, which included the gift of a water clock that chimed the hours by dropping brass balls onto cymbals at the mechanism's base. This sophisticated timepiece was just one of the advances the Arabs had made—advances that left their European counterparts far behind.

The rise of the Abbasids
After the death of the Prophet Muhammad in 632, his successors ruled over a growing Islamic empire (or caliphate). Following the murder in 744 of the caliph al-Walid, a member of the Umayyad family that had ruled from Damascus since 661, civil war broke out, ending only when the Abbasid dynasty came to power in 750. The Abbasids spent their first decade pacifying the empire, with the help of troops from Khurasan in northeastern Iran. These troops, a mixture of Arab-speakers, Persians, and central Asians, had been among the Abbasids' principal backers and had provided them with a power base independent of the Arab tribes based in northern Arabia, Syria, and Iraq who had supported the Umayyads.

It was in part to provide land for his Khurasani soldiers that al-Mansur, the second Abbasid caliph, established the city of Baghdad in 762. He chose the site for its mild climate and its location on the trade routes between Persia, Arabia, and the Mediterranean. It was also just 20 miles to the southeast of the Persian royal seat at Ctesiphon, which it soon eclipsed, enabling the new dynasty to portray themselves as masters of a culture that stretched back to Cyrus the

See also: Siddhartha Gautama preaches Buddhism 40–41 ▪ The palace at Knossos 42–43 ▪ The conquests of Alexander the Great 52–53 ▪ Muhammad receives the divine revelation 78–81 ▪ Mansa Musa's Hajj to Mecca 110–11 ▪ The Arab advance is halted at Tours 132 ▪ The conquests of Akbar the Great 170–71 ▪

Great in the 6th century BCE. The heart of the new capital was a mile-wide, circular enclosure in which sat the caliphal palace and main government offices.

Search for knowledge

The Abbasids laid claim not only to their predecessors' political heritage, but also to their cultural and scientific achievements. Although the Umayyad Empire had included ancient seats of Greek learning such as Alexandria in Egypt, under their rule there had been little sponsorship of scientific endeavor. This changed under the Abbasids, who spent their time consolidating Islamic rule rather than on campaigns of conquest. They sponsored scholars to explore knowledge gained from foreign works, rather than relying solely on the guidance found in the Koran and the *hadiths* (the sayings of the Prophet Muhammad).

The earliest advances were made in medicine. During the mid- to late 6th century, a philosophical school at Gondeshapur in south-

In addition to his profound knowledge of logic and law [al-Mansur... was] very interested in philosophy and observational astronomy.
Said al-Andalusi
Islamic historian (c.1068)

western Iran became a center of medical scholarship. It was staffed mainly by Christians from the Nestorian sect, which had been persecuted in the Byzantine Empire. In 765, al-Mansur is said to have summoned staff member Jurjis ibn Jibril ibn Bukhtishu to Baghdad to diagnose a stomach complaint. So pleased was the caliph with his treatment that he prevailed upon Jurjis to stay on as

his personal physician, and for eight generations until the mid-11th century, members of the Bukhtishu family occupied the position at the Baghdad court, bringing with them knowledge of Greek and Hellenistic texts and medical practices. In 800, Caliph Harun al-Rashid asked Jibril ibn Bukhtishu, Jurjis's grandson, to head the new hospital in Baghdad, the first in the Islamic world.

Al-Mansur established a library in Baghdad to house his collection of manuscripts. This venture was made easier by the Arab adoption of paper as a medium for books, and the establishment in Baghdad in 795 of a paper mill. However, since Arabic speakers had no access to this learning, the library did little to advance an indigenous Arab scientific tradition.

House of Wisdom

To remedy this, Harun al-Rashid (caliph from 786 to 809) and al-Mamun (reigned 813–833) established the *Bayt al Hikma* (House of Wisdom), which not only housed the growing library, but »

Harun al-Rashid

Harun (763–809) succeeded as caliph in 783 after the mysterious death of his older brother al-Hadi, who had reigned for just one year. For the first 20 years of his reign, the Barmakid family, who helped strengthen a powerful central administration, dominated court. Under Harun's rule, Baghdad became the most powerful city in the Islamic world, and flourished as a center of knowledge, culture, invention, and trade. Even so, for almost two decades Harun based himself at Raqqa, closer to the frontiers of the Byzantine Empire, against which he launched a raid

in 806, personally commanding an army of many thousands. Harun's gift of an elephant to Charlemagne in 802 was part of a series of diplomatic exchanges with the Frankish court that were intended to put further pressure on the Byzantines.

Harun's House of Wisdom, a translation bureau, library, and academy for scholars and intellectuals from across the empire, contributed to his nickname al-Rashid ("the Just"). He died in 809 while on an expedition to Khurasan in the northeast of Iran.

also acted as an academy for scholars and a center for the translation of key scientific works into Arabic. Among its leading scholars were Hunayn ibn Ishaq (808–873), a Nestorian Christian from al-Hira in Iraq, who translated more than 100 mostly medical and philosophical works; and Thabit ibn Qurra, a member of a pagan sect known as the Sabaeans, who translated *Elements*, Euclid's great work on geometry, and the *Almagest*, Ptolemy's key work on astronomy.

Translation became a highly prestigious endeavor. One Arab patron paid an extravagant 2,000 dinars a month to ensure his association with a translation of a work by the Greek physician Galen (a dinar, made of pure gold, weighed the same as 72 grains of barley). Within around 150 years, almost all of the key Greek texts

that had been discovered had been rendered into Arabic. Many of them were not available in Western Europe at all, and even if they had been, knowledge of Greek had all but disappeared there. The Muslim world was therefore well set by around 850 to build on the scientific traditions of Classical and Hellenistic Greeks transmitted and developed under the Roman Empire—and to acquire a centuries-long lead over Christian Western Europeans.

Complex calculations

An understanding of mathematics and astronomy is essential to the calculation of the times at which Muslims must observe their five daily prayers (times that varied widely across the vast Islamic Empire), therefore both disciplines were studied assiduously. Another, separate, intellectual tradition contributed to the development

Jews and Christians... translate these scientific books and attribute them to their own people... when they are indeed Muslim works.
Muhammad ibn Ahmad Ibn Abdun
Legal scholar (early 12th century)

of these calculation techniques, arriving in 771 with a delegation of Hindu scholars. The scholars were visiting al-Mansur's court (which in itself illustrates the comparative openness and tolerance of the

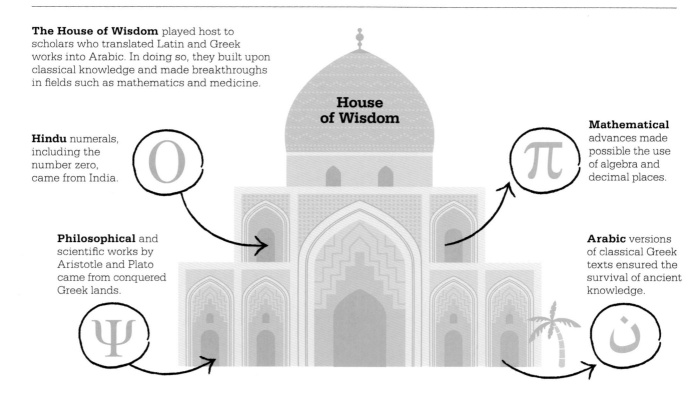

The House of Wisdom played host to scholars who translated Latin and Greek works into Arabic. In doing so, they built upon classical knowledge and made breakthroughs in fields such as mathematics and medicine.

House of Wisdom

Hindu numerals, including the number zero, came from India.

Mathematical advances made possible the use of algebra and decimal places.

Philosophical and scientific works by Aristotle and Plato came from conquered Greek lands.

Arabic versions of classical Greek texts ensured the survival of ancient knowledge.

The *Canon of Medicine* by ibn Sina or Avicenna (980–1037) set the standard for medicine in the Islamic world and medieval Europe, and remained an authority for centuries.

early Abbasids), and brought with them India's relatively advanced mathematics, including the use of trigonometry to help solve algebraic equations. Crucially, the Hindu mathematicians also employed a decimal notation, which one of the members of the House of Wisdom, Al-Khwarizmi (c.780–830), adopted and described in *The Book of Addition and Subtraction According to Hindu Calculation*.

Furthermore, Al-Khwarizmi also explained a method of calculating the square roots of numbers, and pioneered work on algebraic equations. He and his fellow scholars made rapid strides in geometry, taking as their starting point Euclid's and Archimedes's work on spheres and cylinders.

Astronomy and medicine

Al-Khwarizmi compiled the first known tables of daily prayer times at Baghdad, his calculations assisted by direct astronomical observation. The early Islamic astronomers drew from Ptolemy's *Almagest*, adopting his view that the Earth was at the center of the solar system, and that the planets rotated around it along the lines of eight spheres. They also learned from Hindu astronomers, translating and perfecting Indian *zij*, or tables of planetary positions, and continued to refine Ptolemy's system, only occasionally (as in the work of the 10th-century astronomer al-Biruni) toying with a heliocentric system that had the sun at its center. Their calculations were made simpler when in the mid-eighth century they adopted the astrolabe, an instrument in which the celestial sphere was projected onto a flat plane marked with latitude and longitude lines.

By the 13th century, Islamic astronomy was at its zenith, and in 1259 a great observatory was constructed at Maragha in eastern Iran. Here Nasr al-Din al-Tusi and his successors made fine adjustments to account for slight discrepancies in the orbit of the planets, assisted by mechanical clocks that enabled them to record their observations in fine detail. Muslim scholars made advances in many other areas, too, first building on the base of Greek manuscripts translated into Arabic, and then making their own discoveries. They did not accept the theories of the ancients uncritically: al-Haythem (died 1039) produced a key work, the *Book of Optics*, in which he speculated that sight was the result of light traveling from an object to the eye, rather than the other way around as Ptolemy had theorized. Arab physicians continued to make progress, combining their practical observations with theoretical analysis. Al-Razi (died 925) produced the first description of smallpox and measles, as well as compiling a medical compendium that began a tradition of such encyclopedias, culminating in the *Canon of Medicine* by ibn Sina (who was known as Avicenna in the West). Composed around 1015, it included separate sections for diseases that are specific to one body part, and those that afflict the body as a whole.

Islamic science spreads

The Islamic expansion that began in the mid-7th century not only absorbed ancient centers of learning such as Alexandria, but also brought the Muslim world to the fringes of Western Europe through the conquest of Spain (from 711) and Sicily (from 827). A tradition of Islamic learning embedded itself in both areas, and particularly in the Iberian Peninsula, known to the Arabs as al-Andalus. The court established there in 756 by Abd ar-Rahman I, »

The ancient Greek thinker Aristotle teaches Muslim students how to measure the positions of the Sun, Moon, and stars in this imagined scene from an Arabic manuscript.

a refugee Umayyad prince who had escaped the Abbasid revolution, became a magnet for scholars from the East, and its libraries became a repository of precious ancient texts that had been translated into Arabic.

In 967, the French cleric and scholar Gerbert of Aurillac (who in 999 would become Pope Sylvester II) arrived in Spain for a three-year period of study at a monastery in Catalonia. There he had access to manuscripts that had filtered over the border from Muslim-held al-Andalus. He took back to France knowledge of Arabic technology such as the water clock and the

astrolabe, and of a type of abacus that used a decimal system. This was the first example of the system's use in medieval Europe. It was a small beginning, and one paralleled in southern Italy where a medical school was established at Salerno in the 9th century. A few Islamic manuscripts reached the school in the early years, but many more arrived in the late 11th century when Muslim doctor Constantine the African returned from Qairawan in Tunisia. He had gone there to study medicine, and brought back with him works such as the *Complete Art of Medicine* by Ali ibn al-Abbas al-Majusi (known in the West as Haly Abbas), parts of which he then translated into Latin. This translation gave Western doctors and scholars access to comparatively advanced Muslim medical knowledge.

Classical Greek texts arrived directly from the Byzantine Empire to the West (in particular Pisa, which had a trading quarter in Constantinople), including works by the philosopher Aristotle. The main channel for the transmission of Islamic learning into Europe, however, continued to be Spain. As Islamic Spain shrank, pressurized by the Reconquista, the flow of materials accelerated. The Christian reconquest spread increasingly into Muslim emirates until, in 1085, Alfonso VI of Castile captured Toledo. The city became a center for the translation of Arabic works by an international group including the Englishman Herbert of Ketton, Slav Hermann of Carinthia, the Frenchman Raymond of Marseilles, Jewish scholar Abraham ibn Ezra, and Italian Gerhard of Cremona. In the mid-12th century, the group translated many Arabic texts into Latin, including works on mathematics, medicine, and philosophy. Western Europe now had access to Ptolemy's *Almagest*, and to the medical works of Galen, as well as access to new works by Arabic writers who had built on or summarized the work of their ancient predecessors, such as ibn Sina's *Canon of Medicine*. This five-book encyclopedia became one of the most widely used treatises in European medical schools until the 16th century.

Royal patronage

This transmission of knowledge to the West mirrored the process by which the Islamic world had absorbed Greek learning during the great period of translation into Arabic in the 9th and 10th centuries. Noble and royal patrons played similar roles in both phases of the transmission. King Roger II of Sicily (which by 1091 had been reconquered from the Muslims) invited Arab scholar al-Idrisi to his court in 1138 with a commission to construct a map of the world based on Islamic geographical and cartographic works. The result, which took more than 15 years to complete, was by far the most accurate world map yet available to Europeans, and showed areas as far east as Korea. The map was accompanied by the *Book of Pleasant Journeys into Faraway Lands*, in which al-Idrisi's royal patron could have read of wondrous things such as cannibals in Borneo, and the gold trade in Ghana.

A tradition of learning

Roger's grandson Frederick II, Holy Roman Emperor from 1220 until 1250, continued his grandfather's tradition of sponsoring translations of Arabic texts. A remarkable polymath who knew at least four languages, Frederick so impressed his contemporaries with his learning that he became known as Stupor Mundi ("the Marvel of the World"). Among his protégés were the Scottish scholar Michael Scot, who translated key works of Aristotle on zoology, and the Pisan Leonardo Fibonacci, who had been sent by his merchant family to study mathematics at Bougie in Muslim North Africa. There Fibonacci learned of the decimal system, and in 1202 he published the *Book of Calculations*, the most detailed account yet seen in Europe of the Arabic system of numbering.

By the early 13th century, the Abbasid Empire had all but collapsed. The difficulties of ruling such a far-flung empire and the effects of a series of civil wars had led to key provinces such as Spain, Tunisia, and Egypt breaking away to make themselves independent. Even in Baghdad, where the Abbasid caliphs clung on, they were only notionally sovereign. Real power was held by other dynasties such

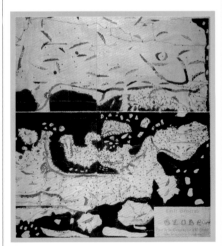

Roger II invited scholar al-Idrisi to create an accurate map of the known world in 1138. Al-Idrisi presented the planisphere, and an accompanying book, to his patron in 1154.

[Roger II of Sicily] is responsible for singular innovations and for marvelous inventions, such as no prince has ever before realized.
Al-Idrisi, c.1138

as the Shia Buyids, and, from 1055, the Seljuqs, a Turkish group originating in central Asia. The final blow was dealt by the Mongols, who surged westward into the Islamic world in the early 13th century. In 1258, the Mongol Great Khan Möngke unleashed an army against Iraq, which laid siege to and then sacked Baghdad, inflicting an appalling massacre on its inhabitants. The last ruling Abbasid caliph al-Musta'sim was executed, and political and cultural leadership of the Islamic world passed first to the Mamluks in Cairo and then, after their conquest of Egypt in 1517, to the Ottoman Turks.

By this time Europeans had rediscovered Greek and Roman learning in almost every field of scholarship through the medium of Arabic texts. It had taken centuries for the new material to be absorbed, and a further wave of interest in classical manuscripts in the 15th century to spark the Renaissance in Europe. The House of Wisdom founded by the Abbasid caliphs had played a key role in ensuring the survival of Greek and Roman science in the Islamic world, allowing its transmission centuries later to Christian Europe. ■

NEVER BEFORE HAS SUCH A TERROR APPEARED IN BRITAIN

THE VIKING RAID ON LINDISFARNE (793)

IN CONTEXT

FOCUS
Viking raiders

BEFORE
550–750 In Sweden, the Vendel period is a time of increased prosperity.

737 Construction of Danevirke fortifications in Denmark shows growing royal authority.

AFTER
841 Vikings establish a permanent settlement in Ireland, which will grow into the city of Dublin.

845 Viking raiders advance along the Seine and sack Paris.

867 Danish Vikings take control of Northumbria in northeast England.

911 Vikings found Duchy of Normandy in northern France.

10th century Swedish "Rus" Vikings are dominant in Kiev and Novgorod in Russia.

O n a calm June day in 793, a party of men landed on the shore of the holy island of Lindisfarne in northern England, and mounted a ferocious attack on its monastery. The invaders murdered some of the monks, dragged others away into slavery, and plundered the church's treasures before slipping away.

This unexpected assault was the first recorded raid by Vikings—pagan seafaring warriors from Denmark, Norway, and Sweden—and news of it sent waves of horror and fear across Christian Europe. Over the next 200 years, Vikings would ravage and loot settlements across large parts of the continent. But they were also colonists and traders with a sophisticated artistic culture who left a lasting imprint on the places they invaded and settled.

An unstoppable force

Within six years of the Lindisfarne raid, bands of Vikings—or "Danes" as they were known in Anglo-Saxon England—were targeting the wealth of other Christian sites in England, Scotland, Ireland, and France. Key to the success of these missions was the Viking longship, a slender vessel with a shallow bottom that enabled its crew to sail far up waterways and alight stealthily on shores. Each ship could carry up to 80 warriors, recruited by a warlord whose authority depended on his military prowess and his success in capturing booty for his followers.

No single motive drove the Vikings to venture across the sea. In parts of Scandinavia, population growth may have forced young men into a piratical lifestyle; in others, perhaps the increasing strength of local clan leaders sparked power struggles that pushed the losers into exile. And the newly rich trading towns in northern Europe

The ravages of heathen men miserably destroyed God's church on Lindisfarne with plunder and slaughter.
Anglo-Saxon Chronicle

See also: The Sack of Rome 68–69 ▪ Belisarius retakes Rome 76–77 ▪ The crowning of Charlemagne 82–83 ▪ Alfred rules Wessex 132 ▪ Christopher Columbus reaches America 142–47

were irresistible targets for a warrior society in which a reputation for valiant deeds was a great asset.

Conquest and settlement

As the Vikings' raiding parties grew in size, many of the men started to settle in the territories they invaded, including those in Britain and France. In the late 9th century, England was divided into a number of kingdoms that offered no coherent resistance to the Viking challenge, while France was consumed by civil war.

This disunited opposition helped the Vikings to conquer northern and central England—where they established a kingdom that lasted almost 100 years—and to occupy land in northern France, where their descendants became French-speaking Normans. In the east, Vikings traded and raided along Russia's rivers, which brought them silver from the Islamic world and contact with the Byzantine Empire.

By the 11th century, most of the Scandinavian kingdoms had adopted Christianity, and turned

Population **pressure** and **political instability** in Scandinavia.

News of **rich targets** across the North Sea attracts **rootless young men** to war leaders.

Attack on Lindisfarne Monastery.

Success of Lindisfarne attack **attracts more warriors** to join new raids.

Raids lead to **permanent** Viking settlements.

from raiding and pillaging to more organized settlement and conquest. Cnut of Denmark created a Viking North Sea empire that included Denmark, Norway, and England. Yet it did not survive his death, and in 1066, an unsuccessful attempt to claim the English throne by the Norwegian King Harald Hardrada, was the final flourish of the Viking age that began with the sack of Lindisfarne. ▪

Viking expansion in the North Atlantic

The Vikings used their knowledge of winds and currents to navigate the seas and discover new lands. Around 800, they colonized the Faroe Islands, and used them as a stepping stone to explore the North Atlantic. By the 870s, their ships had reached Iceland, where settlers founded a colony that grew politically independent.

In 982, Erik the Red, exiled from Iceland for murder, stumbled upon Greenland and established a new colony there. A Norse saga tells how, 18 years later, Erik's son, Leif Eriksson, was driven off course at sea and landed in a region teeming with hardwood forests and wild grapes that he called Vinland (Land of Wine).

Subsequent expeditions to this area, which is located in what is now Newfoundland in eastern Canada, led to a tiny Viking colony, but this was abandoned after attacks by hostile indigenous people. Nevertheless, Leif and his crew had been the first Europeans to set foot on North American soil.

The Vikings were among the most skilled shipbuilders, sailors, and navigators in the Western world of the early Middle Ages.

THE ROMAN CHURCH HAS NEVER ERRED

THE INVESTITURE CONTROVERSY (1077)

IN CONTEXT

FOCUS
The medieval church and the papacy

BEFORE
1048–1053 Pope Leo IX issues decrees against simony and priestly marriage, beginning the reform movement.

1059 A college of cardinals to elect new popes is established.

1075 The Lateran Council decrees that only the pope can appoint bishops.

1076 Gregory VII deposes and excommunicates Henry IV.

AFTER
1084 Henry IV captures Rome, forcing Gregory VII to flee to southern Italy.

1095 The pope calls a Crusade, asserting papal leadership over Christendom.

1122 In the Concordat of Worms, Henry V gives up almost all rights to invest bishops.

Laxness in observing church rules on clerical marriage and investiture of bishops **leads to calls for reform**.

Gregory VII promotes reform, including a **ban on lay investiture**.

Emperor and pope clash over investiture; the emperor is excommunicated.

The **pope's victory** in the investiture struggle **strengthens the reform movement** and papal administration.

For three days in 1077, the Holy Roman Emperor Henry IV stood penitent, barefoot in the snow outside the Italian fortress of Canossa, begging Pope Gregory VII for absolution. This event was the culmination of the Investiture Controversy, a struggle between the two men about the extent of secular authority over the Christian church, and the authority to appoint—or invest—bishops.

Both king and pope were rulers of particular domains, but they also had rival symbolic claims to lead all Christendom. An emperor had to be crowned by the pope before he assumed the imperial title. Pope Gregory VII asserted that the pope's authority was supreme in spiritual matters, and that even in secular affairs it stood far above that of worldly princes.

When at last Gregory signaled forgiveness of the penitent emperor, it marked a bitter blow for imperial prestige and a huge triumph for the independence of the church.

The state of the church

By the early 11th century, the papacy was at a low ebb. It had failed to impose—or had lost—authority over national churches outside Italy,

See also: The Battle of Milvian Bridge 66–67 ▪ The crowning of Charlemagne 82–83 ▪ The fall of Jerusalem 106–07 ▪ Otto I becomes Holy Roman Emperor 132 ▪ Martin Luther's 95 theses 160–63

Henry was refused entry when at last his long trek across the Alps brought him to the castle gates. Only after three days' penitence was the emperor's excommunication lifted.

and monarchs were appointing their own bishops, especially in Germany where the office often came with considerable territorial domains. The feeling that the church had lost touch with its roots was also widespread: monasteries had become storehouses of treasure, bishops were ruling their lands like secular lords, and clerical offices were openly sold. Itinerant preachers started to inveigh against these betrayals, and calls for reform were beginning to be heard from within the church itself.

Gregory vigorously promoted papal authority, and in 1075 a church council declared that only the pope had the power to appoint bishops or move them to a different area. Henry, facing the loss of authority over large tracts of Germany, continued to appoint bishops and called for the pope to step down. Gregory retaliated by excommunicating the king and declaring him deposed. German nobles, already feeling discontented at Henry's attempts to centralize power, felt that this released them from their oath of loyalty, and many rose up in revolt. Caught between the papacy and the nobles, Henry eventually chose to take the road to Canossa in a humiliating retreat.

Final agreement at Worms

But Henry's submission did not last. The issue of investiture was not explicitly settled and underlying dispute caused partisans of pope and emperor to clash repeatedly until 1122, when Henry's son Henry V agreed to the Concordat of Worms. Squeezed between an

> I Henry, by the grace of God august emperor of the Romans… do remit to the holy Catholic church, all investiture through ring and staff and grant that in all the churches there may be free election and consecration.
> **Henry V, 1122**

increasingly assertive insistence on papal supremacy, and the growing independence of the German nobles, the emperor conceded virtually all investiture rights.

Energized by its success, the papal administration (or *curia*) consolidated. A growing thirst for education led to the foundation of universities such as that at Bologna where many students studied canon law. With rising confidence, popes ruthlessly persecuted heretics and swept away lax practices.

The reforms strengthened the church, whose diplomatic stature grew to equal that of any monarch, and it survived in a united form until the Reformation in the 16th century. The blow to the prestige of the Holy Roman Emperors was commensurate. Secular lords seized the opportunity to magnify their own power, fragmenting the empire into a constellation of lordships and competing authorities who paid only lip service to the emperor. ▪

The new monasticism

By the 11th century, many felt that monastic orders had also strayed from their original mission, accumulating wealth and abandoning spirituality. Men such as Bruno of Cologne led calls to return to a purer form of monasticism. Bruno joined a group of hermits near Grenoble in 1084. Their way of life attracted others to found similar groups, which became the core of the Cistercian Order. The Carthusians, established in 1098, had by 1153 nearly

350 houses, yet these enclosed orders did not fully answer the spiritual needs of a society that was becoming increasingly affluent, educated, and mobile. A new wave of mendicant friars appeared in the 13th century: committed to a life of poverty, they traveled and preached among the people. The Franciscans, founded in 1209 by Francis of Assisi, and the Dominicans, established in 1216 by Dominic de Guzman, represented the most successful exponents of this new apostolic form of monastic life.

A MAN DESTINED TO BECOME MASTER OF THE STATE

MINAMOTO YORITOMO BECOMES SHOGUN (1192)

When in 1192 the Japanese clan leader Minamoto Yoritomo became the military commander-in-chief, or shogun, it marked the ascent to power of a Japanese military class, the samurai, and established a line of military rulers who would govern Japan for the next 750 years.

The Japanese imperial court had been dominated since the mid-7th century by regents from the Fujiwara family, who had reduced the emperors to mere figureheads.

At the time of the Gempei Wars the Samurai fought as mounted bowmen, but by the 15th century the sword, in particular the long-bladed *katana*, had become their principal weapon.

The situation became entrenched after the capital moved (following the emperor) to Kyoto in 794. Non-Fujiwara nobles were denied preferment at court, so sought positions in the provinces. The gulf widened between the Kyoto-based bureaucrats and the regional nobility, the samurai, who assumed

See also: The An Lushan revolt 84–85 ▪ Kublai Khan conquers the Song 102–03 ▪ The Mongol invasions of Japan are repulsed 133 ▪ The Battle of Sekigahara 184–85 ▪ The Meiji Restoration 252–53

a dominant role in local government. The Kyoto court appointed the most talented samurai as governors (*zuryō*), both to bind them to the imperial government and to prevent them from building their own power bases. However, the samurai developed loyalty to their extended family, or clan, and its leader rather than to the emperor, and fought one another from their power bases in the provinces. The Minamoto and Taira clans engaged in a series of these struggles which culminated in the Gempei War, during which the Taira were utterly crushed.

The shogunate

Following his victory, clan leader Minamoto Yoritomo established a parallel government based at Kamakura, about 250 miles east of Kyoto. Other clan chieftains became his vassals or *gokenin*, and he dispatched military estate governors to cement his control over the provinces. In 1192, Yoritomo accepted from the emperor the title of shogun, becoming the de facto military ruler of Japan.

Over the following centuries, the emperors made periodic vain attempts to reassert authority over the shogunate, but the shoguns in turn could not maintain control of the samurai and their warlords, who controlled their areas and fought among themselves. Japan dissolved into a patchwork of military warlords or *daimyo*, each with its own power base and retinue of samurai warriors.

Establishment of the office of shogun, which had seemed to offer Japan stability in 1192, ultimately led to the Sengoku, a civil war lasting almost 150 years. This war ended with the reunification of Japan under the new shogunate of Tokugawa in 1603. ▪

Minamoto Yoritomo

A descendant of the royal emperor Seiwa, Yoritomo was the heir of the Minamoto clan, which had been crushed by the Taira clan after a civil war in 1159. After the war, the now orphaned Yoritomo was exiled to Hirugashima, an island in Izu province. Here he remained for 20 years before issuing a call to arms and rising up against the Taira. He established a headquarters in Kamakura, from which he began to organize the warlords and samurai into an independent government.

A decisive victory over the Tiara in 1185 sealed Yoritomo's military success, and he emerged the undisputed leader of Japan.

Yoritomo developed policies to relieve the strain between the military lords and the court aristocrats, and set up an administrative network that soon took over as the central government, but much of the remainder of his life was spent in suppressing those clans who had not accepted Minamoto dominance.

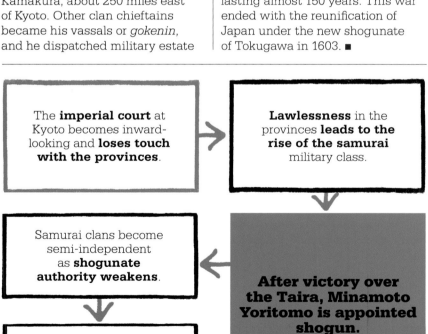

The **imperial court** at Kyoto becomes inward-looking and **loses touch with the provinces**.

Lawlessness in the provinces **leads to the rise of the samurai** military class.

Samurai clans become semi-independent as **shogunate authority weakens**.

After victory over the Taira, Minamoto Yoritomo is appointed shogun.

The shogunate collapses and **power devolves to the *daimyo***.

THAT MEN IN OUR KINGDOM SHALL HAVE AND KEEP ALL THESE LIBERTIES RIGHTS AND CONCESSIONS

THE SIGNING OF THE MAGNA CARTA (1215)

IN CONTEXT

FOCUS
**The development
of subjects' rights**

BEFORE
1100 Henry I's Coronation
Charter promises to abolish
unjust oppression.

1166 The Assize of Clarendon
extends the power of royal
justice at the expense of
baronial courts.

1214 Normandy is lost at the
Battle of Bouvines; barons
bridle at the campaign's cost.

AFTER
1216 The Magna Carta is
reissued on the accession of
Henry III, and again in 1225
in exchange for a tax grant.

1297 The Magna Carta is
again confirmed and written
into statute law by Edward I.

1970 A bill repealing ancient
statute laws leaves untouched
four chapters of the Magna
Carta, including Chapter 39.

On June 15, 1215, King
John of England signed
a charter at Runnymede,
a meadow beside the Thames.
Designed to make peace between
the king and a group of rebel barons,
the Magna Carta, as a form of the
document became known, at first
seemed ineffectual. However, its
assertion of the rights of subjects
against arbitrary actions of the
Crown—the essential principle
of the rule of law—provided a
blueprint which, more than eight
centuries later, is still viewed as a
fundamental guarantee of rights
in the US and elsewhere.

Feudal society

When King John acceded in 1199,
England was a feudal society, a
land-based hierarchy headed by the
king, who owned all the land. The
tenants-in-chief (or barons) received
land from the king in exchange for
loyalty and military service. They
in turn leased the land to their own
armed retainers, who leased to
peasants, or villeins. Yet monarchs,
especially in England, were levying
an ever-increasing series of taxes
and additional financial burdens
on their barons. English kings from
Henry I (1100–1135) onward also

The Magna Carta included clauses
relating to royal forests: the barons
aimed to limit the king's rights under
England's Forest Law, regulate forest
boundaries, and investigate officials.

sought to centralize administration,
partly by establishing a series of
royal courts. These royal courts
raised revenue for the Crown
through fines and charges—but at
the expense of the barons, who had
previously raised those funds from
their own local tribunals.

The exactions of King John

The barons' discontent at these
growing demands intensified under
King John. Ruinously expensive
campaigns against the French in
1200–04 had already resulted in the
loss of Normandy (and earned the

See also: The Norman conquest of England 132 ▪ The Battle of Castillon 156–57 ▪ The execution of Charles I 174–75 ▪ The signing of the Declaration of Independence 204–07 ▪ The storming of the Bastille 208–13

king the mocking nickname "Lackland"). Scutage, a further cash levy that left many barons in debt to money-lenders, was bitterly resented. Not only was the king proving lamentably unsuccessful in war, but he had also broken the unspoken contract between himself and the barons, that allowed them to run their lands as they chose.

Hoping for support from the pope, who had excommunicated John in 1209, the rebellious barons confronted the king. Attempts at diplomacy failed, and by May 1215, the barons had occupied London, forcing John to enter into a treaty with them to avoid a civil war. After careful direction of negotiations by Archbishop Stephen Langton of Canterbury, the agreement—more a truce than a peace—was signed.

Provisions of the charter

The charter was known as the Magna Carta, or Great Charter, to distinguish it from a more restricted Forest Charter issued in 1217. Much of the Magna Carta dealt with redressing baronial grievances,

Centralization of royal administration **reduces barons' power and income**.

Financial demands to fund wars in France **increase**.

Barons revolt and force King John to sign a charter of rights.

Rights of individuals against arbitrary punishment by the Crown are **established**.

The principle that **new taxes** can be raised **only after consultation** with a royal council evolves.

but the section that has exerted the most influence down the ages was Chapter 39. This open-ended clause protected all "free men" from arbitrary actions by the Crown such as arrest or confiscation of land. The charter survived the civil war that broke out soon after the Magna Carta was agreed, and the papal repudiation of the charter's terms in August 1215, which led to the barons' excommunication. Chapter 39 was extended under a 1354 law of Edward III to protect not only "free men" (a small minority in England where most people were technically serfs), but also any man "of whatever estate or condition he may be." It survived longer than most of the other provisions, including the security clause that allowed barons to seize all the king's land if he failed to fulfil his obligations under the agreement.

What had seemed a small concession that day in Runnymede provided a long-lasting rallying cry for opponents of royal tyranny. ▪

Influence of the Magna Carta

The Magna Carta has acquired an almost mythical status as the constitutional bedrock of subjects' rights. It contributed to the development of parliament from the 13th century, and was used by 17th-century rebels to argue against the divine right of kings propounded by the Stuart monarchs Charles I and James II. Several American colonies' charters contained clauses modeled on it, while the design of the Massachusetts seal chosen at the start of the Revolutionary War depicts a militiaman with sword in one hand and the Magna Carta in the other. Revolutionary feeling was fueled by Americans' belief that the Crown had breached the fundamental law enjoyed by all English subjects, and both the United States Constitution, enacted in 1789, and the Bill of Rights adopted two years later, were influenced by the Magna Carta's limitations on the arbitrary powers of a government against its subjects.

THE MOST POTENT MAN AS REGARDS FORCES AND LANDS AND TREASURE THAT EXISTS IN THE WORLD

KUBLAI KHAN CONQUERS THE SONG (1279)

IN CONTEXT

FOCUS
Mongol rule in China

BEFORE
1206 Mongol Empire founded by Genghis Khan.

1215 Genghis Khan sacks Zhongdu (now Beijing), capital of the northern Jin dynasty.

1227 Death of Genghis Khan and fragmentation of the empire into smaller khanates loyal to a single Great Khan.

1260 Kublai declares himself Great Khan.

1266 Kublai orders the reconstruction of Zhongdu, and renames it Khanbalik.

AFTER
1282 Kublai's corrupt chief minister Ahmad killed by Chinese assassins.

1289 Southern extension of the Grand Canal completed.

1368 Mongols driven from China. Replaced by the native Chinese Ming dynasty.

In March 1279, Mongol warriors swept through southern China, capturing the last strongholds of the Chinese Song dynasty. This defeat, which heralded the start of the Yuan dynasty, marked the culmination of the Mongols' rise in under 70 years from an obscure nomadic group from the Central Asian steppes to the masters of a vast empire stretching from China to eastern Europe. One of the major challenges they now faced was to make the transition from roving tribesmen to settled conquerors.

The rise of the Mongols
At the start of the 13th century, the Mongols had consisted of many different warring clans. However, in 1206 Temüjin—later known as Genghis Khan—proclaimed himself the ruler of a united Mongol nation. Shrewd and ruthless, Genghis diverted his people from inter-clan warfare and directed their energies to the more lucrative business of invading—first neighboring tribes in the steppes, then more organized states such as Persia, Russia, and northern China (1219–23). He gave the Mongol hordes a proper military structure and exploited the skills they had learned from their nomadic lifestyle: as expert horsemen, the soldiers were masters of mobile warfare and able to descend with devastating force and lightning speed on their opponents.

The Mongols' rule in China
Genghis's grandson Kublai Khan ruled China from 1260, but the challenges of mediating between the nomadic traditions of the Mongols and the complex culture of the conquered proved difficult.

Paper money was invented by the Chinese c.800. By the Yuan dynasty, banknotes (such as the one above from 1287) were issued by the government.

Kublai Khan

Grandson of Genghis Khan, Kublai Khan (1215–94) governed northern China for his elder brother Möngke, who became Great Khan (the senior Mongol ruler) in 1251. Kublai's restoration of Chinese-style administration displeased many Mongols and he was nearly removed in 1258, but Möngke's death led to Kublai achieving the position of Great Khan himself in 1260. Kublai established a bureaucracy staffed largely by Chinese officials, but he placed Mongol officers (*darughachi*) in key towns to ensure loyalty to the empire. He took measures

to restore the economy, initially encouraged religious tolerance, and welcomed foreigners such as Marco Polo to the Mongol court, aware of the expertise they might bring. After the successes in China, Kublai dispatched armies to Japan, Annam (Vietnam), Myanmar (Burma), and Java; however, these either failed or did not establish a lasting Mongol presence. By his death, Kublai was a disappointed man, who drank to excess, suffered from obesity, and had to be carried to his final campaigns in a litter.

The old informal hierarchies of the steppes no longer sufficed to administer a land that contained great cities, and the immediate rewards of plunder were replaced by the deferred benefits gained by good governance and taxation. As a result, many Mongols missed the old ways. To appease his fellow Mongols, Kublai gave them greater rights and privileges than the native Chinese. Meanwhile, to gain favor with the traditional Chinese elites, he promoted Confucian scholars, funded Taoist temples, and had his

son educated in Buddhist scripture. He also set up schools for peasants and introduced the Mongol postal system of using horses and relay stations to link up the empire, which benefited the merchants.

The end of the empire
The need to restore stability in northern China delayed Kublai's attempts to subjugate the Song in the south until 1268. Although ultimately successful, the 11-year campaign was ruinously costly. To preserve their warrior identity,

the Mongols needed the spoils of conquest to fund their huge army. Kublai's successors failed to work out how to preserve their identity while also keeping their monopoly of power, and the Mongol military gradually declined. After decades of famine, lethal epidemics, and corruption at court, in 1368 the heirs of Kublai were defeated in a rebellion led by Zhu Yuanzhang, founder of the Ming dynasty. After more than a century of occupation, China was back in the hands of the native (Han) Chinese. ▪

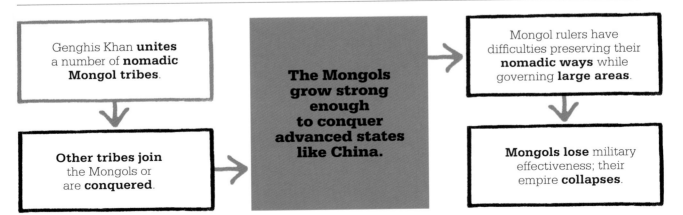

I DID NOT TELL HALF OF WHAT I SAW, FOR I KNEW I WOULD NOT BE BELIEVED
MARCO POLO REACHES SHANGDU (c.1275)

IN CONTEXT

FOCUS
Rise of international trade

BEFORE
106 BCE The first caravan to travel the full length of the Silk Road carries Chinese ambassadors to Parthia.

751 CE Defeat of the Chinese army at the Talas River prevents Chinese expansion west along the Silk Road.

1206 Genghis Khan unites the Mongol tribes, beginning Mongol conquest of Central Asia and China.

AFTER
1340s The Black Death spreads along the Silk Road, reaching Europe in 1347.

1370–1405 Timurlane makes extensive conquests, briefly reviving the Mongol empire and the Silk Road.

1453 The Ottoman conquest of Constantinople blocks Europeans' land route to Asia.

Long-distance trade from China to the Middle East is **damaged** by the collapse of traditional powers.

Mongols conquer lands through which the Silk Road runs, **improving** the route's **security**.

Trade along the route increases, attracting European merchants including Marco Polo.

European powers seek **alternative maritime trade routes** to the east.

The **collapse of Mongol rule** and rise of the Ottoman Empire render the route's territory **less secure**.

Venetian merchant Marco Polo's arrival at Shangdu, the capital of the Great Khan Kublai, in 1275 marked the end of a four-year journey. He had traveled from Italy to the Mongol capital Shangdu along the length of the Silk Road, an ancient network of routes that been carrying precious goods between China and Europe for centuries. The Silk Road had first become a conduit for trade when the Chinese Han Dynasty pushed into Central Asia in the late 2nd century BCE. From then on, goods such as jade and silk were carried west, passed from caravan to caravan by a series of merchants,

See also: Siddhartha Gautama preaches Buddhism 40–41 ▪ Kublai Khan conquers the Song 102–03 ▪ Hongwu founds the Ming dynasty 120–27 ▪ The Treaty of Tordesillas 148–51 ▪ The construction of the Suez Canal 230–35

to be met by caravans of furs, gold, and horses traveling in the opposite direction. Chinese inventions ranging from gunpowder and paper to the magnetic compass were also brought to the west along the route, arriving at Constantinople and the Black Sea ports, the western end of the route where Genoa and Venice chiefly traded.

Mongol revival of the route

By the 13th century, empires that had controlled sections of the Silk Road had fragmented. This left the route less secure for travelers, and so deterred merchants from using it. However, following Mongol conquest of the area between 1205 and 1269, the area was controlled—if loosely—by a single authority, the Great Khan, so a merchant could travel from Khanbalik (Beijing) to Baghdad without leaving Mongol territory. This renewed stability encouraged a revival of trade.

At around this time, European merchants' horizons were also expanding. In the early Middle Ages, traders could work only locally, and transport their goods to points where they might connect to longer-distance trade routes. From the 12th century, Italian city States such as Pisa, Genoa, and Venice, pioneered maritime trade across the eastern Mediterranean, which enabled merchants to connect directly with sea routes that linked West Asia and Egypt to China via the Indian Ocean.

The profits for merchants taking advantage of the "Pax Mongolica," or Mongol peace, could be huge. In the late 13th century, the costs of setting up a caravan might amount to 3,500 florins, but the cargo, once sold in China, could yield seven times that sum, and by 1326 Genoese traders were a common sight in the principal Chinese port of Zaitun.

Decline of land trade

The Silk Road flourished for another century, but the collapse of the Mongol Ilkhanate of Persia in 1335, and the overthrow in 1368 of the Yuan, the Mongol ruling dynasty in China, once again left the route divided between politically weak powers. It was also blocked to European traders at the western end by the growth of the Muslim Ottoman Empire.

A taste of the profits of long-distance trade in luxury goods encouraged European powers to seek alternatives to the now defunct Silk Road, this time by sea. In 1514, Portuguese merchants arrived off the coast of China, near Guangzhou, eager to take up the direct trading links with China that had been pioneered two and a half centuries earlier by their illustrious predecessor, Marco Polo. ▪

All the rare things that come from India are brought to Cambaluc—precious stones and pearls, and other kinds of rarities... a thousand cart-loads of silk enter Cambaluc daily.
Marco Polo, c.1300

Marco Polo

At just 17 years old, Marco Polo (1254–1324) set off from Venice to the court of the Mongol ruler, Kublai Khan. He traveled with his father and uncle, who had previously visited China and been entrusted by Kublai with a message for the pope. Polo was received with great favor at the Mongol court and stayed in China for 17 years. He traveled extensively throughout the country in the Khan's service, leaving for home at last in around 1291.

During a naval battle in 1298, Polo was captured and imprisoned by the Genoese. The stories he told of his sojourn in the lands of the Great Khan attracted the attention of his cell mate, Rustichello, who wrote them down, embellishing them as he went along. The resulting book was translated into many languages and includes much invaluable information about late-13th-century China. After his release, Polo returned to Venice, where he lived for the rest of his life.

THOSE WHO UNTIL NOW HAVE BEEN MERCENARIES FOR A FEW COINS ACHIEVE ETERNAL REWARDS
THE FALL OF JERUSALEM (1099)

IN CONTEXT

FOCUS
The Crusades

BEFORE
639 A Muslim army captures Jerusalem.

1009 Caliph al-Hakim orders Jerusalem's Church of the Holy Sepulchre to be destroyed.

1071 Seljuk Turks defeat and capture Byzantine emperor, Romanus Diogenes.

1095 Byzantine emperor Alexios sends to pope for help.

AFTER
1120 The Order of the Knights Templar is founded.

1145 The Second Crusade is launched.

1187 Muslim leader Saladin captures Jerusalem, and the Third Crusade is launched.

1198 Baltic Crusade begins.

1291 Muslim forces complete the reconquest of Palestine and Syria.

On July 15, 1099, some 15,000 Christian knights surged into Jerusalem after a month-long siege. The victorious crusaders slaughtered Muslim defenders and Jews alike in a bloody act that marked the beginning of 200 years of Muslim–Christian warfare in the Holy Land.

Defending Christianity
Jerusalem had fallen into Muslim hands in 639. Neither the Byzantine emperors in Constantinople nor the Christian kings in Western Europe had the political willpower or the strength to reverse the conquest, although the city was sacred to both.

Victorious crusaders flooded into Jerusalem, and in a ruthless assault seized the city from the Fatimid caliphate, laying the foundations for a new kingdom.

In the 11th century, however, the advances of a new group, the Seljuk Turks, disrupted the pilgrimage routes to Jerusalem, and the Turks' defeat of the Byzantines at Manzikert threatened to push the frontiers of Christianity back to the gates of Constantinople. In 1095, Emperor Alexios I Komnenos sent emissaries to Pope Urban II asking for help to bolster the Byzantine retaliation.

See also: Muhammad receives the divine revelation 78–81 ▪ The founding of Baghdad 86–93 ▪
The Investiture Controversy 96–97 ▪ The fall of Granada 128–29 ▪ The fall of Constantinople 138–41

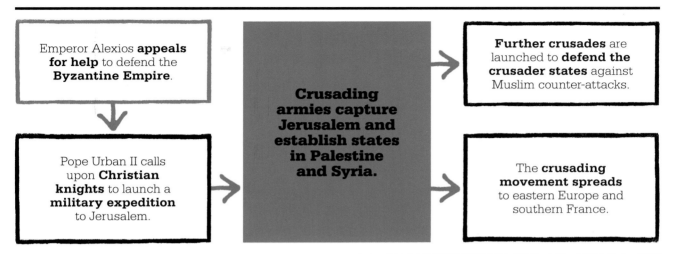

Emperor Alexios **appeals for help** to defend the **Byzantine Empire**.

Pope Urban II calls upon **Christian knights** to launch a **military expedition** to Jerusalem.

Crusading armies capture Jerusalem and establish states in Palestine and Syria.

Further crusades are launched to **defend the crusader states** against Muslim counter-attacks.

The **crusading movement spreads** to eastern Europe and southern France.

The Just War

Pope Urban readily seized a cause that would enhance papal prestige. In a sermon of 1095, he described atrocities against Christians in the Holy Land, calling for an expedition to free them. Christian warriors rallied to the cause, eager to gain both salvation and plunder by joining a so-called Just War in God's name.

Some 100,000 crusading knights, mostly French and Norman, set out in 1096. Progress to Jerusalem was slow: the crusaders suffered several setbacks at the hands of the Seljuk Turks, and the long siege of Antioch severely tested their morale, yet they pressed on and, led by the French knight Godfrey of Bouillon, at last captured the Holy City.

In the area they had conquered, the crusaders established four states, at Edessa, Antioch, Tripoli, and the Kingdom of Jerusalem, known collectively as Outremer. To withstand the vigorous Muslim counter-attacks, the crusaders built a dense network of fortresses such as Beaufort, Margat, and Krak des Chevaliers, which dominated strategic routes into the Holy Land.

As the initial crusading impulse waned, Outremer began to suffer from a shortage of manpower. This was partly resolved by the founding of crusading orders such as the Templars and Hospitaller knights, organizations who swore monastic vows to defend the Holy Land.

Further Crusades

However, even this was not enough, and when Muslim armies captured Edessa in 1144, a Second Crusade was called. This, and the

A race absolutely alien to God has invaded the land of Christians, has reduced the people with sword, rapine, and flame.
Pope Urban II, 1095

Third Crusade mustered in response to the catastrophic loss of Jerusalem in 1187, attracted participation at an even higher level, as monarchs such as Louis VIII of France, Richard I of England, and the Holy Roman Emperor Frederick Barbarossa assumed their leadership.

By 1270 there had been eight further crusades, and the movement had extended to include attacks on Muslims in North Africa; joining the Reconquista (the Christian reconquest of Islamic emirates in Spain); launching expeditions against pagan groups in eastern Europe, and even Christian heretics, such as the Cathars in southern France. In the Middle East, however, the emergence of stronger Muslim states, such as the Mamluks in Egypt, able to mount a strong resistance to crusader pressure, rendered the later expeditions largely ineffectual.

Jerusalem fell to the Muslims for a final time in 1244. The last crusader stronghold in the Holy Land, the city of Acre, was taken by the Mamluks in 1291. ▪

THE WORK OF GIANTS
THE CONSTRUCTION OF ANGKOR WAT (c.1120)

I n the early 12th century, much of mainland Southeast Asia, including Cambodia, and parts of Vietnam, Laos, and Thailand, was controlled by the Khmer Empire from its capital at Angkor (in the northeast of present-day Cambodia)—an impressive urban complex with residential areas, temples, and a network of water reservoirs, built by a succession of god-kings who ruled as the earthly representatives of the Hindu god Shiva.

Around 1120, the Khmer king Suryavarman II commissioned an ambitious new construction project—a 200-hectare (500-acre) temple complex dedicated to the Hindu god Vishnu that would also record the king's achievements. His spectacular Angkor Wat, completed 37 years later, was enclosed by a huge moat, adorned with lotus-shaped towers, and decorated with an 2,600ft-long (800m) gallery of fine bas-reliefs depicting scenes from Hindu mythology and the king as the embodiment of Vishnu.

Angkor Wat is a testament to the remarkable productivity and creativity of one of the greatest powers in Southeast Asia's history, yet its construction also marked the onset of its decline, as later

Buddhas decorate all the immense columns and lintels.
Zhou Daguan
Chinese diplomat

kings faced foreign invasions, shifts in trade, and wars with rival kingdoms that shrunk its territory. The empire's fortunes were revived under Jayavarman VII, who made Mahyanana Buddhism the state religion, and initiated a building spree at Angkor, but his death in 1218 left it fatally weakened.

Outside influences
The Khmer Empire was preeminent among the powerful states that had emerged in present-day Cambodia, Myanmar, and the islands of Java and Sumatra in Indonesia toward the end of the first millennium CE. During the states' formation, their societies had been profoundly

See also: Kublai Khan conquers the Song 102–03 ▪ Marco Polo reaches Shangdu 104–05 ▪
Hongwu founds the Ming dynasty 120–127 ▪ The Gulf of Tonkin Incident 312–313

affected by contact with India and China, made via the major trade route that ran through the Bay of Bengal, then overland across the Malay Peninsula before resuming through the Gulf of Thailand and on to the south of China. As well as enabling the exchange of Southeast Asian commodities such as rare woods, ivory, and gold, this network introduced Indian and Chinese ideas, concepts, and practices—including Hinduism and Buddhism—to the region's civilizations, which adapted them to create original, indigenous varations, particularly in architecture and the arts.

Maritime empires

While the Khmer Empire held sway in mainland Southeast Asia, in the Indonesian archipelago, the empire of Srivijaya, with its base at Palembang in Sumatra, dominated commerce by controlling the two passages between India and China—the straits of Malacca and Sunda. Over time, it had grown rich from its trade in the spices, especially nutmeg, that Europe, India, and

China craved, but by the end of the 12th century it had been reduced to a small kingdom, and was later eclipsed by the Majapahit in Java.

In the late 13th century, Mongol forces under the Chinese emperor Kublai Khan invaded Vietnam, Java, and Myanmar, and although these campaigns failed, in their wake, the Khmer lost control of eastern Thailand. In the early 1400s, the empire contracted further as armies from Champa (now in Vietnam)

After its rediscovery by Europeans in the late 19th century, Angkor Wat suffered decades of looting and unregulated tourism; it was made a UNESCO World Heritage Site in 1992.

and Ayutthaya (now in Thailand)—seized more of its land. In 1431, the latter took Angkor, and the capital was later relocated to the coast, leaving Suryavarman's spiritual masterpiece to be reclaimed by the jungle. ▪

Suryavarman II

One of the Khmer Empire's greatest kings, Suryavarman II ascended to the throne in 1113, after killing his rival, and reunited Cambodia after decades of unrest. He quickly resumed diplomatic relations with China, and in 1128 his kingdom was recognized as a Chinese vassal, which helped deter neighboring states from attacking it. Suryavarman was a warlike leader, waging campaigns in what is now Vietnam against the Dai-Viet between 1123 and 1136, and against the Khmer's traditional enemy to the east, Champa, in 1145. He also pushed

the empire's boundaries deep into Thailand and made advances against the Pagan kingdom of Myanmar.

As well as the awe-inspiring Angkor Wat, which remains the largest religious structure in the world, the king also built other temples in the same style at the capital. His political and military achievements were less enduring, however—when he died in 1150, in the middle of a campaign against Champa, the empire was convulsed by civil war and pushed to the brink of destruction.

HE LEFT NO COURT EMIR NOR ROYAL OFFICE HOLDER WITHOUT THE GIFT OF A LOAD OF GOLD

MANSA MUSA'S *HAJJ* TO MECCA (1324)

IN CONTEXT

FOCUS
Islam and trade in West Africa

BEFORE
c.500 CE The Kingdom of Ghana emerges.

1076 Ghana is conquered by the Almoravids, who establish an Islamic Empire from Spain to the Sahel.

1240 Sundjata establishes the Muslim Malian Empire, capturing Ghana and gaining control of its strategic salt, copper, and gold mines.

AFTER
1433 Mali loses control of Timbuktu, which is incorporated into the Songhai Empire of Gao.

1464 Sonni Ali, king of Songhai, begins the expansion of his empire, as Mali contracts further still.

1502 Mali is defeated by the Songhai Empire.

Islam spreads into West Africa from the 9th century, in the wake of **trans-Saharan trade**.

Mansa Musa's *hajj* **showcases the wealth and power of the Muslim Malian kingdom.**

Muslim scholars from other Islamic countries are **attracted to Mali** and it becomes a great **center of Islamic learning**.

Islam continues to take root throughout West Africa, even after the collapse of Mali.

The Muslim West African kingdom of Mali burst onto the world stage with a flourish in the early 14th century, when its fabulously wealthy ruler, Mansa Musa, made an unusually extravagant *hajj* (pilgrimage) to Mecca, supported by the huge profits made by Mali's control of the trans-Saharan caravan trade. The emperor's year-long expedition became legendary in the Muslim world, and even in Europe, and his subsequent promotion of Islamic culture and learning in his kingdom was symbolic of the faith's gradual infiltration of the trading empires of West Africa.

African trade and Islam
States had begun to form on the fringes of the Sahel region (a semi-arid zone just south of the Sahara) around the 5th century,

See also: Muhammad receives the divine revelation 78–81 ▪ The founding of Baghdad 86–93 ▪ The conquests of Akbar the Great 170–71 ▪ The formation of the Royal African Company 176–79 ▪ The Slave Trade Abolition Act 226–27

[Mansa Musa] flooded Cairo with his benefactions... They exchanged gold until they depressed its value in Egypt and caused its price to fall.
Chihab al-Umari
Arab historian (1300–1384)

beginning with the Kingdom of Ghana, which became known as "the land of gold," a reference to the source of its huge wealth. In the 7th century, the Arab conquest of North Africa gave a new impetus to trans-Saharan trade—the Muslim states had a huge appetite for West African gold and slaves. As this trade grew, Muslim merchants, and with them Islam, were drawn to the area between the headwaters of the Niger and Senegal rivers.

However, peaceful trading was soon followed by conquest. The Almoravids, a Moroccan Berber dynasty, swept south in 1076 and sacked Ghana's capital, shattering its authority over the region.

Ghana's reduced power opened up a vacuum that was gradually filled by Mali, a state founded around the Upper Niger River, which began to expand in the mid-13th century.

Mansa Musa's *hajj* attracted the attention of Europe's cartographers: the emperor is depicted on this Catalan Atlas of 1375, bearing a gold nugget and a golden scepter.

Under Mansa Musa (ruled 1312–37), Mali reached its greatest extent and power, having forged highly lucrative caravan connections with Egypt and other important trade centers in North Africa. Gold, salt, and slaves were taken north in exchange for textiles and manufactured goods.

A center of scholarship

Mansa Musa was not the first West African ruler to make a *hajj* to Mecca, but the huge scale of his entourage—more than 60,000 people, including 500 slaves who bore staffs of pure gold—impressed his observers, and was a potent expression of his wealth.

The expedition had a purpose beyond advertising Mali's prestige however, as the king invited Muslim scholars and a great architect, Abu Ishaq al-Sahili, to make the return journey with him. The latter built West Africa's first mud-brick mosques at Timbuktu and Gao, trading posts recently captured from the neighboring Songhai.

Under Mansa Musa's guidance, Timbuktu became Mali's main commercial hub—boosted by its advantageous location at the junction of the desert trade and the maritime routes down the Niger—and began its rise as the region's intellectual and spiritual capital. A teaching center grew around al-Sahili's Sankore mosque, laying the foundations for the celebrated Sankore University and other *madrasas* (Islamic schools).

After Mansa Musa's death, Mali initially thrived under his son, but thereafter, weak rulers, external aggression, and the need to keep rebellious tribes in check sapped its strength until it was eclipsed by the Songhai Empire of Gao: by 1550 it was no longer a major political entity. Mansa Musa's great empire—one of the most prosperous states in the 14th century—may have been short-lived, but his celebrated *hajj* had longer-lasting effects, helping to spearhead the spread of Islamic civilization in West Africa. ▪

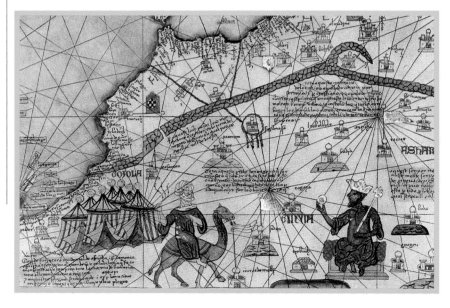

GIVE THE SUN THE BLOOD
OF ENEMIES TO DRINK
THE FOUNDATION OF TENOCHTITLAN (1325)

IN CONTEXT

FOCUS
The Aztec and Inca empires

BEFORE
c.1200 Emergence of the Incas in the Cuzco valley, Peru.

c.1250 Aztecs arrive in the Valley of Mexico.

1300 Aztecs establish settlements on land owned by the lord of Culhuacán.

1325 Aztecs flee south from Culhuacán and enter the land around Lake Texcoco.

AFTER
1376 Acampichtli becomes the first Aztec ruler.

1428 Inca expansion begins. Establishment of the Aztec Triple Alliance.

c.1470 Incas capture Chimor, center of the Chimú culture.

1519 Spanish arrive in Mexico.

1532 Spanish arrive in Peru.

Small, competing states in central Mexico and Peru **attract Aztec and Inca migrants** who fill the power vacuum.

↓

The Aztecs and Incas found capital cities at Tenochtitlan and Cuzco respectively.

↓ ↓

The Aztec Empire expands using military **aggression** and **fear of reprisals** to **retain power**.

The Inca empire expands by **co-opting conquered peoples** and seeking to **integrate** them.

↓ ↓

Neither model of empire can survive the **Spanish invasion**.

In 1325, a band of Central American refugee warriors, known as the Aztecs, saw a sign their patron god Huitzilopochtli had long ago prophesied—an eagle perched on a cactus, marking the spot they had been told to settle. Before long, they had built a temple that became the nucleus of their capital, Tenochtitlan. Within two centuries, the city was the center of the most predominant empire in the history of Mesoamerica—a large region that shared a pre-Columbian culture and extended from modern-day central Mexico southward to Belize, Guatemala, El Salvador, Hondurus, Nicaragua, and northern Costa Rica. This progress was paralleled by the growth at much the same time of Cuzco, the capital of the Incas—an Andean people of humble beginnings, who in just a few decades created the largest state South America had yet seen.

Aztec foundations

The Aztecs may have begun their wanderings in northern Mexico around 1200. For the next 100 years they eked out a miserable existence as mercenaries or barely tolerated squatters, their plight not aided by their reputation as cruel warriors. Frequently, they had to flee after committing violent acts, at times involving human sacrifice; indeed, their flight to Tenochtitlan was prompted by one such incident. The Aztecs had asked their host, the lord of Culhuacán, whether he would give his daughter as a bride for their chief. He agreed, believing she would be greatly honored as queen; however, to his horror they killed and flayed her as a sacrifice to their deity Xipe Totec. Driven out by the lord and his soldiers, the Aztecs fled southward toward the future site of Tenochtitlan.

Although the soil around Lake Texcoco, on which the island of Tenochtitlan was situated, was marshy and there was very little timber available, the capital was easily defensible and the Aztecs used it to consolidate their position. Initially shielded by a treaty with the Tepanec ruler Tezozomoc, who dominated the Valley of Mexico from 1371 to 1426, the Aztecs went on to form a Triple Alliance with the cities of Texcoco and Tlacopan in 1428—a union that kick-started a period of imperial expansion.

Aztec expansion

In the early days, Aztec society had little formal hierarchy. It was based around communities (*calpulli*) that owned land in common and

See also: The Maya Classical period begins 71 ▪ Christopher Columbus reaches America 142–47 ▪
The Treaty of Tordesillas 148–51 ▪ The Columbian Exchange 158–59 ▪ The voyage of the *Mayflower* 172–73 ▪
Bolívar establishes Gran Colombia 216–19

whose chiefs, together with priests, ruled on important decisions. In 1376, the Aztecs chose for the first time an overall leader (*tlatoani*), who came to serve as war leader, judge, and administrator for the burgeoning empire. Under Itzcoatl (1427–40), Moctezuma I (1440–69), Axayactl (1469–81), and Ahuitzotl (1486–1503) Aztec armies subdued their neighbors in the Valley of Mexico and then spread outward, reaching Oaxaca, Veracruz, and to the edges of land controlled by the Mayan people in the east of modern-day Mexico and Guatemala.

As the Aztec Empire expanded, society was transformed. A warrior elite emerged, while at the bottom of society bondsmen (*mayeques*), who owned no land, were bound by labor service to their lords. The militaristic nature of Aztec society was accentuated by an education system in which all males received military training (in separate schools for nobles and commoners). This reinforced the warrior ethos and gave the Aztecs an incalculable advantage over neighboring tribes in Mexico.

The imperial system

Tenochtitlan was adorned by many temples to the gods of the Aztec pantheon. Each god had their own temple, with the Templo Mayor having twin shrines dedicated to Huitzilopochtli and Tlaloc, the rain god. At these temples a stream of human victims was sacrificed—up to 80,000 at the rededication of the

The founding of Tenochtitlan is illustrated in the *Codex Mendoza*: a record of Aztec history and culture created c.1540 by an Aztec artist for presentation to Charles V of Spain.

Templo Mayor in 1487—by burning alive, decapitation, or cutting open the chest and removing the heart.

Many of the Aztec battles were "flower wars": ritual affairs in which opponents were captured (rather than killed) and sacrificed to placate the Aztec gods, who were believed to need blood to sustain them and keep the sun moving across the sky.

Tenochtitlan also exacted tribute from its subjects. Although there was very little in the way of an organized government bureaucracy, there were tax collectors, who crisscrossed the 38 provinces of the Aztec Empire and levied tribute, which included 7,000 tons of maize, 4,000 tons of beans, and hundreds of thousands of cotton blankets »

each year. The empire depended on this tribute to reward the nobility and the warriors, who ensured that the towns subjugated by the Aztecs remained submissive—little mercy being shown to those who revolted.

While the Aztecs provided some security to their subjects, they gave little else. At Tenochtitlan, artificial islands (*chinampas*) were created at great expense to expand the land available to produce food, but no such works were carried out for the subject cities. Defeated states did not provide troops for the Aztec army, and so did not share in the spoils of future victory, and little effort was made to propagate the Aztec language. It was an empire built on fear and in the end it proved brittle: when it was invaded by a small party of Spaniards led by Cortes in 1519, the subject peoples rallied to the newcomers rather than defending the Aztecs, and the empire collapsed within two years.

Inca beginnings

The Incas, whose heartland lay high in the central Andes around Cuzco, in modern-day Peru, had similarly humble origins to the Aztecs, but their rise to imperial

If the land [Peru] had not been divided by the wars… we could not have entered or conquered it unless over a thousand Spaniards had come simultaneously.
Pedro Pizarro
Spanish conquistador (1571)

status was, if anything, even more meteoric. They began as a small, somewhat disregarded tribe and developed their own strategies to co-opt neighboring groups into a successful empire.

The Incas' origin myth told of their emergence from a cave in the high mountains, from where their first leader—Manco Capac—led his people to Cuzco. It is generally believed that the Incas arrived in the region around 1200, and for two centuries they remained a relatively

insignificant farming group, with their society divided up into clans (*ayllus*) of roughly equal status.

Inca expansion

The Incas began to make their mark as a major power around 1438, when the neighboring Chanca people attempted to push the Incas out of the Cuzco valley. By this time, the Incas had a supreme leader (the Sapa Inca), and although the incumbent Viracocha was unequal to the task, his son Pachacuti defeated the invaders, and then led Inca armies to conquer the rest of the Cuzco valley and the southern highlands around Lake Titicaca. Under Pachacuti's son Topa Inca Yupanqui and grandson Huayna Capac, the Incas overcame Chimor (the largest coastal state) in about 1470. They then absorbed the rest of the northern highlands and extended to parts of modern-day Ecuador and Colombia and south to the deserts north of Chile.

Unlike the Aztecs, the Incas recruited troops from among the conquered peoples (placed under the command of Inca officers), thus providing them with the lure of plunder in return for their loyalty.

Inca communication

The empire of the Incas was highly centralized; censuses recorded the number of peasants, who all owed labor service (*mitad*) to the Sapa Inca. This level of organization enabled the construction of public works on a vast scale. Particularly vital was the extensive road network, which extended nearly 25,000 miles (40,000 kilometers) long and was dotted at regular intervals with resthouses that facilitated rapid transit for the army and provided a very efficient system of communication across the far-flung Inca domains. At the

Tlacaelel

As the Aztec Empire expanded and conquered new territories, it became increasingly necessary to create a more complex system of administration. After Itzcoatl became ruler (*tlatoani*) in 1427, he introduced the new post of chief adviser (*cihuacoatl*). The office's first incumbent was Itzcoatl's nephew, Tlacaelel (1397–1487), who held the office until his death. Tlacaelel served through several reigns and he provided invaluable continuity. In addition, he created impetus

for his reforms (mostly benefiting the royal family and nobles) by ordering the destruction of earlier chronicles and the rewriting of Aztec history to establish the basis of Aztec imperial ideology.

He also presided over the formation of the Triple Alliance, solidifying the Aztec position and ensuring a steady stream of sacrificial victims. Given that Tlacaelel was never the Aztec's ruler, his immense influence in Tenochtitlan shows that the Aztec system of authority was not as monolithic as it might at first appear.

Society in the expansionist Aztec Empire was deeply militaristic. A boy had to prove himself a warrior before he could be considered a man. Noble Aztec youths joined warrior societies and progressed through the ranks by taking more captives for sacrifice.

The Shorn Ones Deadly warriors who swore not to step back in battle.

Otomies Named after skilled allies of the Aztecs, Otomies may have been the first warriors to enter battle.

Jaguar Warriors Men had to take four human captives before they could be admitted to the ranks of the Eagle and Jaguar Warriors.

Eagle Warriors Alongside Jaguar Warriors, these may have been the lowest rank of the elite Aztec warrior societies. Their resplendent uniforms resembled their namesakes.

They cut open their chests, drew out the palpitating hearts and offered them to the idols.
Bernal Díaz de Castillo
True History of The Conquest of New Spain (1568)

same time, the Inca domestication of the llama as a beast of burden made it easier to transport heavy loads across the empire.

Unlike the Aztecs, the Incas actively sought to spread their own language (Quechua) and system of religious beliefs, which was initially based around the worship of Inti (the sun god), but which came to feature prominently Viracocha—a supreme creation god and therefore considered a more suitable deity for

a conquering power. They also dispatched colonists (*miqmaq*), shifting troublesome groups into more pacified areas to dilute their resistance and creating networks of loyal settlers on the fringes of the empire. Although definitive population statistics are not known, by the early 16th century the Inca Empire—which the Incas called Tawintusuyu ("The Realm of Four Quarters")—consisted of about 4–6 million people in total, operating to the advantage of the Inca minority and their subjects.

Despite its many strengths, the highly centralized nature of the Inca Empire proved fatal in the early 1530s, when Spanish invaders led by Pizarro captured the Sapa Inca Atahuallpa; without their leader, the Incas rapidly collapsed.

The new colonizers
The Aztecs and the Incas built the first true empires in their regions of the Americas. They were able to do so by creating food surpluses through irrigation projects, thus

releasing a large proportion of their population to fight in the armies that conducted their campaigns of expansion. They also reorganized the traditional tribal structure to favor a warrior and noble elite. In both cases, the momentum of conquest demanded further wars to reward the warrior caste or to provide an incentive for newly conquered peoples to remain loyal and thus to gain the rewards of participation in new campaigns.

Neither the Aztecs nor the Incas survived long enough to govern after their expansion slowed down. Had they done so, they might have developed strategies to bring long-term stability to their empires, or might have declined to the status of competing city-states fighting to control limited resources. Instead, the Spanish conquest of the Aztecs in 1521 and their defeat of the last Incas by 1572 put paid to the ambitions of both empires and left the Spanish firmly established as colonial rulers in the region for the next 300 years. ■

SCARCE THE TENTH PERSON OF ANY SORT WAS LEFT ALIVE

THE OUTBREAK OF THE BLACK DEATH IN EUROPE (1347)

IN CONTEXT

FOCUS
The Black Death

BEFORE
1315–1319 Famines strike western Europe: 15 percent of Dutch city-dwellers die.

1316 Edward II of England fixes staple food prices as shortages drive them upward.

Late 1330s Bubonic plague spreads gradually westward from western China.

AFTER
1349 Accused of starting the plague, Jews are murdered in the thousands in Germany.

1349 Pope bans the flagellant "Brothers of the Cross."

1351 Statute of Labourers is passed in England.

1381 Peasants' Revolt stirs political rebellion across large parts of England.

1424 Dance of Death painted on the cloister walls of the Cimetière des Innocents, Paris.

In late November 1347, a galley entered the Italian port of Genoa, having fled a Tatar siege of Kaffa in the Crimea. It bore a deadly cargo: the bubonic plague. Within a mere two years, this lethal pestilence had killed more than a third of the population of Europe and the Middle East, and altered the regions' economic, social, and religious makeup forever.

Spread of the Black Death
Having probably originated in Central Asia or western China in the 1330s, the plague's initial progress westward was slow, but after it reached Crimea and Constantinople in 1347 it spread rapidly along maritime trade routes. Having hit Genoa, it appeared quickly in Sicily and Marseilles; by 1348 it had struck Spain, Portugal, and England, and it reached Germany and Scandinavia by 1349.

The epidemic's main vector was infected fleas and the rats that harbored them, both of which flourished in the unsanitary conditions of the time. The main symptoms of the disease were swellings, known as buboes, that appeared in the groin, neck, or armpits. These were followed by black blotches on the skin (hence "Black Death") and then, in around three-quarters of cases, by death.

Contemporaries ascribed the causes of the pestilence variously to divine punishment for immorality, adverse conjunctions of the planets, earthquakes, or bad vapors. There was no cure, but preventive advice included abstinence from hard-to-digest food, the use of aromatic herbs to purify the air, and—the only effective measure—avoiding the company of others.

More than a hundred million people may have died of the plague; estimates put the world population

Employees are refusing to work unless they are paid an excessive salary.
The Ordinance of Labourers, 1349

See also: The crowning of Charlemagne 82–83 ▪ Marco Polo reaches Shangdu 104–05 ▪ The Columbian Exchange 158–59 ▪ The opening of Ellis Island 250–51 ▪ Global population exceeds 7 billion 334–39

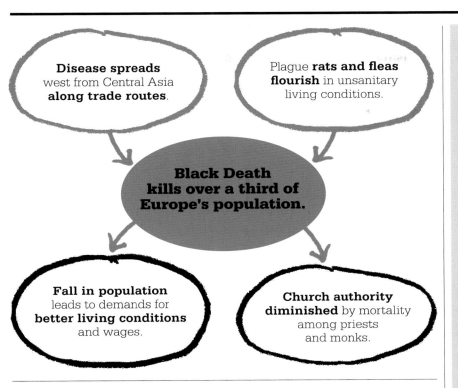

> **Disease spreads** west from Central Asia **along trade routes**.

> Plague **rats and fleas flourish** in unsanitary living conditions.

> **Black Death kills over a third of Europe's population.**

> **Fall in population** leads to demands for **better living conditions** and wages.

> **Church authority diminished** by mortality among priests and monks.

at 450 million before it arrived, and 350 million afterward. Its effects were more deadly in some areas than in others—in Egypt, about 40 percent of the population are thought to have died. Populations did not reach pre-plague levels again for nearly three centuries.

Reactions to the plague

Survivors reacted in varying ways. Jewish communities in Germany were accused of causing the plague by poisoning wells, and many were attacked. In Strasbourg alone, 2,000 Jews were killed.

With the population diminished, landholdings fell vacant, labor became scarce, and peasants' bargaining power increased. By 1350, English laborers could demand five times the wages they had asked in 1347, and tenants were

paying rent in cash rather than with compulsory labor. Governments tried to clamp down on wages—the 1351 Statute of Labourers aimed to freeze rates at 1346 levels—but peasants responded with outbursts such as the Jacquerie in France in 1358, and the Peasants' Revolt in England in 1381.

By the time it ended, the Black Death had killed proportionately as many clergy as laity, and some clergy deserted their posts. As a result the church's authority, like that of the nobility, was greatly weakened. The plague had loosened the ties that had previously bound medieval society together, leaving a freer and more volatile population to face the challenges posed by the Renaissance, Reformation, and the economic expansion of the 16th and 17th centuries. ▪

Shattered society

The plague's catastrophic toll cast a long shadow over contemporary social attitudes. A landscape of mass graves, abandoned villages, and an all-pervading fear of death deepened the sense that God had abandoned his people, and diluted the claims of traditional morality. Crime rose: the incidence of murder in England doubled in two decades from 1349. Flagellants roamed the countryside, scourging themselves with knotted ropes, until a Papal bull banned the practice in 1349. Bequests to charitable foundations—hospitals in particular—rose as the rich gave thanks for their survival. Artistic production tended to the morbid: depictions of the Dance of Death appeared, showing Death cavorting among the living; and writers such as Boccaccio, who chronicled the plague in his *Decameron*, stressed the briefness and fragility of life.

Death selects his victims indiscriminately from among the social orders in the allegorical *Danse Macabre* or Dance of Death.

I HAVE WORKED TO DISCHARGE

HEAVEN'S WILL

HONGWU FOUNDS THE MING DYNASTY (1368)

IN CONTEXT

FOCUS
Ming China

BEFORE
1279 Kublai Khan overthrows the Song and establishes the Mongol Yuan dynasty.

1344 In central China, the Yellow River begins to shift course, leading to droughts and a subsequent upsurge in peasant rebellions.

1351 Outbreak of Red Turban revolt against the Yuan.

AFTER
1380 Hongwu takes on the role of chief minister, laying the basis of an authoritarian political culture.

1415 Yongle revives and extends the Grand Canal, enabling it to carry goods from southern China to Beijing.

1520 The first Portuguese trading missions to China.

c.1592 Publication of *Journey to the West,* one of the masterworks of Chinese classical writing.

1644 Chongzhen commits suicide, ending the Ming era.

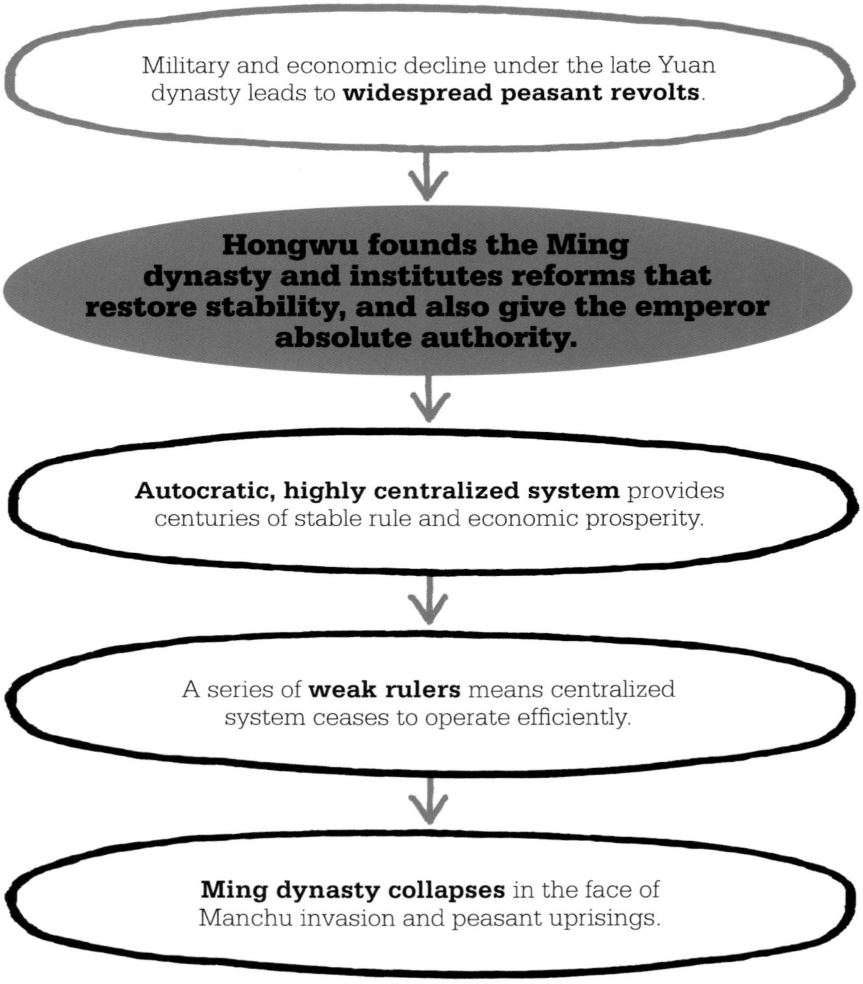

Military and economic decline under the late Yuan dynasty leads to **widespread peasant revolts**.

Hongwu founds the Ming dynasty and institutes reforms that restore stability, and also give the emperor absolute authority.

Autocratic, highly centralized system provides centuries of stable rule and economic prosperity.

A series of **weak rulers** means centralized system ceases to operate efficiently.

Ming dynasty collapses in the face of Manchu invasion and peasant uprisings.

S urrounded by officials at the imperial palace in Nanjing, Zhu Yuzhuang, the son of poor peasant farmers, offered sacrifices to Heaven and Earth as he was proclaimed first emperor of China's Ming ("brilliant") dynasty.

It was the culmination of a remarkable rise to power by the monk turned rebel general, who had ousted the despised Yuan dynasty—founded by Kublai Khan, the Mongol conqueror of China—the country's rulers since 1279. Zhu reigned as emperor Hongwu ("Vastly Martial"—a reference to his military prowess) from 1368 until his death in 1398, by which time he had firmly established one of China's most influential, but also most authoritarian, dynasties. He and his successors brought three centuries of prosperity and stability to the country, establishing its government and bureaucracy in a form that would endure, with slight modifications, until the demise of the imperial system in 1911, and broadening the base of its economy.

Driving out the Mongols
Zhu's new dynasty arose from the chaos that accompanied the decline of the Yuan. In the 1340s and 50s, factionalism in the Mongol court, rampant government corruption, and a series of natural disasters, including plagues and epidemics, resulted in wholesale breakdown in law and order and administration as peasant groups rose up against their faltering

See also: The First Emperor unifies China 54–57 ▪ Kublai Khan conquers the Song 102–03 ▪ Marco Polo reaches Shangdu 104–05 ▪ The Revolt of the Three Feudatories 186–87

foreign overlords. Zhu himself lost most of his family in an outbreak of plague in 1344, and after a few years spent as a mendicant monk, begging for food, he joined the Red Turbans, one of a constellation of native Han Chinese peasant secret societies in rebellion against the Yuan. Determined, ruthless, and an able general, the young rebel climbed the ranks to the leadership of the Red Turbans, and later overcame his rivals to become the national leader against the Yuan.

Zhu took control of much of southern and northern China and declared himself emperor before pushing the Mongols out of their capital at Dadu (Beijing) in 1368. The rest of the country was then subdued, although the Mongols resisted in the far north until the early 1370s, and the unification of China was not achieved until the defeat of the last Mongol forces in the south in 1382.

Reform and despotism
Zhu's first priority as emperor Hongwu was to establish order—decades of conflict had ravaged China and impoverished its rural population. His humble beginnings may have influenced some of his early policies: responsibility for tax assessment was entrusted to rural communities, sweeping away the problem of rapacious tax collectors who had preyed on poorer areas; slavery was abolished; many large estates were confiscated; and lands owned by the state in the under-

The tribulations of Hongwu's early life led him to improve the lot of China's rural poor, but they also created a cruel and irrational man who murdered all those he suspected of disloyalty.

populated north of the country were handed to landless peasants, to encourage them to settle there.

From 1380, Hongwu instituted government reforms that gave him personal control over all matters of state. After executing his prime minister, who had been implicated in a plot to overthrow him, he abolished the prime ministership and the central secretariat and had the heads of the next layer of

government, the six ministeries, report directly to him, ensuring he oversaw even minor decisions.

From then on, Hongwu acted as his own prime minister. His workload was almost unbearable—in a single week-long stint, he had to scrutinize and approve some 1,600 documents—and as a result, the state became incapable of responding swiftly to crises. Although in time a new »

The Forbidden City—the imperial palace in Beijing—adhered to hierarchic Confucian ideology: the higher one's social status, the further one could enter into the city.

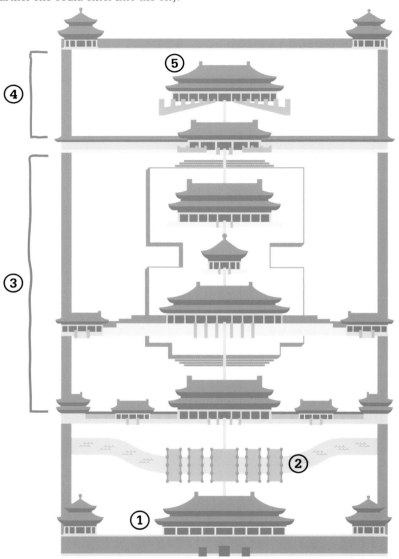

1 Meridian Gate The grand entrance had five gates. The central one was always reserved for the emperor.

2 The Golden Water Bridge Crossing points like the bridges were arranged in odd numbers. Only the emperor could use the central passage, with the next highest rank able to use the neighbouring paths.

3 Outer Court This area was reserved for state affairs and ceremonial purposes.

4 Inner Court Only the emperor and his family could enter the Inner Court.

5 The Palace of Heavenly Purity To fool assassins, the palace had nine bedrooms: the emperor slept in a different one each night.

grand secretariat emerged—an advisory board through which the emperor responded to the six ministries and other government agencies—the Ming retained a more autocratic and highly centralized structure than that of previous Chinese dynasties. This was reflected in the protocol of the Ming court, too: under the Song dynasty (960–1279), the emperor's advisers had stood before him to discuss matters of state, but under the Ming they were required to *kowtow*—kneel and knock their heads to the floor—before him, a reverential acknowledgment of his absolute power and superiority.

Curbing the military
In the later years of the Yuan dynasty, the state had been torn apart by competing power bases outside the central court, and in a bid to avoid this scenario, Hongwu diluted the strength of the army. Although he adopted the Yuan military system—establishing garrisons in key cities, particularly along the northern frontier, where the threat of nomad incursions was ever-present, and creating a hereditary caste of soldiers that supported itself on land granted by the government—he also ensured that military units were periodically rotated through the capital for training, and that a group of centrally selected officers shared authority in the army with the garrison commanders, thus preventing the rise of influential warlords with a strong local base.

Perfecting the civil service
Hongwu also had a deep mistrust of the elite scholar class that had been at the heart of government for centuries. However, he was aware that they played a vital role in the efficient running of the state, and

so he promoted education and trained scholars specifically for the bureaucracy. In 1373, he suspended the traditional examinations used to recruit civil servants and ordered the establishment of local county and prefectural schools. From these, the best candidates would be called for further study at a national university in the capital, where eventually 10,000 students from the original intake were enrolled. The civil service examinations were restored in 1385, when the emperor considered the well-trained graduates of the university ready to take them, and were so competitive that soldiers were stationed outside the cubicles where the examinees sat to avoid any collaboration or illicit use of reference materials.

The pool of potential recruits into the administration was thus widened, but civil servants still received a very conservative education based on the Four Books and Five Classics of Confucianism and a selection of neo-Confucian works that expounded the virtues of loyalty to the emperor and adherence to Chinese tradition. Innovation was discouraged and

Some people in the morning are esteemed [by the Hongwu emperor], and in the evening they are executed.
Memorial of the official Hsieh Chin, 1388

bureaucrats became set in their ways. Those who were perceived as having stepped outside their brief were publicly flogged, sometimes to death.

This maltreatment of public servants was a sign of the cruel side of Hongwu's personality. He was also violently paranoid, and vicious in his suppression of dissent. In 1382, he established a secret police, the Embroidered Brocade Guard, whose 16,000 officers stamped out all signs of resistance. The Guard's reach and

influence was wide, and as a result, until the very last years of its rule, the Ming dynasty experienced no significant rebellions by either the military or the aristocracy.

International diplomacy

The dynasty's self-confidence appeared to grow even further under Hongwu's successor, Yongle (reigned 1402–24), who moved the capital from Nanjing to Beijing, and embarked on an ambitious program of reconstruction and public works, including measures to improve the navigability of the Grand Canal. He also built the extravagant Forbidden City, which housed an imperial palace complex containing more than 9,000 rooms.

Yongle's initially aggressive foreign policy led to four campaigns against Mongolia and an attack on Annam (Vietnam) in 1417 that resulted in its incorporation into the Ming Empire. He also sought recognition from the rulers of faraway states: between 1405 and 1433, he launched six large-scale maritime expeditions to Southeast Asia, East Africa, and Arabia. Led by the great fleet admiral Zheng He, their purpose was to confirm »

This silk scroll records one of the most celebrated tribute gifts from Zheng He's voyages: a giraffe brought back from Africa in 1414.

The voyages of Zheng He

A Muslim of Mongol descent, Zheng He was captured by the Ming as a boy, castrated, and sent into the army, where he acquired military and diplomatic skills and distinguished himself as a junior officer. He went on to become an influential eunuch in the imperial court, and in 1405, Yongle chose him to lead a grandly conceived maritime expedition around the rim of the Indian Ocean, as both fleet admiral and diplomatic agent. Over the next 28 years, Zheng He commanded one of the largest naval forces in history: the first mission had 63 vessels, including 440 ft (130 m) long "treasure ships" carrying more than 27,000 crew.

Although these voyages were dramatic in their conduct and scope—the last three sailed as far south as Mombasa on the east coast of Africa—they were not in any real sense commercial or exploratory ventures. Their intention was strictly diplomatic, designed to enhance China's prestige abroad and to extract declarations of loyalty and exotic tributary gifts for Yongle.

China's domination over the area by exacting tribute and other gestures of homage to the emperor.

The later Ming

However, the enormous cost of Zheng He's ambitious ventures put great strain on the treasury, and to ensure they would never be repeated, all records relating to them were destroyed. Official ideology regarded China as the center of the world, and the later Ming saw no reason to encourage further maritime contact. The Chinese did not regard relations with foreign powers as possible on an equal basis: where diplomatic relations were conducted, the foreigners were considered (by the Ming, at least) as tributaries. The confidence and stability of the Ming bureaucracy also created a sense of self-sufficiency, with little use for external influences.

Ocean-going vessels were made to report all the cargo they landed, and private maritime trade was periodically banned (until it was legalized again in 1567 for all except trade with Japan). In Beijing, a shopkeeper's unauthorized contact with foreigners could result in the confiscation of his stock.

Diplomatic isolation was reinforced by military uncertainty: Annam became independent once more in 1428, while huge resources were devoted to containing the threat posed by the Mongol tribes on China's northern borders. In 1449, Emperor Zhengtong personally led a disastrous expedition against the Mongol leader Esen Khan in which the majority of the 500,000 Chinese soldiers died of hunger, were picked off by the enemy, or perished in a final battle as they retreated.

Extending the Great Wall

In the 1470s, the building of the final stages of the Great Wall—begun by the Qin dynasty in the 3rd century BCE—was not only a bid to prevent a similar disaster, but also to compensate for the Ming's waning energy. Like their predecessors, they were unable to absorb the lands of the nomadic groups to the north of the border, or to send out expeditions that had any lasting effect on discouraging their raids. Therefore, a fixed, strongly garrisoned border defense was the best compromise.

During the 16th century, a succession of short-lived emperors who were dominated by their

On taking the throne, Hongwu issued his own traditional bronze coinage, although a shortage of metal led to the reinstatement of paper money, made of mulberry bark.

consorts, mothers, or by eunuch (castrated) advisers, was capped by the long reign of Wanli (1573–1620), who simply withdrew from public life entirely: for the last decades of his reign, he refused even to meet with his ministers. The dynasty began to decline: the machinery of government faltered and the army had little strength to respond to the serious threat posed by the Jurchen in Manchuria (now in northeast China). In 1619, this tribal people, who later renamed themselves Manchu, began to encroach on China's northern borders.

Global trade

Economically, however, Ming China's great productivity was a magnet for European maritime states seeking new commercial connections in East Asia, and in the early 16th century, European traders finally reached the coast of China. In 1514, a Portuguese fleet

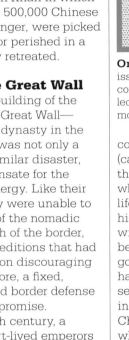

Hongwu's final resting place, the Xiaoling Mausoleum, lies at the foot of the Purple Mountain in Nanjing, and is guarded by an avenue of stone statues of pairs of animals, including camels.

appeared off Canton (now Guangzhou) in the south, and by 1557, Portugal had established a permanent base at Macao. Spanish and Portuguese merchants (the former operating from Nagasaki in Japan and Manila in the Philippines)—and from 1601, the Dutch—secured an important share in trade with China.

Even though Ming policy discouraged foreign maritime trade, individual Chinese merchants had participated actively in the revived economy. Before long there were flourishing Chinese colonies in Manila and on Java in Indonesia, near the Dutch-controlled trading city of Batavia, and Chinese merchants controlled a large share of local trade in Southeast Asia. The technical sophistication of the Chinese porcelain industry under the Ming led for the first time to the mass production of ceramics for export to European markets.

The effects, though, of this growth in trade were not wholly positive: while a huge influx of silver from the Americas and Japan, used by the Europeans to pay for Chinese goods such as silk, lacquerware, and porcelain, stimulated economic growth, it also caused inflation.

Technological change

Ming China had inherited a legacy of scientific and technological innovation from the Song dynasty, which had left the country at the forefront of many scientific fields, including navigation and the military applications of gunpowder—a substance discovered during the Tang era whose use had spread to Europe from China in the 13th century. Under the Ming, though, the pace of progress slowed and by the later part of the dynasty, ideas had begun to flow in from Europe.

> Today the great civil and military officers, the numerous officials, and the masses join in urging us to ascend the throne.
> **Proclamation Document of the Hongwu Emperor, 1368**

The Chinese military began to use artillery of European manufacture, and knowledge of European mathematics and astronomy was introduced to the country through Jesuit missionaries, including Matteo Ricci, who lived in Beijing from 1601 to 1610. He translated the ancient Greek mathematician Euclid's *Geometry* into Chinese, as well as a treatise on the astrolabe (an astronomical instrument used for taking the altitude of the sun or stars). In 1626, the German Jesuit Johann Adam Schall von Bell wrote the first treatise in Chinese on the telescope, bringing Heliocentrism (an astronomical model in which the sun lies at the center of the universe) to a Chinese audience.

The Ming collapse

The late Ming began to suffer many of the same issues that had led to the fall of the Yuan. Crop failures reduced the productivity of China's vast agriculture, and famines and floods led to widespread unrest in rural areas. The army's pay began to fall into arrears, leading to discipline problems and desertions, while localized peasant uprisings

coalesced into more general revolts. Meanwhile, on the northeastern frontier, the Manchus had built a state along Chinese lines at Mukden in Manchuria—calling their regime the Qing dynasty in 1636—and were now poised to take advantage of the Ming's imminent collapse. They were aided in this by a revolt led by Li Zicheng, a rebel leader whose forces entered Beijing in 1644 unopposed, prompting the emperor to commit suicide. In desperation, the Ming military called on the Manchus for help. The tribesmen swept into the capital and drove out the rebels, but then seized the throne, and proclaimed the Qing dynasty in China.

An enduring legacy

Although the Ming had fallen victim to an agrarian crisis that coincided with renewed nomadic activity on its frontiers, this was a combination that had also brought down dynasties before it. The bureaucracy that had given China centuries of constancy and reduced the possibility, or even the need, for internal dissent, was slow to adapt itself to times of fast-moving crisis.

Yet even so, the Ming era had brought great wealth and success to China. The population expanded from around 60 million at the start of its rule, to around three times that number by 1600. Much of this growth was centered in medium-sized market towns, rather than in large cities, and an increase in agricultural production led to the rise of an affluent merchant class in the provinces. Many of the elements of orderly government that Hongwu had inaugurated were carried over into the succeeding Qing dynasty, providing China with a degree of unity, stability, and prosperity that the European states of that period could only envy and admire. ∎

CAST DOWN THE ADVERSARIES OF MY CHRISTIAN PEOPLE
THE FALL OF GRANADA (1492)

At midnight on January 2, 1492, Abu 'Abd Allah, the Muslim Emir of Granada, handed over the keys of his city to King Ferdinand and Queen Isabella, joint rulers of the Christian Spanish states of Aragón and Castile. This act marked the end of nearly 800 years of Muslim rule in the Iberian peninsula and the eclipse of a great civilization renowned for its architectural splendors and a rich tradition of scholarship. At the same time, it signaled the birth of a self-confident, united Spain that would soon divert its energies away from crusading against its Muslim neighbors, turning them instead towards building an overseas empire in the New World.

A kingdom of so many cities and towns, of such a multitude of places. What was this, if not that God wanted to deliver it and place it in their hands?
Andrés Bernáldez
Archbishop of Seville (1450)

Christian conquests

Muslim Spain (or al-Andalus) dated from the Islamic conquest of the Visigothic kingdom in 711. Christian resistance survived in Asturias, in the far north, but it took centuries for the kingdoms of Castile, Aragón, León, and Navarre to gain the strength to push slowly southward into Muslim lands. This gradual reconquest, known as the Reconquista, gathered pace during the 11th century, when the Muslim regions broke up into numerous competing emirates ("*taifas*") and lost the strategically important city of Toledo in central Spain, in 1085.

The growth of the crusading spirit in western Europe also accelerated the progress of the Reconquista. Formal crusades against the Spanish Muslims (or Moors) were declared several times from the mid-14th century and a military culture emerged, in which raids into al-Andalus acquired the air of righteous expeditions. From

See also: Founding of Baghdad 86–93 ▪ The fall of Jerusalem 106–07 ▪ The fall of Constantinople 138–41 ▪ Christopher Columbus reaches America 142–47 ▪ The Treaty of Tordesillas 148–51

| **Muslims weakened** by the **break-up** of the centralized caliphate. | **Christians** amass **wealth** after seizing **land and assets** from Muslims. | **Union of kingdoms** of Aragón and Castile **ends Christian infighting**. |

Reconquista escalates as Christians benefit from greater resources and unity, culminating in the fall of Granada to the Castilian-Aragonese army.

| **Jews and Muslims** are **expelled** from Spain. | United Spanish kingdom allocates resources to **overseas expansion** in the **New World**. |

---

the 12th century, military orders, such as Santiago and Alcántara, were founded. They frequently spearheaded independent thrusts into Muslim territory, amassing great wealth in the process, which enabled them to sustain extended campaigns and ransom Christians taken prisoner in the wars. They also repopulated land conquered from the Muslims with Christians.

The end of Muslim Spain

In Portugal, the Reconquista was completed with the conquest of the Algarve in 1249, while in Spain the Muslims clung on to power in the south. However, this was not to last. In 1474, Queen Isabella ascended to the throne of Castile, in northern Spain. Her husband Ferdinand was already king of the neighboring state of Aragón, and they resolved to permanently expel the Muslims from the south. The union of the two crowns enabled them to devote more resources to

completing the Reconquista. It also put an end to centuries of Christian infighting, and this unity coincided with a period of Muslim division. From 1482, the monarchs undertook a series of military campaigns to conquer Granada—the last Muslim

Known as the Catholic Monarchs, Ferdinand and Isabella joined forces and used military might to restore Christianity in Spain, suppress other religions, and colonize the Americas.

emirate in the Iberian peninsula. The cities were put under siege and fell one by one, until finally the major city of Granada surrendered in 1492.

Despite an agreement reached at the capitulation of Granada, which contained guarantees for freedom of worship, in 1502 the monarchs decreed that any Muslims over the age of 14 who refused to convert to Christianity must leave Spain within 11 weeks. This edict, combined with the expulsion of the large Jewish community in Granada 10 years earlier, left Spain a more homogeneous and less tolerant place, and the crusading impulse, now shorn of obvious targets, would have to find other channels.

Christopher Columbus's expedition to the New World in 1492—the same year as the fall of Granada—provided the Spanish with just such an outlet, leading to their colonization of the Americas and Spain's subsequent emergence as the first global superpower. ▪

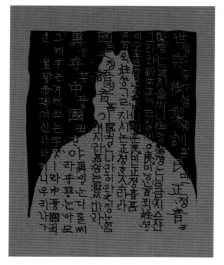

I HAVE NEWLY DEVISED 28 LETTERS
KING SEJONG INTRODUCES A NEW SCRIPT (1443)

IN CONTEXT

FOCUS
Chosŏn Korea

BEFORE
918 The Goryeo dynasty is founded.

1270 Goryeo comes under the structural, military, and administrative influence of the Mongol Yuan dynasty.

1392 Yi Sŏngyye founds the Chosŏn dynasty.

1420 King Sejong founds the Chiphyŏn-jŏn research institution.

AFTER
1445 A 365-volume medical encyclopedia is published.

1447 The first work printed in Han'gŭl is published.

1542 The first sŏwŏn private academy opens. The academies become centers of debate and house neo-Confucian texts.

1910 Japan annexes Korea and deposes the last Chosŏn ruler.

In 1443, the Korean court of King Sejong announced the creation of Han'gŭl, a national alphabet for the Korean language, and launched a program of publications in the new script. The measure was one of a number of strategies encouraged by Korea's king that were designed to stabilize Korea and improve prosperity, and enabled his Chosŏn (or Yi) dynasty to survive for another 450 years.

Rise of the Yi dynasty

The Mongol Yuan dynasty had interfered in the Korean Peninsula from the late 12th century until 1368, when it was overthrown by the Ming dynasty. Korea was left in chaos as its Koryŏ kings tried to reverse the effects of a century's authoritarian domination. The redistribution of land and the sacking of pro-Mongol ministers led almost to civil war, but in 1392 Sejong's grandfather Yi Sŏngyye, a former general, stepped in, deposed the last Koryŏ king, and assumed the throne as King T'aejo.

King T'aejo's immediate priority was to secure stability, and the installation of a state ideology based on neo-Confucianism was key to achieving that. This ideology

King Sejong of Chosŏn, also known as Sejong the Great, revolutionized government by making it possible for people other than the social elite to become civil servants.

sought to re-establish proper relations between the ruler and his people, and conferred privileged status on a bureaucratic class that would act as guardian of the social hierarchy. Buddhism had been the dominant ideology under the Koryŏ dynasty, but T'aejo undermined its hold in the region by breaking up large estates controlled by Buddhist temples and redistributing the land, some to Confucian shrines.

See also: The An Lushan revolt 84–85 ▪ Kublai Khan conquers the Song 102–03 ▪ Hongwu founds
the Ming dynasty 120–27 ▪ The Meiji Restoration 252–53

Hyanggyo were Confucian schools,
built throughout provincial Korea
and used for both ceremonial and
educational purposes.

Neo-Confucianism

The neo-Confucianism that
became dominant in Korea under
the Chosŏn dynasty had evolved
in China during the 11th and 12th
centuries as a means to revive
Confucianism, which had declined
in favor of Taoism and Buddhism
under the Tang and early Song.
A more rationalist and secular
form of Confucianism, the new
philosophy rejected superstitious
and mystical elements that had
influenced Confucianism during
and after the Han dynasty. Writers
such as Confucian scholar Zhu Xi
stressed the importance of
morality, respect for social
harmony, and education as
means of understanding the
Supreme Ultimate (*tai qi*), the
underlying principle of the
universe. In practice, however,
neo-Confucian virtues such
as loyalty, determination, and
the belief that a supreme
monarch should rule the state
to parallel the Supreme Ultimate
that governed the universe,
tended to favor a hierarchical,
bureaucratic state staffed
by scholars who jealously
maintained the status quo.

Neo-Confucianism emphasised
the importance of education as a
way of producing a class of literati
capable of ensuring the harmonious
running of the state. T'aejo's
grandson, King Sejong (reigned
1418–1450) raised this principle to
new heights, founding in 1420 the
Chiphyŏn-jŏn (Hall of Worthies), an
elite group of 20 scholars tasked
with research that would promote
the better running of the kingdom.

Encouragement of wider literacy
was an important neo-Confucian
ideal, and T'aejo had already ordered
the foundation of government-
sponsored schools. At the time,
however, Korean was written in
Chinese characters, which were
not well adapted to express the
sounds of the language. Sejong
himself is said to have developed
the simplified script, the Han'gŭl,
whose principles were explained
in *Proper Sounds for the Education
of the People*, a book published in
1445. Having only 28 characters—
later reduced to 24—the script
was far easier than Chinese was
to learn, but its introduction faced
bitter resistance from traditionalist
nobles. They feared it might open
civil service examinations to people
from other social classes, which
would risk diluting their power. As
a result Han'gŭl faded from use,
relegated as the "vulgar letters"
of the lower orders, until its
rediscovery in the 19th century,
since when it has thrived as a
vehicle for Korean nationalism.

The reforms of T'aejo and
Sejong, however, broadly survived,
creating a class of yangban—elite
government officials dedicated to
the perpetuation of the state. The
yangban also acted as a break on
any tendency to autocracy among
the Yi monarchs, which helped
the resulting dynasty to endure for
more than five centuries. ▪

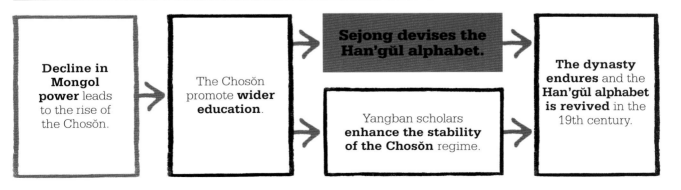

**Decline in
Mongol
power** leads
to the rise of
the Chosŏn.

The Chosŏn
promote **wider
education**.

**Sejong devises the
Han'gŭl alphabet.**

Yangban scholars
**enhance the stability
of the Chosŏn** regime.

**The dynasty
endures** and the
**Han'gŭl alphabet
is revived** in the
19th century.

FURTHER EVENTS

THE ARAB ADVANCE IS HALTED AT TOURS
(732)

By the 8th century, the Islamic people of the Arabian peninsula had conquered much of North Africa and crossed into Europe, occupying Spain and moving into southern France. Their northward expansion seemed unstoppable—until 732, when they met combined Frankish and Burgundian troops at Tours. The Franks and Burgundians won the battle, and the Arab leader, Abdul Rahman Al Ghafiqi, was killed. Although there was another invasion in 735–39, the Arabs never got further than Tours. The Franks kept their power in Western Europe, Christianity was preserved as the continent's dominant faith, and only Spain remained under Muslim rule.

ALFRED RULES WESSEX
(871–99)

Alfred was an able ruler and military leader who successfully defended his kingdom from Danish invaders. He expanded his territory from his base in Wessex (central southern England), uniting a large part of southern England under him. He built fortifications, founded a navy, encouraged education, and promoted Old English as a literary language via translations of Latin books. Alfred became known as "King of the English" and, although the Danes still held the northeast, is seen as the monarch who first embraced the idea of a united England with a distinctive culture based on Christianity and the English language.

THE SPREAD OF THE MISSISSIPPI CULTURE
(c.900)

There was a long tradition, lasting several millennia, of native North American groups based around large earth mounds that had been built for use in rituals or to house the dwellings of the ruling class. These communities were mostly confined to local areas, from Ohio to Mississippi, but the Mississippi culture spread widely through eastern North America. They grew maize intensively, worked copper, and developed hierarchical societies. Recognition of this complex culture has been a key element in debunking the idea that American Indian peoples were primitive and in forming a clearer understanding of their civilization.

OTTO I BECOMES HOLY ROMAN EMPEROR
(962)

German ruler Otto I suppressed revolts, united the Germanic tribes, and defeated outside aggressors such as the Magyars. In addition, he changed the relationship between the ruler and the Catholic church by exercising tight control over the clergy and using his close links to the church to increase royal power. He also extended his rule into northern Italy, creating what became the Holy Roman Empire. This major political power—whose emperors claimed to be the secular leaders of Christian Europe, vying with the Popes for power—dominated much of Europe for more than 900 years.

THE GREAT SCHISM
(1054)

During the late centuries of the 1st millennium CE, the Eastern and Western parts of the Christian church had several disagreements about authority (with the pope claiming seniority over the Eastern patriarchs, but the latter disputing this), the wording of the Creed, and liturgical matters. These disputes came to a head in 1054, when Pope Leo IX and Patriarch Michael I excommunicated one another, creating a split called the Great Schism. This division between what are now the Catholic and Orthodox churches has never been healed.

THE NORMAN CONQUEST OF ENGLAND
(1066)

In 1066, the English king Edward the Confessor died childless, and a dispute arose over who should succeed him. One of the claimants to the throne was Duke William of Normandy, who invaded England, defeated the English at the Battle of Hastings, and was crowned king. This event forged a long-standing link between England and mainland Europe, in which England's rulers held French lands and spoke French. The Normans introduced a new ruling class, built castles and cathedrals, and transformed the English language with many new French-based words, all of which are legacies that still endure.

THE HUNDRED YEARS' WAR
(1337–1453)

The Hundred Years' War was a series of conflicts fought between England and France that began when Edward III asserted his right to the French throne, a claim that the French Valois dynasty disputed. By the end of the war, English possessions in France had been reduced to the coastal town of Calais and its immediate environs. This result transformed England from a power that aspired to be part of a larger European empire to an island nation separate from Europe. France, inspired especially by the leadership of Joan of Arc, gained a stronger sense of national identity.

THE BATTLE OF GRUNWALD
(1410)

At the Battle of Grunwald, a combined Polish and Lithuanian force crushed the army of the Teutonic Knights. This military order, set up to assist crusaders and pilgrims, controlled large territories in Eastern Europe, including Prussia and Estonia, and campaigned against Slavs and pagans in the Baltic. This decisive battle put an end to the Knights' military power, arrested German eastward expansion,

and left the Polish–Lithuanian alliance as the strongest power in Eastern Europe.

THE MONGOL INVASIONS OF JAPAN ARE REPULSED
(1274, 1281)

In the late 13th century, the Mongols were at the height of their power under their leader Kublai Khan. From their base in central Asia, they had moved east to take control of China. In 1271, they sent troops by sea to conquer Japan. The attack was unsuccessful, in part because the Mongol ships were caught in a typhoon, referred to by the Japanese as a *kamikaze* (divine wind). The Mongol defeat was decisive in checking their advance and shaping the idea of a strong, independent Japan, free from outside intervention or influence. This concept of Japanese nationhood lasted for centuries.

SCOTLAND UPHOLDS INDEPENDENCE AT BANNOCKBURN
(1314)

The Battle of Bannockburn, Scotland, was a major clash in an ongoing war between England and Scotland. Despite being vastly outnumbered, the Scots, under King Robert Bruce, inflicted a heavy defeat on the English and their ruler Edward II. This left Bruce in full control of Scotland, from where he continued to lead raids on northern England. The war went on for decades, and Scotland remained independent until 1707. The battle was such a sweeping victory that it is still remembered as a key event in Scottish history,

symbolizing the independence from the rest of Britain to which many Scots still aspire.

THE CONQUESTS OF TAMERLANE
(1370–1405)

Timur, also known as Tamerlane, was the last of the great nomadic Mongol conquerors. In an attempt to revive the great empire of Kublai Khan, he roamed widely across Europe and Asia, from northern India to Anatolia and Russia. By the end of the 14th century, he had conquered Persia, Iraq, Syria, Afghanistan, and eastern Russia, destroying Delhi in 1398, and pushing on toward China in 1405 but dying en route. His empire did not endure, and Mongol horse-based fighting techniques were no match for the firearms that increasingly drove warfare in the 15th century.

THE HUSSITE REVOLT
(1415–34)

The Hussites, followers of the religious reformer Jan Hus, were precursors of the Protestants who lived in Bohemia (modern Czech Republic, then part of the Austrian Habsburg empire) and fought their Catholic rulers for the freedom to worship in their own way. Hus was executed for heresy in 1415, sparking a series of wars that eventually led to the defeat of the Hussites. The area remained under Catholic Hapsburg rule, but most of the people of Bohemia stayed true to their Protestant beliefs. Their revolt against their Catholic rulers in 1618 triggered the Thirty Years' War, when the Bohemian Protestants were again defeated.

THE EAR
MODERN
1420–1795

Y
ERA

Brunelleschi designs the groundbreaking dome of Florence cathedral, signaling the **beginning of the Renaissance**.

Christopher Columbus reaches America, starting an era of **European trade and colonization**, and transforming the **ecology** of the Americas.

Martin Luther writes 95 theses against the Catholic Church, leading to the **Reformation** and the rise of **Protestantism**.

The Battle of Sekigahara ushers in the **Edo Period** in Japan—a time of **unity, stability**, and **artistic** achievements.

1420   **1492**   **1517**   **1603**

1453   **1494**   **1556**   **1618**

The **Ottoman Turks** conquer **Constantinople**, marking the end of the Eastern Roman Empire and creating a new **Muslim capital**.

Spain and Portugal sign the **Treaty of Tordesillas**, dividing the newly conquered **lands** in the **Americas** between them.

Abu Akbar becomes ruler of the **Mughal Empire in India**; Persian and Indian art forms merge to create a unique style.

Religious tensions between **Protestants and Catholics** come to a head at the Defenestration of Prague, leading to the **Thirty Years' War**.

The course of world events always looks different in retrospect from the way that it appears at the time, but the contrast in perspective is rarely as extreme as in the Early Modern Era, which spanned the 15th, 16th, 17th, and 18th centuries. This period is often viewed today as the age during which Europe climbed toward world domination, but to Europeans living at the time it often seemed to be full of unprecedented disasters. The unity of Christendom was split by the Reformation, and sectarian conflict between Catholics and Protestants, combined with power struggles between competing royal dynasties, made Europe a place of frequent warfare—a continent tearing itself apart. Meanwhile, the Muslim armies of the Ottoman Empire threatened the heartland of Europe, seizing the Byzantine city of Constantinople and twice penetrating as far as Vienna.

Yet historical retrospect certainly recognizes changes underway that were to make European nations the founders of the modern world. The flowering of arts and ideas in the Renaissance meant that Europe ceased to be a cultural backwater. Printing and paper, both originally invented in China, were used by Europeans to create mass-produced books that went on to revolutionize the dissemination of information. Gunpowder weapons, also invented by the Chinese, were deployed most effectively by European armies and navies. Above all, explorers and sailors from Europe's western seaboard established oceanic trade routes that laid the foundations for the first global economy.

The start of colonialism
The importance of Christopher Columbus's transatlantic voyage in 1492 cannot be exaggerated. It established a permanent link between two entire ecosystems that had evolved in isolation from each another for almost 10,000 years. The initial impact on the inhabitants of the Americas was catastrophic. Eurasian diseases and the infamous brutality of the Spanish conquistadors decimated the population. A remarkably small number of European invaders conquered the most sophisticated American states with startling ease, laying potentially the entire New World open to European exploitation and colonization.

However, the arrival of European sailors in Asia did not have the same dramatic impact. Powerful

English **religious separatists** (pilgrims) set sail in the *Mayflower* to seek a new life; they found a **colony** in **North America**.

The Royal African Company is established in England; **slaves are taken** from the West African coast for sale in the Americas.

Tsar Peter the Great founds **St Petersburg** on the Baltic coast to encourage **trade** and **modernize** Russia along European lines.

The Battle of Quebec **ends French rule** in **Canada**; it was part of the Seven Years' War, which involved most major European nations.

1620   **1660**   **1703**   **1759**

1649   **1687**   **1751**   **1768**

The English Civil War culminates in the **execution** of King Charles I; England becomes a **republic for** the next 11 years.

Isaac Newton publishes his theories about **gravity** based on **mathematics and logic**, paving the way for the Enlightenment.

The first volume of Diderot's three-part *Encyclopédie* is published, distilling the **rational ideas** of the **Enlightenment**.

Captain Cook **sets sail** on his first voyage; he maps the **New Zealand** coast and claims southeastern **Australia** for Britain.

countries, including India, Imperial China, the Mughal Empire, and the Japanese shogunate at first merely tolerated the Europeans as traders, allowing them to control only a few islands or enclaves along the coast, as long as they did not interfere or become too troublesome.

Economic growth

From the second half of the 17th century, signs of economic growth accelerated in Europe. Productivity of labor in trades and agriculture increased notably in areas like the Dutch Netherlands. New financial institutions, such as central banks and joint stock companies, laid the foundations of modern capitalism. Complex patterns of maritime trade linked European colonies in the Americas to Europe, Africa, and Asia. Slaves, mostly bought by

European traders in West Africa, were transported in vast numbers to work on colonial plantations, so that in some parts of the New World people of African descent greatly outnumbered both Europeans and the native population. At home, Europeans consumed luxury goods from China and India, and products such as sugar and coffee from plantations in the Caribbean and Brazil. North America, the West Indies, and India were all regions of colonial contention—the precipitous decline of the Mughal Empire having opened up parts of India to European territorial conquest.

Intellectual movements

Even at this stage, the degree of European ascendancy should not be exaggerated. China had gone through difficult times in the mid-

17th century transition from the Ming to the Qing dynasty, but in the 18th century, imperial China was enjoying a golden age of power and prosperity. The population of Europe had begun a steep increase to unprecedented levels—a result of improved food production and declining epidemic diseases—but China also experienced rapid population growth.

What really marked out Europe as unique at this time was the development of knowledge and thought. The 17th-century scientific revolution began a transformation of our understanding of the universe. The rationalist movement known as the Enlightenment challenged all preconceptions, traditions, and conventions. The modern world was under construction in the European mind. ■

AS MY CITY FALLS, I SHALL FALL WITH IT

THE FALL OF CONSTANTINOPLE (1453)

IN CONTEXT

FOCUS
The Ottoman Empire

BEFORE
1071 Turkish forces inflict a significant defeat on the Byzantine Empire at the Battle of Manzikert.

1389 The Ottomans defeat the Serbs at Kosovo, making possible Ottoman advance into Europe.

1421 Murad II comes to the Ottoman throne and plans extensive conquests.

AFTER
1517 The Ottomans conquer Mameluke Egypt.

1571 The Ottoman navy suffers a crushing defeat at Lepanto.

1922 The empire ends with the foundation of modern Turkey.

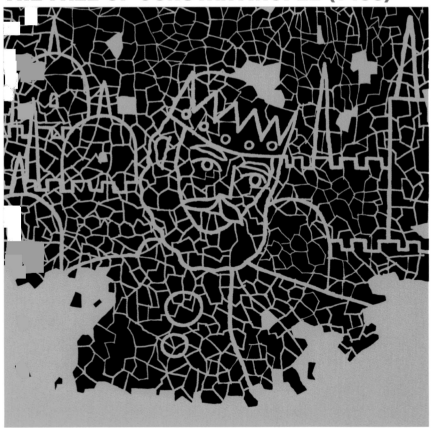

In 1453, the Ottoman Turks attacked and took the city of Constantinople, the capital of the Byzantine Empire. The loss of this millennium-old Christian empire, which had once stretched virtually all the way around the Mediterranean, was a profound shock to the Christian world. As if to symbolize the Muslim victory Sancta Sophia, one of the greatest cathedrals in Christendom, was converted into a mosque.

The Ottoman Turks had already conquered much of the surrounding territory before Sultan Mehmet II (1432–1481) laid siege to the city and bombarded it with heavy artillery. Having breached its walls, his army of more than 80,000 men then overwhelmed the small force

See also: Belisarius retakes Rome 76–77 ▪ Muhammad receives the divine revelation 78–81 ▪
The founding of Baghdad 86–93 ▪ The fall of Jerusalem 106–07 ▪ The Young Turk Revolution 260–61

inside. Constantine XI, the last Byzantine emperor, was killed, and with the fall of the city, his empire ended. Constantinople then became the capital of the Ottoman Empire, which lasted until 1922.

A weakening empire

The Byzantine Empire was already in terminal decline by the time Constantinople was taken. It had shrunk to include only the capital city, some land to its west, and the southern part of Greece. The decline began at the Battle of Manzikert (1071), during which the army of the Turkish Seljuk dynasty drove the Byzantines out of their crucial territory in Anatolia. From this point, rival claims for the Byzantine crown, disputes over tax, loss of trade revenue, and poor military leadership all contributed to the contraction of the empire.

In 1203, the Fourth Crusade— a western European expedition originally intended to conquer Jerusalem—became entangled in the empire's politics. Some of the crusade leaders pledged to help restore the deposed Byzantine Emperor Isaac II Angelos in return

[Blood flowed] like rainwater in the gutters after a sudden storm.
Nicolò Barbaro
Eyewitness to the fall of Constantinople (1453)

for support for their expedition. They were initially successful: Angelos' son was crowned as co-emperor but, in 1204, he in turn was deposed by a popular uprising. The Byzantine senate elected a young noble, Nicolas Canabus, as emperor, and he refused to back the crusaders. Denied their promised payments, the crusaders and their allies, the Venetians, responded with a ruthless attack on the city. They raped and killed civilians, looted churches, and demolished priceless works of art. Constantinople was all but destroyed.

Rise of the Ottomans

Before capturing Constantinople, the Ottoman Empire had already expanded from Anatolia into the Balkans. Afterward, in the 16th century, it expanded into the eastern Mediterranean, along the banks of the Red Sea, and into North Africa. The defeat of the Mamelukes in Egypt in 1517, and

When lighted tapers were put to the "innumerable machines" ranged along a four-mile section of the city walls, the world's first concerted artillery barrage exploded into life.

wars against the Safavids, one of Persia's most significant ruling dynasties, gave the Ottomans control of a whole swathe of the Arab Middle East.

The Ottoman Empire was a Muslim State and the sultans saw it as their duty to promote the spread of Islam. Nevertheless it tolerated Christians and Jews in a subsidiary status and made extensive use of slaves. Many languages were spoken and faiths followed within its domains, but it dealt with the potentially conflicting religious and political differences by setting up vassal (subordinate) states in some regions. Territories such as Transylvania and the Crimea paid tribute (made regular payments) to the emperor, but they were not » ruled by him directly, and acted as buffer zones between Muslim and Christian areas. Some vassal states, including Bulgaria, Serbia, and Bosnia, were eventually absorbed into the larger empire; others retained their vassal status.

Government and military

The Ottomans evolved a strong system of government that combined local administration with central control. The sultan—whose brothers were customarily murdered at his accession—was supreme ruler. He had a council of advisers, later a deputy, who ruled on his behalf. Local areas were ruled by military governors (beys) under the emperor's overall control, but local councils kept the beys' authority in check.

Non-Muslim communities within the empire were allowed a degree of self-rule through a system of separate courts called millets. The millets allowed Armenian, Jewish, and Orthodox Christian communities to rule according to their own laws in cases that did not involve Muslims. This balanced combination of central and local control enabled the Ottomans to

Janissaries wore distinctive uniforms and, unlike other military units, were paid salaries and lived in barracks. They were the first corps to make extensive use of firearms.

hold together a large and diverse empire for much longer than would have been possible with a more wholly centralized system.

The Ottoman army was also crucial to the empire's success. It was technically advanced—employing cannon from the siege of Constantinople onward—and tactically sophisticated. Its high-speed cavalry units could turn what looked like a retreat into a devastatingly effective flanking attack, surrounding the enemy in a crescent-shaped formation that would take them by surprise.

At the heart of the army were the Janissaries, a unit of infantry that began as the imperial guard and expanded to become the most feared elite force of the period. Initially, the unit was made up of men who, as children, had been abducted from Christian families in the Balkans. Under the *devsirme* system, which was also known as the "blood tax" or "tribute in blood," boys aged from eight to 18 were taken by Ottoman military, forcibly converted to Islam, and sent to live with Turkish families, where they learned the Turkish language and customs. They were then given rigorous military training, and any who showed particular talent were selected for specialized roles ranging from archers to engineers.

Mehmet II

Mehmet (1432–1481), the son of the Ottoman emperor Murad II, was born in Edirne, Turkey. As was usual for an heir to the Ottoman throne, Mehmet had an Islamic education, and at 11 years old was appointed governor of a province, Amasya, to gain experience of leadership. A year later, Murad abdicated in favor of his son, but shortly afterward was called back from his retirement in Anatolia to lend military support. "If you are the Sultan," Mehmet wrote, "come and lead your armies. If I am the Sultan, I order you to come and lead my armies."

Mehmet's second, main, rule was from 1451 to 1481. His victory at Constantinople was followed by a string of further conquests: the Morea (southern Greece), Serbia, the coast of the Black Sea, Wallachia, Bosnia, and part of the Crimea. He rebuilt Constantinople as his capital and founded mosques there, while also allowing Christians and Jews to worship freely. Known for his ruthless military leadership, he also welcomed humanists to the capital, encouraged culture, and founded a university.

Naturalistic motifs in cobalt blues and chrome greens surround Islamic calligraphy in these Iznik wall tiles, commissioned for the Topkapi Palace during the classical age of Turkish art.

Janissaries were not permitted to marry until they retired from active duty, but they received special benefits and privileges designed to secure their sole allegiance to the ruler. Although they made up only a small proportion in the Ottoman army, they had a leading role and played a key part in many victories, including those over the Egyptians, Hungarians, and Constantinople.

The Ottoman heyday
The empire reached its peak under Emperor Suleiman the Magnificent. He forged an alliance with the French against the Habsburg rulers of the Holy Roman Empire, and signed a treaty with the Safavid rulers of Persia that divided Armenia and Georgia between the two powers and put most of Iraq into Ottoman hands. Suleiman conquered much of Hungary, and even laid siege to Vienna, although he did not succeed in taking it.

The Ottomans took their Islamic faith to their territories, building mosques everywhere—and with the mosques came scholarship and education. Ottoman cities were impressive. Constantinople itself was virtually rebuilt: the Ottomans reinforced its fortifications as well as adding many mosques, bazaars, and water fountains. The city's dazzling centerpiece was the royal palace of Topkapi, commissioned by Sultan Mehmet II in around the 1460s. Masons, stonecutters, and carpenters were summoned from far and wide to ensure the complex would be an enduring monument. It contained mosques, a hospital, bakeries, and a mint among much else, and attached to it were imperial societies of artists and craftsmen who produced some of the finest work in the empire.

Gradual decline
This cultural flowering continued after Suleiman's death, but the empire faced serious challenges in other arenas. A rising population was putting pressure on available land; there were military threats and internal revolts; and defeat by a coalition of Catholic forces at the sea battle of Lepanto in 1571 prevented the empire's expansion further along the European side of the Mediterranean.

The Ottoman empire steadily lost prestige and influence until its decline earned it the title "the sick man of Europe." Incapable of responding to the convulsions of the 19th century, it lost territory and struggled against a rising tide of nationalism among its conquered peoples. Its long history finally ended with defeat in World War I and the foundation of the modern Turkish state by Kemal Attatürk. ∎

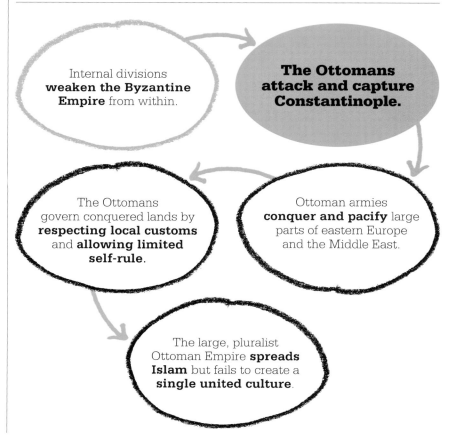

Internal divisions **weaken the Byzantine Empire** from within.

The Ottomans attack and capture Constantinople.

The Ottomans govern conquered lands by **respecting local customs** and **allowing limited self-rule**.

Ottoman armies **conquer and pacify** large parts of eastern Europe and the Middle East.

The large, pluralist Ottoman Empire **spreads Islam** but fails to create a **single united culture**.

FOLLOWING THE LIGHT OF THE SUN

WE LEFT THE OLD WORLD

CHRISTOPHER COLUMBUS REACHES AMERICA (1492)

IN CONTEXT

FOCUS
Voyages of discovery

BEFORE
1431 Portuguese navigator Gonçalo Velho sails on a voyage of exploration to the Azores.

1488 Bartolomeu Dias rounds the Cape of Good Hope, discovering the passage around southern Africa.

1492 King Ferdinand and Queen Isabella of Spain agree to sponsor Columbus's voyage.

AFTER
1498 Vasco da Gama's fleet arrives in Calicut, India.

c.1499 Italian explorer Amerigo Vespucci discovers the mouth of the Amazon.

1522 Ferdinand Magellan's Spanish expedition to the East Indies, from 1519 to 1522 results in the first circumnavigation of the Earth.

Europeans develop **a taste for Asian spices** and luxury goods.

Land routes to Asia are hazardous and **blocked by the Ottoman Empire**.

The **Portuguese** explore **Indian Ocean routes**.

After the fall of Granada, **Spanish religious zeal** turns outward.

The **Spanish Crown supports** the exploration of a potential route **to Asia across the Atlantic Ocean**.

Columbus sets sail westward across the Atlantic to Asia, but instead reaches America.

Christopher Columbus (c.1451–1506), an Italian-born navigator and trader from Genoa, made a journey in 1492 that initiated a lasting contact between America and Europe, and changed the world.

When he set out, Columbus was expecting to reach Asia, since no Europeans at the time knew that an entire continent blocked this route. When he reached an island in the Bahamas after sailing for over two months, he believed that he had arrived at the outer reaches of Indonesia. From there, Columbus continued to explore the Caribbean, visiting Cuba, Hispaniola, and several of the smaller islands. He met with a mostly peaceful response from the native people, whom he observed might make good servants or slaves. He also noticed their gold jewelry, and took a sample of local gold, as well as some native prisoners, back to Europe.

Columbus was to return to the Caribbean on three later voyages, bringing in his wake countless European visitors and settlers.

Motivation to explore

The rulers and merchants of Western Europe wanted to explore the Atlantic for primarily economic reasons. Spices that would not grow in Europe's climate, such as cinnamon, cloves, ginger, nutmeg, and pepper, were prized not only for their taste but also because they could help to preserve foods. There was also an enthusiastic market for luxury goods such as silk and precious stones, commodities that came primarily from the islands of the Indonesian group, such as the Moluccas, which were known in Europe as the Spice Islands.

Bringing such commodities across Asia by land was difficult and dangerous because of local wars and instabilities along the route; it was also costly, since during their journey goods would pass through many different merchants' hands. There were certainly excellent economic reasons to develop sea routes: anyone who could find a more direct way of importing these goods to Western Europe would become very rich.

See also: The Viking raid on Lindisfarne 94–95 ▪ The Treaty of Tordesillas 148–51 ▪ The Columbian Exchange 158–59 ▪ The voyage of the *Mayflower* 172–74 ▪ The formation of the Royal African Company 176–79

Another reason why Europeans started to explore sea routes in the late Middle Ages was to investigate the possibility of establishing European colonies in Asia. These could act not only as trading posts, but also as bases for missionaries, who could convert the locals to Christianity. This they believed would help to reduce the perceived threat of Islam.

By the 14th and 15th centuries, the Spanish, Portuguese, English, and Dutch had developed ocean-going ships, and trained sailors who could navigate over long distances. Explorers used various types of vessels, among the most successful of which was the caravel—a fast, lightweight, and extremely maneuverable ship that was usually equipped with a mix of square and lateen (triangular) sails. The lateen sails made it possible to sail to windward (into the wind), which allowed explorers to make progress even in variable wind conditions. Explorers also used the carrack, or nau, a larger vessel that was similarly rigged. On his first

transatlantic voyage, Columbus took two caravels, each probably of 50–70 tons, and one carrack of about 100 tons, the extra capacity being useful for carrying stores.

Skills and technology quickly developed in both shipbuilding and navigation. Sailors used the cross-staff—a basic sighting device—or later a mariner's astrolabe, to calculate a vessel's latitude. They achieved this by measuring angles, such as the angle of the sun to the horizon. They used a magnetic compass to gauge direction, and theircharts and knowledge of prevailing winds and currents improved with each voyage.

Portuguese navigators

European navigators had been striking out into the Atlantic for many decades. Sailors from Bristol, England, for example, were sailing in the 1470s in search of a mythical island called "Brasil," thought to be west of Ireland. The Portuguese established trading colonies on Madeira, and Prince Henry the Navigator, son of Portugal's King

> I intend to go and see if I can find the island of Japan.
> **Christopher Columbus, 1492**

John I, commissioned numerous journeys of exploration to the Azores in the 15th century. Henry had started the first school for oceanic navigation, with an astronomical observatory at Sagres, Portugal in about 1418. Here he promoted the study of navigation, map-making, and science. Henry sent ships down the west coast of Africa, to which he was particularly attracted by the potential to trade in slaves and gold. His ships pushed southward, setting up trading posts along the »

Christopher Columbus

Born in Genoa, Christopher Columbus became a business agent for several prominent Genoese families and undertook trading voyages in Europe and along the African coast.

Columbus followed his voyage to America with a second in 1493, during which he explored the Lesser and Greater Antilles, and set up a colony at La Isabela in what is now the Dominican Republic. His third voyage (1498–1500) took him to the Caribbean island of Hispaniola and on to Trinidad, where he found the coast of South America and

guessed, from the size of the Orinoco River, that he had found a huge land mass. During this time, settlers complained to the Crown about the way he ran his Caribbean colony, and he was dismissed as governor.

On his last voyage (1502–04) he sailed along the Central American coast, hoping to find a strait to the Indian Ocean. He returned to Spain in poor health and an increasingly disturbed state of mind, feeling he had not received the recognition and benefits he had been promised. Columbus died in 1506.

way. Subsequent rulers continued to sponsor voyages and, in 1488, Portuguese captain Bartolomeu Dias rounded the southern tip of Africa. Soon another Portuguese navigator, Vasco da Gama, led the push to round the Cape and pressed on across the Indian Ocean, linking Europe and Asia for the first time by ocean route.

Since Portugal dominated the sea route along the African coast, Portugal's European neighbor and rival Spain needed to find an alternative route, if it was to gain access to the riches of the East. Although educated people knew by this time that the Earth was round, they did not know about the existence of the Americas. An alternative way to the East seemed, therefore, to be to sail west across the Atlantic. This route seemed especially attractive to the many seamen—including Christopher Columbus—who believed the planet's diameter to be rather smaller than it actually is.

Seeking sponsorship

In 1485, Columbus presented to John II, king of Portugal, a plan to sail across the Atlantic to the Spice Islands. John refused to invest in the scheme, however. This was partly because Portugal was already exploring the West African coast with some success, and partly because the experts John consulted about the proposal were skeptical about the distances involved.

Columbus cast his net more widely, seeking backing from the powerful maritime cities of Genoa and Venice, and sending his brother to England to do the same—but still he received no encouragement. He therefore turned to Ferdinand of

Such inhumanities and Barbarisms were committed... acts so foreign to human nature that I now tremble as I write.
Bartolome De Las Casas
Spanish historian (c.1527)

Aragon and Isabella of Castile, the "Catholic Monarchs" who jointly ruled Spain. At first they turned him down, their navigational consultants also skeptical about the length of his proposed route, but eventually,

---

Columbus's voyage was a bold undertaking. Despite a general understanding that the world was spherical, many believed the westward journey was doomed to fail, fearing the crew would die of thirst before ever reaching land.

Start

The voyage to America and back lasted seven months, from August 3rd, 1492—March 15th 1493.

On August 3rd, 1492, Columbus departed Spain with three ships: the Niña, Pinta, and Santa Maria.

The crew consisted of 87 men—20 on the Niña, 26 on the Pinta, and 41 on the Santa Maria.

Provisions on board the ships included vinegar, olive oil, wine, salted flour, biscuits, dry legumes, and salted sardines.

Columbus calculated that Asia was 2,400 miles away from Spain. In fact it is around 12,200 miles away.

On October 12th, 1492, the ships finally reached the Bahamas.

Finish

Columbus discovered Hispaniola in 1492 when his flagship ran aground on its shores. La Isabela, founded there in 1496, is the oldest permanent European settlement in the Americas.

after protracted negotiations, they agreed to sponsor the voyage. Securing a new trade route would certainly bring material rewards, but Isabella also saw the voyage in terms of a religious mission that could bring the light of Christianity to the East.

Columbus sails west

Having been granted viceroyship and governorship of any lands he could claim for Spain, plus other benefits including 10 percent of any revenues they yielded, Columbus set sail westward in 1492. He called at Gran Canaria before sailing west, sighting land five weeks later. In early 1493 he returned to Europe with two ships, the third having been wrecked off the coast of present-day Haiti, and was duly appointed Governor of the Indies.

Columbus's second expedition was organized just a few months later. This involved 17 ships loaded with some 1,200 people who would found Spanish colonies in the Caribbean. As well as farmers and soldiers, the colonists included

I should not proceed by land to the East, as is customary, but by a Westerly route.
Christopher Columbus, 1492

priests, who were specifically charged with converting local people to Christianity. Religious conversion became a key part of European colonization, illustrating the colonist's ambition to impose their own culture and exert control over newly colonized peoples.

Columbus's achievement in 1492 is often described as the European "discovery" of America. This is a problematic claim not only because Columbus thought he had reached Asia, but also because Vikings from Scandinavia had reached North America some 500 years earlier—archaeological

remains at L'Anse aux Meadows in Newfoundland reveal that they even settled there. However, the Viking settlement was not long-lived, and was unknown to Columbus and his contemporaries.

Nevertheless, Columbus's 1492 journey did inaugurate a lasting contact between the Americas and Europe. The pitiless destruction he and his men wrought upon the indigenous peoples of the West Indies, whom he encountered when he first arrived in the Americas, also began a process of decimation of American Indian populations that would continue for a century. ∎

THIS LINE SHALL BE CONSIDERED AS A PERPETUAL MARK AND BOUND

THE TREATY OF TORDESILLAS (1494)

Spain and Portugal signed a treaty on June 7, 1494, at Tordesillas in Spain, that resolved the countries' disputes about the possession of newly discovered territory. The rulers settled on a meridian 370 leagues west of the Cape Verde Islands as a line of demarcation. All the lands to the west of this line would belong to Spain; all those to the east would belong to Portugal. The line was chosen because of its location: it lies roughly halfway between the Cape Verde Islands, which already belonged to Portugal, and the Caribbean islands, which Christopher Columbus had claimed for Spain in 1492.

See also: Marco Polo reaches Shangdu 104–05 ▪ The foundation of Tenochtitlan 112–17 ▪ Christopher Columbus reaches America 142–47 ▪ The Columbian Exchange 158–59 ▪ The formation of the Royal African Company 176–79

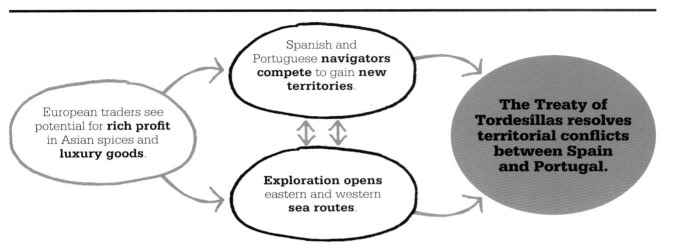

European traders see potential for **rich profit** in Asian spices and **luxury goods**.

Spanish and Portuguese **navigators compete** to gain **new territories**.

Exploration opens eastern and western **sea routes**.

The Treaty of Tordesillas resolves territorial conflicts between Spain and Portugal.

By the 1490s both countries were discovering substantial territories, including lands in the New World, although at this point the size and extent of the Americas was unclear to Europeans. In spite of the fact that the Spanish Crown had funded Columbus's voyages, Spain's claim to his discoveries was not clear-cut. The 1479 Alcaçovas treaty between the Catholic Monarchs of Spain and the rulers of Portugal gave all newly discovered lands south of the Canary Islands to Portugal. When Columbus landed at Lisbon after his first voyage, he told John II, king of Portugal, that he was claiming Hispaniola and Cuba for his Spanish backers. John wrote to Spain's rulers immediately to say that he was preparing to send his own ships to claim the Caribbean for Portugal.

Legalizing possession
To prevent such disputes erupting each time a navigator made a fresh discovery, the leaders of both countries decided to review the terms of the Alcaçovas treaty. The papacy had been involved in the 1479 treaty, and now Pope

Alexander VI (a Spaniard) proposed a combined north–south and east–west dividing line, suggesting that any lands west and south of a line 100 leagues west and south of the Azores and Cape Verde Islands be allocated to Spain. John rejected the proposition, considering it to be biased in favor of his rivals, and eventually all parties agreed on the meridian between the Cape Verde Islands and the Caribbean. The resulting treaty set the agenda for future colonization, and influenced the fate of vast swathes of the world.

I and my companions suffer from a disease of the heart that can be cured only with gold.
Hernán Cortés, 1519

Portuguese colonies
By the time the Tordesillas treaty was signed, Portugal had already made headway in exploring Africa and southern Asia. Working south from a North African base at Ceuta, explorers established a series of trading posts on the West African coast, gradually pushing south until, in 1498, Vasco da Gama rounded the Cape of Good Hope and sailed into the Indian Ocean. In the 16th century, Portugal had settlements in India, the Moluccas, Sumatra, Burma, and Thailand, and by 1557 they had established their long-standing enclave in Macau, which became a hub for their trade with many Asian communities.

The treaty line passed through South America, allocating a north-western portion to the Portuguese. In 1500, explorer Pedro Álvares Cabral landed on the coast of Brazil and claimed it for Portugal. The conquistadors exploited their new colony, forcing indigenous peoples to cultivate sugar cane, and later to grow coffee, and mine gold. The laborers died in huge numbers, both from diseases introduced by » the colonists, and as a result of their ruthless treatment, and slaves were brought in from Africa to replace them. Brazil, ruled from the mid-16th century by Portuguese governors-general, remained a colony until the early 19th century.

The Spanish in America

Following Columbus's transatlantic voyages and the settlement of the treaty, Spain turned increasingly to America, sponsoring expeditions that combined exploration with conquest and colonization. The first of these, led by Hernán Cortés, was to Mexico, which was then home to

The siege of Tenochtitlan, the Aztec capital, was decisive in the Spanish conquest of Mexico, and brought the Spaniards a step closer to their goal of colonizing the Americas.

the small but rich Aztec Empire. The empire's large, central capital was at Tenochtitlan (modern Mexico City). With just a small force of about 600 men, Cortés overthrew the million-strong empire, eventually killing its ruler, Moctezuma. Another Spanish leader, Francisco Pizarro, conquered the Inca Empire, which centered on Peru but also included Chile, Ecuador, and large parts of Bolivia and northwestern Argentina. Again with just a small force (180 men), Pizarro laid the foundations of another Spanish stronghold and source of great wealth in precious metals. Peruvian silver became the main source of Spain's income from its colonies.

Several factors contributed to Cortés' and Pizarro's astounding conquests. The Aztecs were overwhelmed by a kind of battle

unknown to them, involving firearms and the single-minded slaughter of opponents—Aztec practice was to capture prisoners, whom they would later kill in ritual sacrifice. The Spanish were also helped by alliances they made with local peoples who were hostile to the Aztecs. The result for Spain was a flow of wealth across the Atlantic and a secure base for building on their involvement in the Americas.

Further Spanish colonization followed, including that of Colombia, known to the Spanish as New Granada. By the end of the 17th century, much of western and central South America was in Spanish hands. Conquered areas, and the people who lived in them, were parceled out to the Spanish conquistadors, who

>
>
> Those regions which we found and explored with the fleet… we may rightly call a New World.
> **Amerigo Vespucci, 1503**

undertook to convert the locals to Christianity. They did convert them, but they also made them perform forced labor, especially in the silver mines. Laborers fell victim to disease and exploitation—like their counterparts in Brazil, but on a lesser scale—and slaves from Africa were brought in to supplement their numbers.

The Spanish Crown tried to control this large empire, appointing viceroys to rule over the settlers and the native American peoples, and taking a fifth of the profits from silver mining. Settlers increasingly resisted this external interference, however, and by the 19th century the empire was diminishing as areas from Colombia to Chile won their independence.

Circumnavigation

The Treaty of Tordesillas set the seal of approval on Spain's activity in America, but this deterred neither Spain nor Portugal from looking for a westward route to eastern Asia, a potential source of spices, luxury goods, and great wealth for traders from Europe. Amerigo Vespucci, an Italian navigator working for the Portuguese Crown, was one of the first to take this exploration further. He explored the coast of South America, and is remembered because America is named after him. The Portuguese navigator Ferdinand Magellan was next to explore this route, this time on behalf of Spain. He believed that the Spice Islands could be less than halfway around the world when sailing west from the treaty line, which would give Spain a claim to them. In 1519 he set out with five ships in an ambitious attempt to make the first circumnavigation of the globe. Although Magellan himself died en route, some of the expedition survivors completed the voyage, giving Spain a basis for its claim to land in Southeast Asia.

In 1529 the rival Crowns signed another treaty at Zaragoza. This agreement assigned the Philippines to Spain and the Moluccas to Portugal.

The treaty's heritage

European countries not party to the Tordesillas agreement simply ignored it, and soon began to move in to develop their own empires. Britain colonized North America, for example, the Dutch moved into the Spice Islands, and several European countries set up colonies in the Caribbean. The treaty did, however, influence a significant proportion of the world. It underlined a development that was already beginning in Europe in which wealth and influence were passing from the old central European powers (based in the Holy Roman Empire) to the coastal, maritime powers that looked to build empires in new territories. These empires brought both Spain and Portugal enormous riches, and their overseas empires left a significant cultural legacy: much of South and Central America is Spanish-speaking, and there is a major Portuguese heritage in parts of Africa and Asia, the greatest of all being in Brazil. ■

Ferdinand Magellan

Born into a noble Portuguese family, Magellan (1480–1521) was orphaned as a boy, and sent to the Portuguese royal court to act as page.

As a young man, he became a naval officer. He served in Portugal's colonies in India and took part in the conquest of the Moluccas, but after a disagreement with the Portuguese king, he went to Spain to look for support for his venture westward. By 1518 he had the backing of the Spanish king Charles I, and set off the following year with five ships.

After losing one ship to the weather and another to a desertion, Magellan navigated the narrow sea route (named the Strait of Magellan in his honor) between what is now mainland South America and Tierra del Fuego. He emerged in an ocean he named Pacific, because of its calmness. He crossed this expanse of water, stopping at Guam, and then in the Philippines, where he was killed. Only one ship, under Juan Sebastien del Cano, made it back to Europe in 1522, having achieved the first circumnavigation of the globe.

THE ANCIENTS NEVER RAISED THEIR BUILDINGS SO HIGH

BRUNELLESCHI DESIGNS THE DOME OF FLORENCE CATHEDRAL (1420)

IN CONTEXT

FOCUS
The Renaissance

BEFORE
1296 Building work begins on the Santa Maria del Fiore cathedral (Il Duomo), Florence.

1305 Giotto completes his frescoes at the Arena (Scrovegni) Chapel in Padua.

1397 The Medici bank is founded in Florence; becomes the largest bank in Europe.

AFTER
1434 Cosimo de' Medici becomes de facto ruler of Florence and supports the arts.

1447 Francesco Sforza comes to power in Milan. His court becomes a center of culture.

1503 Leonardo da Vinci starts work on the *Mona Lisa*.

1508 Michelangelo begins to paint the Sistine Chapel ceiling in the Vatican.

In 1418, the wealthy Guild of Wool Merchants of Florence launched a competition to find a design for a dome to complete their unfinished cathedral—the Cattedrale di Santa Maria del Fiore, commonly known as Il Duomo. The city of Florence was one of the richest in Italy, a center of banking and trade and it was on the basis of this wealth that the city could afford to commission a cathedral dome of unprecedented size.

This lavish spending on art and architecture would soon be echoed across Italy, as the region's growing prosperity meant that rulers and rich citizens could spend money to beautify their towns and enhance

This enormous construction towering above the skies, vast enough to cover the entire population of Tuscany with its shadow.
Leon Battista Alberti
On Painting and Sculpture
(1435)

their prestige. The strong economy and deep civic pride in Italy laid the foundations for one of the most significant intellectual movements in history: the Renaissance.

Il Duomo

At the time of the competition, Florence's cathedral featured a vast octagonal space toward its eastern end, but since work on the building began in 1296 no one had worked out how to make a dome to cover it. The dome would have to be the largest cupola constructed since the late Roman period and the guild specified that it should be built without external buttresses, favored by their political rivals in France, Germany, and Milan and also considered old-fashioned. This seemed an impossible task. The young goldsmith and clockmaker-

Dominating the skyline of Florence, Brunelleschi's groundbreaking dome remains the tallest building in the city, rising majestically from the surrounding red-tiled roofs at 374 ft (114 m) high.

turned- architect Filippo Brunelleschi won the competition with his daring plan for a huge eight-sided brick dome, but many doubted that he would be able to construct it.

The main problem was being able to support the structure in such a way that it did not spread and collapse under its own weight. Brunelleschi's ingenious solution was to construct two concentric domes—an inner supporting dome and a larger outer one. The domes were then joined together with huge brick arches and a complex interlocking system of "chains" made from rings of stone and wooden beams that were attached by iron clamps to prevent the dome from expanding outwards

The result—which was finally completed in 1436—remains the largest masonry dome in the world. Combining the style of antiquity with new engineering techniques, it exhibited the blend of ancient wisdom and modern knowledge that typified the Renaissance.

The Renaissance in Italy

Meaning "rebirth," the Renaissance was a movement that started in Italy and began to spread across Europe from the mid-15th century. Its roots lay in the rediscovery of the culture of ancient Greece and Rome and it influenced all the arts, as well as science and scholarship. Painters, sculptors, and architects broke free from the traditions of medieval art. They visited the monuments of ancient Rome, looking at classical statues and the carvings on Roman buildings, and created works of art in the classical style. This new movement inspired architects, such as Leon Battista Alberti and Brunelleschi, and a wave of great artists, including Michelangelo and Leonardo da Vinci. Most of these figures were active in many fields— Brunelleschi was a sculptor and engineer as well as an architect; Michelangelo painted, sculpted, and wrote poetry; while da Vinci's achievements spanned both the arts and the sciences. »

Michelangelo's painted ceiling at the Sistine Chapel in the Vatican combines the Renaissance interest in physical beauty and realism with religious subject matter.

Renaissance painters and sculptors sought to represent the physical world in a more realistic way than their Medieval predecessors: they valued anatomical accuracy and developed scientific methods of illustrating perspective. As in classical art, there was more focus on human beauty and the nude.

There was also a revival of interest in classical learning, which was influenced by Greek scholars from the Byzantine Empire, who settled in Italy when Constantinople (the empire's capital) fell in 1453. The *émigrés* brought with them ancient Greek literary, historical, and philosophical texts, which had been lost to the West, and taught the Italians Greek so they could read and translate the works. This led to the emergence of Renaissance Humanism in Italy, which involved studying the humanities—grammar, rhetoric, history, philosophy, and poetry—and, more broadly, a high regard for the dignity and potential of the human race.

At the time of the Renaissance, life, business, and politics in Italy were dominated by a number of powerful city-states—mainly Florence, Milan, Ferrara, and Venice—together with Rome, from where the pope could exercise great secular ("temporal") power as well as being the spiritual head of the Catholic Church. The city-states generated a lot of wealth from trade and—as in the case of Florence—banking. Their ruling families, such as the Gonzaga in

The idea of the Renaissance Man, whose expertise and curiosity extends to a range of diverse subjects, reflects the great thinkers of the era: polymaths such as Leonardo da Vinci, who mastered disciplines from art to science.

Humanism placed mankind at the center of the universe. It gave the credit for human accomplishments to people instead of God.

The rediscovery of classical texts inspired thinkers to emulate and even surpass the work of philosophers such as Aristotle.

Science and a growing knowledge of how the world works contributed to fields as diverse as architecture and medicine.

Renaissance artists made several great achievements, which were inspired by the discovery of lifelike Greek and Roman sculpture, and aided by a new understanding of perspective.

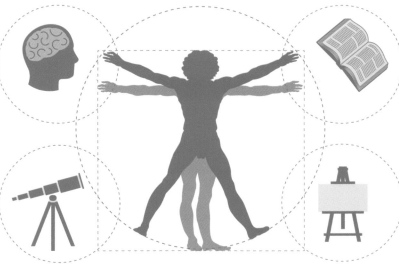

Mantua, the d'Este in Ferrara, the Sforza in Milan, and the Medici in Florence, spent lavishly on palaces, churches, and works of art, and became patrons of many great Renaissance artists. These wealthy families also encouraged the revival of classical learning by employing scholars as tutors for their children. In addition, several members of the Medici family became popes.

Spread of the Renaissance

From the end of the 15th century, the Renaissance spread from Italy to other parts of Europe and a Northern Renaissance emerged. Northern countries, particularly the Netherlands and Germany, produced their own great artists, such as Albrecht Dürer (1471–1528) and Hans Holbein the Younger (1497–1543)—both gifted realists. Renaissance Humanism also spread northward, but northern writers and philosophers, most notably Erasmus of Rotterdam (1466–1536), tended to place more emphasis on Christianity, education, and reform than their Italian counterparts.

The invention of printing using movable type by Johannes Gutenberg in Germany in the 1430s enabled Renaissance ideas to spread even more quickly. Before Gutenberg, the only way printed

For the wise man there is nothing invisible.
Filippo Brunelleschi

text was possible was for each page to be carved by hand into a block of wood, but as this was so laborious books were invariably written out by hand. Gutenberg's method involved arranging individual metal letters and punctuation symbols in lines and pages; when many copies of a page had been printed, the type could be taken apart and reused. He combined this new idea with the existing technology of paper-making and the kind of press used in wine production, and the result was the printing of multiple copies of books for the first time.

Gutenberg's invention had a major impact. It meant that books, which had previously been costly and took months to produce, were now easily available and much more affordable, so ideas and information could circulate quickly and reach more people. While the church had used mostly Latin as its universal language, writers now wrote in their local tongues, and as a result literature in French, English, German, and other languages flourished. In addition, copies of the ancient classics were reproduced in quantity, thus helping to spread ideas that were central to both the Renaissance and Humanism.

The Renaissance's impact

By the mid-16th century, the influence of the Renaissance was waning in southern Europe, but it lasted slightly longer in the north. However, many great Renaissance works endured and they continued to inspire future generations of painters and architects. Indeed, the longstanding popularity of oil paintings and classical style of architecture, and the rise of Humanism, would all have been impossible without the movement that began with Brunelleschi in Florence in the 15th century. ∎

Filippo Brunelleschi

Born in Florence, Filippo Brunelleschi (1377–1446) was the son of a civil servant, who educated Filippo in the hope that he would follow in his footsteps. However, Filippo was artistically talented and instead trained as a goldsmith and a clockmaker before becoming an architect. When he was around 25, he traveled to Rome with his friend, the sculptor Donatello, where he studied the remains of ancient Roman buildings and read the treatise *On Architecture* by the Roman writer Vitruvius. In 1419, he won his first major commission—the design of an orphanage, the Ospedale degli Innocenti in Florence, which, with its arched loggia, is one of the first great Renaissance buildings. A number of other fine works, including chapels in Florentine churches and fortifications for the city, cemented his reputation, but the stunning dome of Il Duomo is his masterpiece. In addition to his buildings, Brunelleschi did important work on the theory of linear perspective, and designed machinery to produce special effects in theatrical productions.

WAR HAS BECOME VERY DIFFERENT

THE BATTLE OF CASTILLON (1453)

Feudal system declines as royal power rises.

More **efficient firearms** are invented.

The role played by artillery at Castillon highlights advantages of hiring professional forces over levying troops from nobles.

Royal power becomes more centralized as **nobles lose military and political strength**.

Armored **knights and bowmen** are gradually replaced by infantry armed with **pikes and firearms**.

IN CONTEXT

FOCUS
Military revolution

BEFORE
1044 The first surviving formula for gunpowder appears in a Chinese military compendium.

1346 Edward II uses cannons at the Battle of Crécy.

1439 Jean Bureau is made master gunner of the French artillery.

1445 Charles VII creates a French standing army.

1453 Constantinople falls to an Ottoman army employing heavy cannons.

AFTER
1520s The Italian Wars demonstrate the effectiveness of infantry with firearms.

1529 Michaelangelo designs a star fort for Florence.

c.1540 Some German cavalry adopt wheel-lock pistols as their main armament.

In July 1453, John Talbot, Earl of Shrewsbury, marched out of Bordeaux with some 6,000 men towards the English-held town of Castillon, which the French were preparing to besiege. The French had constructed a fortified camp big enough to contain 10,000 men, and were armed with some 300 guns under the command of artillery expert Jean Bureau. Expecting reinforcements, Talbot signaled an attack but as the English approached, they found themselves outnumbered by a well-prepared army. The French artillery

See also: The signing of the Magna Carta 100–01 ▪ The outbreak of the Black Death in Europe 118–19 ▪ The fall of Constantinople 138–41 ▪ Christopher Columbus reaches America 142–47 ▪ The Defenestration of Prague 164–69

fired, their bowmen followed suit, and the English were mowed down en masse. It was the first field battle in European history to be decided by gunpowder.

Hundred Years' War ends

The Battle of Castillon was the climax of the Hundred Years' War, fought since 1337 by England and France, countries that had long been closely linked by their ruling families. By the time of Castillon, great changes had taken place in the fabric of European life, which profoundly altered the armies with which the French and English monarchs fought.

The Europe of the 15th century was principally a money economy, and everyone, including soldiers, expected to be paid. Kings were thus increasingly reliant on mercenaries who fought for pay. This was a sharp contrast to the feudal system that had existed previously, in which fighting men were provided by the nobility in exchange for land. Eventually, rulers began employing mercenaries on a permanent basis: a standing army. But it wasn't until the later 17th century that this model became the norm.

> There is no wall, whatever its thickness that, artillery will not destroy in only a few days.
> **Machiavelli, 1519**

French troops (left) engage with the English over wooden defenses in this 15th-century illustration of the Battle of Castillon, from a French chronicle of King Charles VII's life.

Cannons and guns

The kings who fought for control of France relied increasingly on large armies and expensive artillery. Cannons, like those that secured the French victory at Castillon, transformed warfare. The stout walls of medieval castles provided little defense against a cannonball. To better resist artillery rulers began, from the 16th century, to build a new type of fortification, the star fort. These forts had walls sunk into ditches to strengthen them against direct fire and also used cannons themselves in an active defense.

At the same time, hand firearms that fired projectiles that smashed through the armor of mounted knights and required little skill to wield, gradually replaced the bow. Drilled infantry—wielding pikes and firearms—replaced massed ranks of archers, and formed the core of the new line of battle.

To pay for their new armies, rulers steadily began to centralize their domains. More efficient taxation systems and bureaucracies were established, curbing the power of an aristocracy whose influence was already diminished by the decline of the feudal system.

Victory at Castillon, guaranteed by gunpowder, ensured the survival of an independent France that was becoming more like a centralized state and less like a feudal country. As a result of the French triumph, France was able to consolidate the territory under its control and the map of this part of western Europe began to take on its modern form. England, bereft of its European possessions, also became more centralized, and its rulers turned away from contintental Europe, leveraging the country's resources to begin maritime exploration of the Atlantic and North America. ▪

AS DIFFERENT FROM OURS AS DAY AND NIGHT
THE COLUMBIAN EXCHANGE (1492 ONWARDS)

IN CONTEXT

FOCUS
Ecological change

BEFORE
Pre-1492 American and Eurasian ecosystems exist in complete isolation.

AFTER
1518 Charles V of Spain grants a license to sell African slaves in America's Spanish colonies.

1519 Spanish conquistadors bring horses to Mexico.

c.1520 Spanish settlers introduce wheat to Mexico.

c.1528 Spanish traders introduce tobacco to the Old World.

c.1570 Spanish ships bring the first potatoes to Europe.

1619 Dutch traders bring Africans from a captured Spanish slave ship to Jamestown, Virginia.

1620 The Pilgrims bring livestock such as chickens and pigs to Massachusetts.

The arrival in the 1490s of the first Europeans in North and Central America reconnected ecosystems that had developed in isolation from one another for thousands of years. In the so-called Columbian Exchange, lives and economies that had altered only gradually over centuries were suddenly transformed by the influx of new crops, animals, technology, and pathogens. Many of the effects were unforeseen and misunderstood by both Europeans and American Indians at the time, but once the first landing had been made, there was no turning back.

Food and farming

When Europeans began to settle in the Americas, they brought with them their own domesticated animals and foods. The enormous range included citrus fruits, grapes, and bananas; coffee, sugar cane, rice, oats, and wheat; and cattle, sheep, pigs, and horses. To cultivate their crops and pasture their animals, the settlers cleared huge areas of woodland, destroying the habitats of some native wild species in the process, and unintentionally contaminating American fields with the seed of weeds such as

[The lands are] very suitable for planting and cultivating, for raising all sorts of livestock herds.
Christopher Columbus

dandelion and sow thistle. The exchange in the other direction brought potatoes, tomatoes, sweet corn, beans, pumpkins, squash, and tobacco to the Old World, as well as turkeys and guinea pigs.

The introduction of new staple crops transformed lives on both sides of the Atlantic. Potatoes and maize, carbohydrate-rich and easily grown, helped overcome chronic food shortages in Europe and, along with manioc and sweet potatoes, spread on to Africa and Asia. In the New World wheat, which thrived in the temperate latitudes of North and South America and in the

See also: Christopher Columbus reaches America 142–47 ▪ The Treaty of Tordesillas 148–51 ▪ The voyage of the *Mayflower* 172–73 ▪ The Slave Trade Abolition Act 226–27

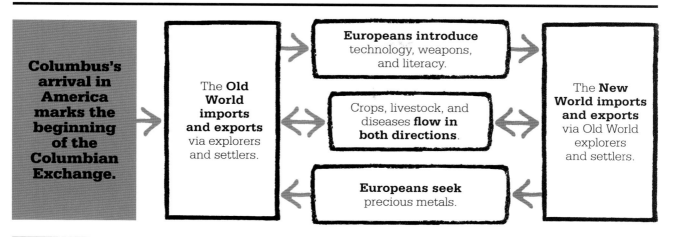

Columbus's arrival in America marks the beginning of the Columbian Exchange.

The **Old World imports and exports** via explorers and settlers.

Europeans introduce technology, weapons, and literacy.

Crops, livestock, and diseases **flow in both directions**.

Europeans seek precious metals.

The **New World imports and exports** via Old World explorers and settlers.

---

highlands of Mexico, eventually became a fundamental food crop for tens of millions of settlers. The arrival of horses in the New World was also revolutionary, permitting more effective and selective hunting, as well as facilitating travel and transport.

Biological catastrophe

The most immediately devastating impact of the Columbian Exchange followed the introduction of new diseases into the Americas. The settlers and the chickens, cattle, black rats, and mosquitoes that accompanied them introduced contagious diseases to a people who had no biological defense against them. American Indians' immune systems were not adapted to cope with alien diseases such as smallpox, measles, chickenpox, influenza, malaria, and yellow fever. Once they were exposed to them, they began to die in the hundreds of thousands. Half the Cherokee nation died in a smallpox epidemic in 1738, and some other tribes were wiped out entirely. European explorers encountered and brought back American illnesses such as Chagas Disease, but the effect on Old World populations was negligible compared with the consequences of Old World pathogens in the New World.

Exchange economics

From the start, the Columbian Exchange had a strong economic driver. Commodities ranging from gold and silver to coffee, tobacco, and cane sugar were transported on a vast scale, mostly to the benefit of European traders and plantation owners.

Very soon, slave trading became a key part of this network too. The movement of people from continent to continent in vast numbers provided a continual supply of labor for expanding new economies at the cost of unspeakable oppression, misery, and early death to many generations. The dramatic and irrevocable changes brought about on both sides of the Atlantic by the Columbian Exchange continued to shape lives for centuries. ▪

Cultural exchange

New World peoples were using stone-age tools, had no wheeled vehicles, and few domesticated animals when they encountered Old World societies, who used guns and alphabets, farmed pigs, sheep, and cattle, and kept bees. The huge cultural changes that ensued, especially in the Americas, were complicated by the two societies' very different attitudes to the "ownership" of nature and property; attitudes that would have significant consequences for future American Indian–European relations. The arrival of the horse led to the emergence of a new, nomadic American Indian tribe that came to dominate the southern Great Plains. Christianity started to spread in the New World, some elements of which fused with pre-Columbian beliefs in the old Inca and Aztec territories. West African religion also arrived, while introductions such as literacy and metal tools and machines, brought advances in education, agriculture, and the evolution of warfare.

MY CONSCIENCE IS CAPTIVE TO THE WORD OF GOD

MARTIN LUTHER'S 95 THESES (1517)

IN CONTEXT

FOCUS
**Reformation and
Counter-Reformation**

BEFORE
1379 English reformer John
Wycliffe criticizes church
practices in *De Ecclesia*.

1415 Czech reformer Jan Hus
is burned at the stake.

1512 During a stay in Rome,
Martin Luther's eyes are
opened to church corruption.

AFTER
1520 Lutheran services are
held regularly in Copenhagen.

1534 Henry VIII of England
breaks from Rome and
becomes head of the church
in England.

1536 John Calvin begins his
church reforms in Switzerland.

1545–63 The Council of Trent
reaffirms Catholic doctrines,
beginning the Counter-
Reformation movement.

I n the autumn of 1517, Martin
Luther, a monk and teacher of
theology at the University of
Wittenberg in Germany, set off a
chain reaction that would transform
Europe. Deeply concerned by what
he saw as corrupt practices in the
Catholic Church, he wrote a series
of 95 theses—arguments—against
them, which he then circulated
within the university. According to
some reports, he also nailed them
to the door of the Castle Church in
Wittenberg. The theses were soon
published more widely, prompting
Pope Leo X to charge Luther
with heresy. Luther responded by
breaking with the Catholic faith,

See also: The Investiture Controversy 96–97 ▪ The beginning of the Italian Renaissance 152–55 ▪ The Defenestration of Prague 164–169 ▪ The execution of Charles I 174–75 ▪ Henry VIII breaks with Rome 198

Injury is done to
the Word of God when,
in the same sermon, an equal
or larger amount of time is
devoted to indulgences
than to the Word.
Martin Luther, 1517

so initiating the Reformation—the rise of churches based on reformed practices, and a focus on scripture rather than on priestly authority. Because of the churches' origin in protests against Catholic practices and beliefs, they became known as Protestant churches.

Spread of the Reformation

Luther was not alone in seeking religious reform. Swiss preacher Ulrich Zwingli (1484–1531) led a Protestant church based in Zurich, and Frenchman John Calvin broke from the Catholic church in around 1530. Forced to flee France, he went to Geneva, Switzerland, where he supported the reform movement, eventually helping to shape Protestant doctrine.

Reformers' beliefs did not necessarily concur. Calvinists were markedly different from Lutherans,

At the Diet of Worms in 1521, Luther refused to recant: "Unless I am convicted of error by the testimony of Scripture... I cannot and will not retract.... Here I stand. God help me!"

and Anabaptists were persecuted by Protestants as well as Catholics for their radical views. Luther himself supported the brutal suppression of the Anabaptist-led Peasants' Revolt in the 1520s. What the Protestants had in common was that their views brought them into fundamental theological conflict with the Catholic church.

Reformers' ideas spread via the relatively new technology of the printed word. Before movable type and presses made printed books possible in the 1450s, books were all written by hand in Latin, the international language of the church. Print allowed information to be reproduced cheaply and quickly, and demand rapidly grew for books written in the vernacular. Luther wrote his theses in Latin, but before long they had been translated and printed in German, French, English, and other languages. Books and pamphlets describing church abuses and outlining Protestant theology soon followed, and were printed in large numbers.

Importance of The Word

A central idea in Protestant theology was that authority came not from the priesthood, but from scripture itself. For this reason, access to the Bible was essential both for the reformers and their followers. Bibles printed in native European languages were appearing by the 16th century, Luther's German translation of the New Testament was published in 1522, and a translated version of the whole Bible including the Apocrypha followed in 1534. A year later, Miles Coverdale (1488–1569), sometime friar, preacher, and Bishop of Exeter, produced the first complete Bible in English. A French translation by theologian Jacques Lefèvre d'Étaples (c.1450–1536) appeared between 1528 and 1532.

By the mid-16th century, Reformation ideas had been widely disseminated. Lutheranism spread across Germany and Scandinavia; Calvinism took hold in much of Switzerland, and made significant inroads in Scotland. There were »

also Calvinists in France, where they were called Huguenots, although that country was split between Catholics and Protestants, who fought in the Wars of Religion of the second half of the 16th century. Spain, Portugal, and Italy remained Catholic.

In England, the seeds of reform were sown early. Many people objected to abuses such as the use of church funds to pay for clerics—including the Pope and foreign bishops—to lead a life of luxury. However, Protestant ideas were not yet widely enough held for the faith to take hold. Things changed when Henry VIII of England broke with Rome in 1534, rejecting papal authority and proclaiming himself head of the church in England. As supreme ecclesiastical leader, he exercised his sole right to authorize the publication of the English Bible, the Coverdale Bible, but English religious practice and doctrine remained Catholic. A moderate

Cartoon images of the pope as a bestial monstrosity communicated to an international audience, literate and not, a common Protestant idea that the papacy was the institution of the devil.

form of Protestantism was later established in England under Henry's daughter Elizabeth I.

Reformers risked their lives by speaking out at a time when heresy was punishable by death. Czech reformer Jan Hus had been burned at the stake in 1415, Zwingli died in a battle between Protestant and Catholic forces in 1531, and English Bible translator William Tyndale was executed in 1536. Luther, urged to recant by Pope Leo X in 1520, threw the written request on a bonfire, so church authorities handed him over to Frederick the Wise, Elector of Saxony and founder of the University of Wittenberg, for punishment. Frederick convened a formal enquiry or "Diet" at Worms, at which Emperor Charles V presided. The emperor rejected Luther's arguments and banned his views in the empire, but Luther refused to recant. He was outlawed and excommunicated, but Frederick saved him from execution by faking his abduction, then hiding him at the Wartburg castle. Luther continued to write and organize, garnering increasing support.

Powerful allies

Support from people in positions of power assisted the spread of the Reformation. Like Henry VIII in England, the princes of Germany resented church wealth, taxation, and its independent law courts, and were also eager to strengthen their own power. Throughout the Middle Ages, popes had made alliances with kings and emperors, and intervened in secular affairs. Many German princes wanted to prevent such alliances by cutting ties with Rome and removing bishops from their princedoms, so their support for the reformers was motivated by political expedience as well as personal piety.

I do not accept the authority of popes and councils, for they have contradicted each other.
Martin Luther, 1517

In what would become the first in a long list of religiously motivated conflicts between Catholics and Protestants, the Holy Roman Emperor Charles V invaded Lutheran territory in an effort to stamp out the movement. Lutherans united against him and, despite his triumph at the Battle of Mühlberg in 1547, he was unable to suppress them. A temporary compromise was eventually reached at Augsburg in 1555 when the emperor conceded that each prince within the empire could choose how to worship in his own domain. The peace was not to last, however; bitter divisions drawn by the Reformation would cause people across Europe to take up arms again, and the continent was ravaged by more than a century of religiously-motivated conflict.

Reform from within

Even before Luther wrote his 95 theses, a movement for reform had begun within the church. Inspired partly by Renaissance Humanism, it brought on a resurgence of scholarship and philosophy, and motivated churchmen such as Spaniard Francisco Ximenes, who produced a Bible with texts in Hebrew, Greek, Latin, and Aramaic.

> Has the Catholic Church been dead for a thousand years to be revived only by Martin?
> **Cardinal Girolamo Aleandro, 1521**

However, Luther's clear theological challenges prompted the papacy to prepare a more widely considered response. In 1545, Paul III called together the Council of Trent at which bishops and cardinals reaffirmed Catholic doctrines, from the importance of the priesthood and sacraments to the legitimacy of indulgences. But the council also introduced reforms: it forbade abuses such as the holding of multiple offices by one priest, set up training seminaries for priests, and, in an attempt to slow the spread of Protestant doctrine, established a commission to specify which books Catholics were forbidden to read. In addition, a number of popes from Paul III onward lived austerely, appointed like-minded bishops, and reviewed papal finances.

Counter-Reformation

The council met periodically for 18 years, and provoked a renewal and resurgence of Catholicism from within the church that is usually called the Counter-Reformation. The new Society of Jesus (also known as the order of Jesuits), founded by Spanish knight Ignatius Loyola in 1534, was approved by the pope in 1540 as an answer to the Reformation, and it spread a powerful Counter-Reformation message across Europe. The contemporary revival of Christian art, which coincided with the flowering of the baroque style in Italy, added a vibrant emphasis.

The Ecstasy of Saint Teresa, a white marble altarpiece and one of the masterpieces of High Roman Baroque, by Gian Lorenzo Bernini, the leading sculptor of his day.

Baroque churches were imposing and ornate, filled with affecting sculptures, paintings, and strikingly posed biblical scenes. This potent propaganda served to underline the difference between Catholic churches and their Protestant counterparts, which were usually plain and undecorated. Baroque art, together with the zeal of reforming popes and Jesuit priests, helped to ensure that the Catholic church survived and flourished in countries such as Italy and Spain, even while the Protestant movement was gathering strength elsewhere. Europe, which had once been united under the pope in the Roman Catholic Church, was now irrevocably split into Catholic and Protestant states. The seeds were sown for over a century of conflict as subjects took up arms against their rulers, kings and princes clashed, and nations attacked nations in the name of religion. ∎

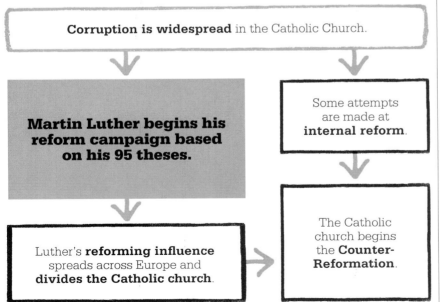

Corruption is widespread in the Catholic Church.

Martin Luther begins his reform campaign based on his 95 theses.

Some attempts are made at **internal reform**.

Luther's **reforming influence** spreads across Europe and **divides the Catholic church**.

The Catholic church begins the **Counter-Reformation**.

HE BEGAN WAR IN BOHEMIA WHICH HE SUBJUGATED AND FORCED INTO HIS RELIGION

THE DEFENESTRATION OF PRAGUE (1618)

IN CONTEXT

FOCUS
The Wars of Religion

BEFORE
1562 The French Wars of Religion begin a 36-year period of conflict in France.

1566 The sack of the monastery at Steenvoorde, Flanders, leads to the Dutch Revolt.

AFTER
1631 Gustavus Adolphus' victory at Breitenfeld protects German states from forcible reconversion to Catholicism.

1648 The Peace of Westphalia, a series of peace treaties, ends the Thirty Years' War (1618–1648) in the Holy Roman Empire, and the Eighty Years' War (1568–1648) between Spain and the Dutch Republic.

1685 Revocation of the Edict of Nantes leads to renewed persecution of French Protestants.

Protestant nobles threw the imperial regents from the council room window, signaling the start of a revolt against the Habsburg emperor and one of the opening phases of the Thirty Years' War.

I n May 1618, a group of Protestant leaders in Prague met a number of councillors in an upper room in Prague Castle. The councillors were Catholics, working as regents for Ferdinand, the new king of Bohemia (now part of the Czech Republic); the Protestants wanted to be sure that the king and regents would not remove the religious freedoms that their former rulers had granted them. When the regents refused to give this assurance, the Protestants threw two of them, together with their clerk, out of the castle window.

The trio landed some 65 ft (20 m) below in a dung heap stacked against the castle walls. Known as the Defenestration of Prague, this event began the Thirty Years' War, a series of conflicts that devastated huge areas of Europe.

Religious differences

The Defenestration took place in the wake of long-standing disputes between Catholics and Protestants about whether people should be allowed to worship freely in their own way. These differences affected much of Europe, and before war ignited Bohemia, there were violent religious conflicts in several other parts of the continent.

The disputes also involved rivalries for power between royal and aristocratic families who favored the different sides and used the conflicts to promote their own interests. The Netherlands, for example, were home to many Protestants, but were ruled by Catholic Spain, whose ruler Philip II

wanted to eliminate Protestantism. The largely Protestant Seven Provinces in the northern Low Countries revolted against the king's rule. Religious clashes escalated into violence against the perceived repression of the Habsburg Crown, leading to the formation of the independent Dutch Republic in the north of the region.

Philip also planned to conquer England, which was moderately Protestant under Elizabeth I, and wanted to place a Catholic monarch on the English throne. In 1588, he sent his famous Armada to invade the country, but a combination of superior English

I would rather lose all my lands and a hundred lives than be king over heretics.
Philip II of Spain, 1566

See also: The fall of Granada 128–29 ▪ Christopher Columbus reaches America 142–47 ▪
Martin Luther's 95 theses 160–63 ▪ The opening of the Amsterdam Stock Exchange 180–83

naval tactics and stormy weather foiled the attempt, and England remained independent.

These religious differences proved particularly devastating in 16th-century France, where the substantial Protestant minority generally known as the Huguenots were widely persecuted. Many Protestants, especially Calvinist ministers, had their tongues cut out, or were burned at the stake. In the so-called St. Bartholomew's Day Massacre of 1572, a group of targeted assassinations followed by a wave of mob violence against the Huguenots lasted several weeks and left thousands dead.

There followed a series of so-called Wars of Religion that lasted some 36 years. After eight periods of fighting, punctuated by uneasy truces and broken agreements, the wars came to an end in 1598 when the French king Henry IV, who had been a Protestant leader before taking the throne, promulgated the Edict of Nantes. This agreement gave the Huguenots certain rights, including freedom of religion in particular geographical areas. It also maintained Catholicism as the established religion in France, and obliged Protestants to observe Catholic holidays and pay church taxes. Disputes between the two sides still flared from time to time, however, and many Huguenots left France to seek safety in other countries such as England and the Netherlands.

Thirty Years' War

The religious wars and disputes in France, the Netherlands, and England formed a troubled backdrop to the Thirty Years' War in Europe. Most people in Bohemia were

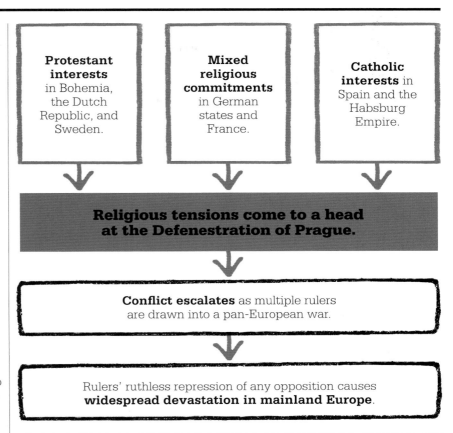

| Protestant interests in Bohemia, the Dutch Republic, and Sweden. | Mixed religious commitments in German states and France. | Catholic interests in Spain and the Habsburg Empire. |

Religious tensions come to a head at the Defenestration of Prague.

Conflict escalates as multiple rulers are drawn into a pan-European war.

Rulers' ruthless repression of any opposition causes **widespread devastation in mainland Europe**.

Protestants, but the area was part of the large Holy Roman Empire, which also included Germany, Austria, and Hungary, and was ruled by Catholic Habsburg emperors. The emperors acted as overlords to local kings, princes, and dukes. Some of them, notably Matthias, who was on the throne when the Defenestration took place, granted their Protestant subjects the right to worship as they wished. Matthias achieved this by ratifying the Letter of Majesty, a charter that had been signed by the previous emperor, Rudolf II, which guaranteed Protestants religious freedom and certain other basic rights. However, Matthias' successor, the ardently Catholic

Ferdinand, felt no obligation to honor the Letter of Majesty. He suppressed Protestant churches and appointed Catholics to high positions. This reignited a dispute that had existed in Bohemia since the first stirrings of the Protestant Reformation in the 15th century.

After the Defenestration, both sides began preparing for war, but the process was accelerated when, in 1619, Matthias died. Ferdinand, who was already King of Bohemia, then also became Holy Roman Emperor. Bohemia's Protestant leaders tried to reduce the Catholic emperor's local power by deposing him as King of Bohemia and inviting their » own candidate, the Protestant Frederick V, Elector Palatine, to rule in his stead.

Frederick's credentials as a Protestant were excellent, not only because of his own faith, but also by marriage: his wife was Elizabeth Stuart, daughter of England's Protestant king James I. However, in order to make Frederick king, the Bohemians had to depose a monarch who had been legally crowned, a move that deprived them of support from a number of their potential allies.

In 1620, the forces of Bohemia gathered to face those of the Holy Roman Empire at White Mountain, outside Prague. The forces seemed evenly matched: the Protestants under Frederick and Christian of Anhalt had a larger force, but the empire's soldiers were experienced and well led by the Spanish–Flemish nobleman Field Marshall Tilly, and renowned general Albrecht von Wallenstein. After only one hour, Bohemian forces were crushed— 4,000 dead or taken prisoner

compared to 700 of the empire's forces—and Tilly entered Prague. Frederick fled, and many of the Protestant leaders were executed; ordinary Protestants were ordered to leave or convert to Catholicism; and Bohemia was left devastated, depopulated, and almost powerless. The area remained overwhelmingly Catholic into the 20th century.

A destabilizing reform

What happened in Bohemia was a symptom of the instability of the wider Holy Roman Empire. In its history there had often been power struggles between emperors and local rulers, but a general balance of power had emerged in which the emperor resolved to respect the rights of the individual states that made up the empire. This balance was upset by the changes of the Reformation, when Protestant beliefs strengthened in some places (such as Saxony), and Catholicism prevailed in others (such as Bavaria). A series of struggles then escalated into armed conflict.

> The [Protestant] wound is degenerated into gangrene; it requires fire and sword.
> **Fernando Álvarez, c.1560s**

Most of the battles were in the German and central European lands. In a few years the Habsburg imperial army, raised for Ferdinand and led by skilled military leader Albrecht Wallenstein, had crushed its rivals in Germany, and gone on to overwhelm Denmark. By 1629, Ferdinand was in a position to reclaim the lands that had passed into Protestant hands.

However, the Protestants still had two powerful allies. One was Sweden, under King Gustavus Adolphus, an able military leader; the other was France, a Catholic country, but one that wanted to curtail imperial power. In 1630, Gustavus arrived in Germany with a large army and won a significant victory at Breitenfeld in 1631, with financial assistance from France.

In the mid-1630s the Habsburgs fought back, with the help of Spain. The conflict had now become an all-encompassing war involving virtually every one of Europe's major countries in a struggle for power. The emperor wanted to win

Gustavus achieved his decisive victory at Breitenfeld with a new, combined-arms approach in which infantry, artillery, and cavalry worked together in self-supporting units.

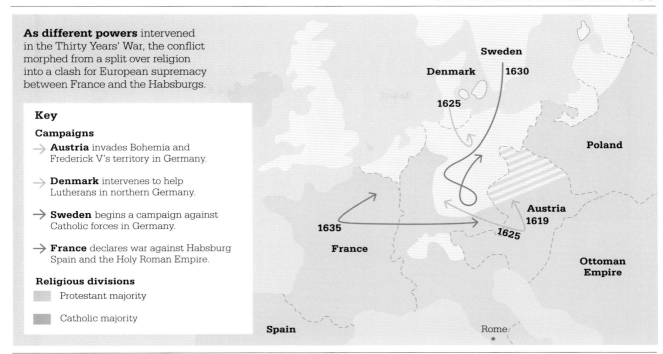

As different powers intervened in the Thirty Years' War, the conflict morphed from a split over religion into a clash for European supremacy between France and the Habsburgs.

Key

Campaigns

→ **Austria** invades Bohemia and Frederick V's territory in Germany.

→ **Denmark** intervenes to help Lutherans in northern Germany.

→ **Sweden** begins a campaign against Catholic forces in Germany.

→ **France** declares war against Habsburg Spain and the Holy Roman Empire.

Religious divisions

Protestant majority

Catholic majority

back his lands in Germany, while the Spanish wanted their allies the Habsburgs in power so that they could cross Europe with ease in their hoped-for attack on the Netherlands. France, fearful of being surrounded by the Habsburgs and their allies, continued to try to reduce imperial power.

The end and the aftermath

By the 1640s, anti-imperial forces were regaining the upper hand. France defeated Spain at Rocroi in the Oise valley in 1643, while in 1645, Sweden met the imperial army at Junkau, southeast of Prague. Around half the 16,000-strong imperial army was killed in this bloody battle, and it looked as if the Swedes would march on Prague or Vienna. However by this point, both sides were exhausted, and no advance was made on either city.

The battles of the Thirty Years' War were conducted on a large scale. Forces of thousands came together in cavalry charges backed up by firearms, and large numbers of mercenaries were employed. The battles were fought with professional speed and ruthlessness, but what came after was sometimes even worse. Vast armies committed infamous atrocities as they pillaged huge areas of country to find food, and removed anything that might be useful to their enemies. Rural areas suffered particularly badly at the hands of the scavenging troops—Germany lost around 20 percent of its population—but trade and manufacturing were also affected by the damage and devastation left behind. Central Europe took decades to recover from the war, although countries with strong trade networks and sea power, such as England and the Netherlands, fared better.

Repeated artillery battles also wore down both armies. Exhausted, the sides eventually came together to make peace.

Representatives of the empire, Spain, France, Sweden, and the Dutch Republic, as well as rulers of German princedoms and cities, and other interested parties, assembled in 1648 in two north-German cities, Osnabrück and Münster, to agree the Peace of Westphalia. The talks could not resolve basic differences between political and religious interests; they did however produce an agreement to end the war, and the Peace established an overall balance of power among a number of independent nations.

Although Europe was now permanently divided into states that were predominantly Catholic or predominantly Protestant, they had agreed to learn to coexist with one another. The Peace set the precedent of creating agreements between nations by means of high-level diplomatic meetings, the like of which have played a key part in international relations ever since. ∎

ROYALTY IS A REMEDY FOR THE SPIRIT OF REBELLION

THE CONQUESTS OF AKBAR THE GREAT (1556)

While in exile in Persia, Akbar's father, Humayun, develops close ties with the **Safavid court**, who help him recover some of his **territories in India**.

Akbar wins the Second Battle of Panipat and the Mughal dynasty goes on to becomes the dominant power on the Indian subcontinent.

Akbar strengthens the cultural, commercial, and political **bonds between Persia and India**.

Persian artists and scholars are lured to India by the **Mughal court's brilliance**.

Persian culture influences northern Indian literary, architectural, and artistic traditions, resulting in **a distinctive Mughal style**.

IN CONTEXT

FOCUS
Islamic empires

BEFORE
1501 The Safavid dynasty unites Persia; they make Shi'a Islam the state religion, and suppress all other religions and other forms of Islam.

1526 At the first Battle of Panipat, Babur, a descendant of Mongol rulers Timur and Genghis Khan, conquers Delhi and founds the Mughal Empire.

1540 Babur's successor, Humayun, rapidly loses much of the empire and is exiled.

AFTER
1632 The Taj Mahal, the crowning glory of Mughal architecture, is commissioned.

1658–1707 The Mughal Empire reaches its greatest extent under Aurangzeb, but his harsh rule leads to revolt.

1858 The last Mughul emperor is removed by the British.

In February 1556, Abu Akbar became the new ruler of the Muslim Mughal dynasty in northern India, founded 30 years earlier by Turkic-Mongol invaders from Central Asia. The emperor's forces immediately confronted the army of Hemu, a rival claimant to the throne of Delhi, at the Second Battle of Panipat. The Mughals inflicted a crushing defeat on Hemu, and regained territory lost by Akbar's father, Humayun. Akbar then gradually consolidated and extended his authority, annexing all of northern and part of central India. Rulers were deposed and killed and citizens massacred as once-independent kingdoms became provinces of his empire.

See also: Muhammad receives the divine revelation 78–81 ▪ The founding of Baghdad 86–93 ▪ The fall of Granada 128–29 ▪ The fall of Constantinople 138–41 ▪ The founding of the Safavid dynasty, Persia 198

Support and survival

Akbar maintained the political unity of his sprawling realm by building an administration capable of expansion as new territories were incorporated. He created a network of highly paid nobles who served as provincial governors, or were employed as commanders of field armies or as part of the central military—the backbone of the empire. He also recruited talented men from across India (and Persia), both Muslim and Hindu, into his government, remunerating them with money or land.

This system rewarded individual merit and loyalty, but kept the administration from becoming too centralized—a distinct advantage in an empire that was difficult to hold together from a single center. The emperor

In this miniature painting, the Mughals are seen battling their Hindu enemies at Panipat. As later conquests added money, men, and weapons to the imperial army, it became supreme.

himself was often on the move, traveling with his court and harem in well-appointed tents.

Another unifying factor was the spread of Islam, together with its arts and culture; however, Akbar believed in religious freedom and allowed the empire's non-Muslim populations, which included a large Hindu majority, to live by their own faiths, laws, and customs.

Interaction with Persia

Babur, the founder of the Mughal dynasty, and Humayun, Akbar's father, had developed diplomatic, cultural, and political links with another Islamic empire in the region, Safavid Persia, which stimulated Mughal interest in Persian fine arts such as miniature painting and the "art of the book." Akbar set up studios to produce illustrated books in the cities of Fatehpur Sikri and Lahore (now in Pakistan), and Persian architects and artisans were brought to India to design and construct palaces, forts, mosques, and public buildings, including Humayun's tomb in Delhi. This domed structure inspired major architectural innovations, and a unique Persian-influenced building style developed across the Indian subcontinent.

The Mughal Empire continued to prosper under Akbar's son Jahangir, but later in the 17th century it declined amid religious conflict and economic problems. The emperors were defeated by Afghan invaders, then came under the control of the Marathas, Hindu warriors who dominated Indian affairs in the second half of the 18th century, and finally were taken over by the British after Britain defeated the Marathas in 1818. ▪

Akbar

Akbar was just 13 years old when he inherited the Mughal throne, and initially ruled under a regent, Bairam Khan, who assisted him in forcibly unifying India's collection of regional kingdoms into a single, centralized political system within which the emperor was the supreme source of authority.

Under Akbar, the dynasty became an artistic as well as military power. Painting and literature blossomed under the emperor's patronage—although he himself was illiterate, he acquired a library of 24,000 books. His capital at Fatehpur Sikri also became a center for religious debate and his court a place of culture and learning. Although he never renounced Islam, Akbar was open to the ideas of other faiths, and he invited Hindu, Christian, and Buddhist philosophers to debate with the Muslim theologians at his court. He even conceived a new religion, which combined elements of all these faiths, with himself as the deity.

THEY CHERISHED A GREAT HOPE AND INWARD ZEAL

THE VOYAGE OF THE *MAYFLOWER* (1620)

In 1620, a group of English people who could not legally worship as they wished to in England set sail across the Atlantic to begin a new life in America. This group later became known as the Pilgrims. They set off on two ships, but one proved unseaworthy so they had to continue in just one, the *Mayflower*. Winter storms ravaged the 66-day crossing and the ship's main beam fractured. While still aboard, the Pilgrims drew up the *Mayflower Compact*, which pledged their loyalty to the Crown but also asserted their right to make their own laws within the English legal framework. They settled at Plymouth and, although many died that first winter, their community endured.

Early colonization

At that time, England, like other countries, was competing to establish colonies in North America. Jamestown had been founded thirteen years before the Pilgrims landed at Plymouth, but it was not a religious community. The Colony of Virginia, centered around Jamestown, had been established by English colonists in 1607 under

English **Protestants** seeking **religious freedom** sail to **North America** on the *Mayflower*.

More **religious separatists** follow, swelling the **colony's population**.

Other English colonies are founded by **companies** granted **royal charters** from the **Crown**.

The colonists develop a form of government based on the pursuit of religious freedom, following the English parliamentary model.

See also: Christopher Columbus reaches America 142–47 ▪ The opening of the Amsterdam Stock Exchange 180–83 ▪ The signing of the Declaration of Independence 204–07 ▪ The opening of Ellis Island 250–51

The *Mayflower* **attempted** to depart England on three occasions: from Southampton and then Dartmouth in August, and finally from Plymouth on September 6, 1620.

a charter from the Crown, and was their first permanent settlement in the Americas. French explorers had established fur trading posts up the rivers of Canada; Dutch and Swedish colonists arrived in North America in the early 17th century, and in 1613 the Dutch established a trading post on the western shore of Manhattan Island.

Government and trade
Both Plymouth and Jamestown developed representative institutions in which colonists elected officials to govern their own affairs. Inspired by the English parliamentary model, and growing out of the assertion of rights articulated in the *Mayflower Compact*, these early developments established a model of self-rule that came to characterize English colonization in North America.

Each colony had a governor, appointed by the British monarch, and a legislature, elected by the colonists. There was often tension between the two, because the

legislature had to work within the framework of existing English law. However, the king and government in London, working with the governor, saw the colonies as a resource, rich in raw materials, that they could exploit to their advantage.

To ensure America remained a ready market for British industry, colonial trade was restricted by the Navigation Acts, which required that all commodity trade take place in British ships crewed by British sailors. The colonists came to see these measures as a willful suppression of their trade and manufacturing. Tensions arose on both sides of the Atlantic as British and colonial merchants sought to protect their interests.

Colonial growth
Relations between the colonists and the indigenous peoples of the East Coast were also starting to strain. The increasing colonial population put pressure on land and resources, pushing people west to settle on land belonging to American Indians.

The groups struggled to coexist harmoniously. An uneasy peace, punctuated by violence, typified relations between settlers and Native Americans for many years. ▪

Religious persecution

In the early 17th century, the English were legally obliged to worship as prescribed by the Church of England. Although the English church had already broken from the Catholic Church, many people still felt that its hierarchical priesthood and set rituals, hymns, and prayers were Catholic features that should be swept away.

Puritans, so-called because of their desire for religious purity, hoped to reform the church from within. Other groups, known as Separatists, set up their own "separate" congregations, but when their leaders were imprisoned or even executed, they moved to the more tolerant Netherlands. Here they could adopt the simpler form of worship they preferred, but it was very hard to earn a living because the country's professional guilds were closed to them. This is part of the reason that the Pilgrims, and later others, decided to seek a new life in North America.

WE WILL CUT OFF HIS HEAD WITH THE CROWN UPON IT

THE EXECUTION OF CHARLES I (1649)

IN CONTEXT

FOCUS
English Civil War

BEFORE
1639 English and Scottish forces clash in the first "Bishops' War."

1642 The Civil War begins at Edgehill, Warwickshire.

1645 Oliver Cromwell's "New Model Army" scores victories at Naseby and Langport.

1646 Charles is forced to surrender to his opponents.

AFTER
1649 The Commonwealth of England (a republic) is formed.

1653 Cromwell takes the title Lord Protector for Life, giving him the power to call or dissolve parliaments.

1658 Cromwell dies and is succeeded as Protector by his son, Richard.

1660 The monarchy is restored: Charles II becomes King of England.

King Charles I asserts his **divine right to rule**.

The king needs to **raise taxes** to pay for wars.

Parliament attempts to **limit the king's authority**. A **civil war** erupts between Crown and parliament for the **right to rule**.

Parliamentary forces, led by Cromwell, **win the war**.

The king is executed and an English republic is instituted.

During the 1640s, England was plunged into a series of wars, fought to decide the future of the country and known collectively as the English Civil War. On one side were the Royalists—predominantly landed gentry and aristocrats who supported King Charles I and his right to rule independently of parliament. On the other were the Parliamentarians—mainly smaller landowners and tradesmen, many of whom held Puritan beliefs and disliked Charles's autocratic stance. By 1648, the Parliamentarians had beaten Charles on the battlefield and Oliver Cromwell, their leader, ejected from parliament all those who were prepared to negotiate

See also: The signing of the Magna Carta 100–01 ▪ Martin Luther's 95 theses 160–63 ▪
The Defenestration of Prague 164–69 ▪ The opening of the Amsterdam Stock Exchange 180–83

with the king, leaving the remainder (known as the Rump Parliament) to vote to end the monarchy. Charles was tried for treason against England and was beheaded in 1649, after which England began an 11-year period as a republic.

The causes of war

King Charles I and parliament were natural opponents. Charles was sympathetic to Catholics while parliament was Protestant, and he believed in the divine right of kings—the idea that the monarch's appointment is approved by God and so he or she has absolute power.

The clash first came to a head over the king's repeated attempts to raise money for a war in France. Parliament tried to curb his power to do so by introducing a Petition of Right in 1628, making it a necessity for its members to approve taxation. However, Charles got around this by levying taxes using antiquated medieval laws, selling trading monopolies to raise cash, and ruling without parliament. In 1640, the king was forced to call parliament

for the first time in 11 years to raise money to quell a Scottish revolt. Once called, parliament tried to bring in further measures to limit his power, such as making it illegal for the king to dissolve parliament, but he responded by trying to arrest five MPs. The dispute escalated into the First Civil War in 1642.

The war and its effects

Initially, the Royalists gained the upper hand but in 1644 the Parliamentarians reorganized their troops under Oliver Cromwell. With their disciplined, professional approach, this "New Model Army" forced Charles to surrender in 1646. However, the king restarted the war two years later, and this Second Civil War—which ended in a Royalist defeat at the Battle of Preston in 1648—began the chain of events that led to his execution in 1649 and the formation of a republic under Cromwell called the Commonwealth of England.

Like Charles, Cromwell found relations with parliament difficult, but he tried to bring in reforms.

He ruled with stern Puritan authority, imposing it ruthlessly on the Scots and the Irish. Soon after he died, the country—perhaps tired of Puritan austerity—welcomed Charles I's exiled son home to reign. Charles II agreed to limitations on royal power and to uphold the Protestant faith, but his heir—his Catholic brother James II—clashed with Anglican bishops and offended Protestants by offering prominent positions to Catholics.

Fears of having another Catholic king mounted until, in 1688, in what became known as the Glorious Revolution, James was deposed. The king was sent into exile and replaced by his Protestant daughter Mary, who ruled with her Dutch husband William of Orange. In 1689, William and Mary accepted a Bill of Rights, which ensured their subjects had basic civil liberties, such as trial by jury, and making the monarchy subject to the law of the land. Britain has remained a constitutional monarchy, in which no king or queen could defy Parliament as Charles I did, ever since. ▪

King Charles I of England

The son of Stuart King James I of England (King James VI of Scotland) and Anne of Denmark, Charles was born in 1600 and became king in 1625. From the start, he alienated both subjects and parliament with his demands for taxation (mostly to fund wars in France) and his assertion of his divine right to rule. He also clashed with the church because of his sympathies with Catholicism (he was married to the French Catholic princess, Henrietta Maria). In addition, he was unpopular in Scotland, where he tried to replace the prevailing

presbyterian system of church governance (without bishops) with the more hierarchical episcopal system (with bishops, following the Anglican model), which led to political and military conflict in 1639 and 1640 (known as the Bishops' War). During the English Civil War, he took an active part in leading the Royalist armies until he was captured; initially, he was put under house arrest, then he was imprisoned before his execution in 1649. He continued to assert his divine right to rule during his trial.

THE VERY BEING OF THE PLANTATIONS DEPENDS UPON THE SUPPLY OF NEGRO SERVANTS

THE FORMATION OF THE ROYAL AFRICAN COMPANY (1660)

IN CONTEXT

FOCUS
Slaves and colonies

BEFORE
1532 The Portuguese found their first settlement in Brazil.

1562 British slave trading in Africa begins with the voyage of John Hawkins.

1625 The British claim Barbados on behalf of James I.

1655 The British capture Jamaica from Spanish colonists.

AFTER
1672 The company is reconstituted as the Royal African Company.

1698 African trade is legally opened to all English merchants, provided they pay a ten percent levy to the company on all goods exported from Africa.

I n 1660, the Company of Royal Adventurers Trading to Africa was established in England. Its charter, endorsed by the king, gave its ships the exclusive right to trade on the West African coast, and permitted its members to set up forts there, in exchange for giving the English Crown half the resulting profits. Twelve years later, the company was reorganized as the Royal African Company and given still greater powers: to build forts and "factories" (where slaves were held before being shipped over the Atlantic), and employ its own troops. The company's particular significance is due to its crucial role in facilitating and developing the

See also: Christopher Columbus reaches America 142–47 ▪ The Treaty of Tordesillas 148–51 ▪ The Columbian Exchange 158–59 ▪ The Slave Trade Abolition Act 226–27

The Atlantic slave trade was banned from 1807, but continued for decades. This engraving shows captives aboard an American ship, the *Wildfire*, bound for Cuba in around 1860.

slave trade. It transported many thousands of Africans to a life of slavery, working with West African leaders to build a trade that lasted even after the company disbanded in 1752, and that would eventually see millions of Africans displaced to lives of toil in the Americas.

Foundation of the company

Soon after its foundation, the company became involved in the Second Dutch War, a trade conflict between the Netherlands and England during which the Dutch took many English forts, excluding them from the slave trade during the war. Involvement in the war almost brought the Company of Royal Adventurers to bankruptcy, but in 1672, with a new charter from the king, the company re-emerged, renamed, restructured, and granted the right to carry slaves for sale in the Americas. It prospered, transporting some 100,000 slaves between that year and 1698 when, royal power having been restricted by the Bill of Rights, the company lost its

monopoly over the trade. After 1698, other merchants were allowed to join the trade but had to pay a levy to the company of 10 percent on all their African exports. The involvement of other merchants strengthened the trade to the point

that it became part of the fabric of British mercantile life, continuing throughout the 18th century.

The slave trade itself was much older than the Royal African Company. Portuguese traders in the late 14th century were the first »

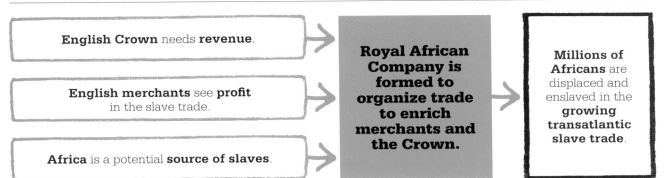

English Crown needs **revenue**.		
English merchants see **profit** in the slave trade.	**Royal African Company is formed to organize trade to enrich merchants and the Crown.**	**Millions of Africans** are displaced and enslaved in the **growing transatlantic slave trade**.
Africa is a potential **source of slaves**.		

> I herded them as
> if they had been cattle
> toward the boats.
> **Diogo Gomes,**
> **Portuguese explorer (1458)**

Europeans to ship slaves from West Africa. By the 16th century, the Portugese were bringing slaves in huge numbers to Brazil to work on sugar cane plantations. Brazil remained the biggest destination for the import of African slaves until the outlawing of the trade. The first English slaving expeditions took place in the 1560s, in which merchants bought captured slaves from African rulers. During the 17th century, with the increase of English colonization, the market for African slaves grew and the Royal African Company took full advantage of it.

Triangular trade
The transatlantic slave trade soon became part of a larger triangular trading network, in which ships took slaves from Africa to the Americas; refilled the holds with goods to transport to Europe; then took European manufactured goods on to Africa for sale, completing the triangle. Ships carried commodities such as sugar, molasses, and coffee from the Caribbean to England; rice, indigo, cotton, and tobacco from the southern colonies in North America; and furs, timber, and rum from the northeast. On the England-to-Africa leg, they carried a range

of items including cloth, guns, iron, and beer. Goods such as ivory and gold were carried directly from Africa to Europe, not as part of the triangular trade but still bolstering the system.

The trade network brought huge profits to plantation owners in the Americas, and to English manufacturers, as well as to the merchants who dealt in the slaves and other goods. Port operators, West African leaders who sold slaves, bankers who loaned money for expeditions, and even English factory workers whose jobs depended on raw materials imported from abroad, all benefited.

As a key part of this trading network, the slave trade made possible the rapid rise of Western capitalism in the 18th century. Even factories some distance away from England's trading ports became involved. A notable example was the business of arms manufacture, which was based in the English Midlands at population centers such as Birmingham, conveniently close to supplies of iron. Some 150,000 guns, mostly made in these Midland factories, were exported to West Africa every year; almost all of them

> The shrieks of the women,
> and the groans of the
> dying, rendered the whole
> scene of horror almost
> inconceivable.
> **Olaudah Equiano,**
> **African writer and freed slave (1789)**

Tobacco from Virginia was in great demand in Europe. Planters shipped their products directly to their home countries and used the profits to buy African labor and European goods.

were then exchanged with African merchants for slaves. English cutlery from Birmingham and Sheffield was also traded in the same way. So many people had vested interests in the triangular trade that it became difficult for European politicians even to criticize the system, let alone abolish it.

The number of people who were enslaved and traded was vast. It has been estimated that by the time the slave trade was outlawed in Britain in 1807, British merchants had forced some 3 million Africans into lives of slavery in the Americas. Unknown numbers of people did not even reach America, but died en route in the appalling conditions on board the slave ships. It is likely that even more were carried by Portuguese traders bound for Brazil; ships from other nations carried smaller numbers. Some historians have estimated the total number at around 10 million; others put the figure still higher.

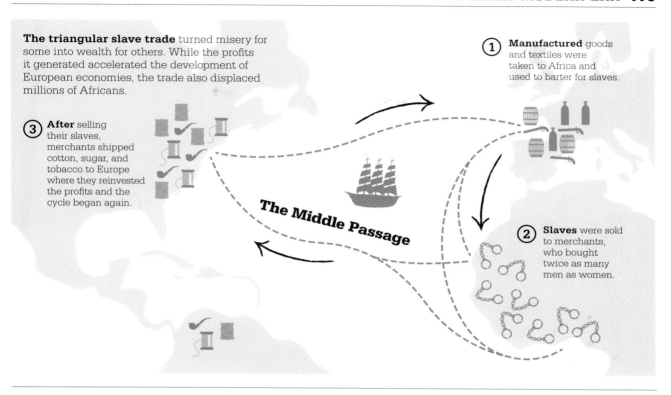

The triangular slave trade turned misery for some into wealth for others. While the profits it generated accelerated the development of European economies, the trade also displaced millions of Africans.

(1) **Manufactured** goods and textiles were taken to Africa and used to barter for slaves.

(3) **After** selling their slaves, merchants shipped cotton, sugar, and tobacco to Europe where they reinvested the profits and the cycle began again.

The Middle Passage

(2) **Slaves** were sold to merchants, who bought twice as many men as women.

European colonies

Spanish, Dutch, and French settlers pioneered the plantation system in the Caribbean, producing crops such as sugar and coffee on huge farms, or plantations. Their principal Caribbean colonies included Cuba (a colony of Spain), Haiti (France), and the Dutch Antilles (the Netherlands). The use of slave labor on these plantations generated substantial profits for owners. The British presence in the area increased in the 17th century, when Britain's most successful colony was Barbados, where there were 46,000 slaves by the 1680s. In the 18th century there was also a sugar boom in Jamaica.

Most of the native populations were wiped out in the European conquests, and European workers did not fare well in the local conditions, so plantation owners increasingly relied on merciless exploitation of slaves. Slavery was also prevalent in the colonies of North America, especially the southern areas where crops such as tobacco were grown on plantations. Slaves were often treated as non-human objects, forced into labor and subjected to cruelties such as beating, branding, and worse.

Slavery beyond the triangle

Colonists from Europe also practiced slavery beyond the Atlantic trading triangle. The Dutch pioneered slave trading in Southeast Asia, and also traded across the Indian Ocean with areas such as Madagascar and Mauritius. Most of this trading was conducted under the auspices of the Dutch East India Company, which had its eastern headquarters in the city of Jakarta, known to the Dutch as Batavia, as well as a base in Sri Lanka. From these points they sent slaves around the Indian Ocean, from eastern Indonesia to southern Africa. Once the Portuguese and English had set up bases, there was also further slave trading along the Indian coast.

The slave trade was not solely carried out by Europeans. Muslim merchants also transported slaves from East Africa for sale elsewhere in the Muslim world.

However, the triangular trade was a crucial element in the creation of a global economy run by Europeans and their colonial offshoots for their own profits. It permitted a phenomenal growth in the wealth of countries that ran the trade. In Britain, for example, the value of foreign trade rose from £10 million at the beginning of the 18th century to £40 million at the end. But the human cost of the trade in slaves, which influenced patterns of thought and behavior for centuries to come, remains incalculable today. ∎

THERE IS NO CORNER WHERE ONE DOES NOT TALK OF SHARES

THE OPENING OF THE AMSTERDAM STOCK EXCHANGE (1602)

IN CONTEXT

FOCUS
The Dutch Golden Age

BEFORE
1585 The founding of the Dutch Republic; Protestants in the south move northward.

1595 Cornelis de Houtman leads an expedition to Asia, starting the Dutch spice trade.

AFTER
1609 The Bank of Amsterdam is founded.

1610–1630 Land is reclaimed; the Dutch Republic increases in area by one-third and agricultural output increases.

1637 A single tulip bulb sells for up to 10 times the annual income of a skilled craftsman.

1650 Half the Republic's population lives in urban areas; the Netherlands is the most urbanized region in Europe.

The Amsterdam Stock Exchange—the world's first permanent market for stocks and shares—opened in 1602 under the auspices of the Dutch East India Company (known in the Netherlands as VOC). The company was a vast enterprise—in effect, the first international corporation—and it was created to facilitate trading expeditions to Asia. Unusually, the Dutch government had granted the company the power not only to trade, but also to build fortifications, establish settlements, raise armies, and enter into treaties with foreign rulers. Since the organization had a huge network of ships, ports, and personnel, it required considerable funding and many investors. The Amsterdam Stock Exchange was

See also: Christopher Columbus reaches America 142–47 ▪ The Treaty of Tordesillas 148–51 ▪ The Defenestration of Prague 164–69 ▪ Stephenson's *Rocket* enters service 220–25 ▪ The construction of the Suez Canal 230–35

The Dutch East India Company ran its own shipyards, the largest being in Amsterdam, shown here. Very powerful in the 17th century, the company went bankrupt and was dissolved in 1800.

originally set up to enable investors to trade their shares in the Dutch East India Company, but it then developed to become a vibrant market in financial assets and one of the drivers of a growing capitalist economy in the Dutch Republic.

An expanding economy

In the 17th century, the Netherlands was growing economically despite being involved in a long war with Spain. The northern part of the region (the Dutch Republic, which was Protestant) had split from the southern half (Flanders, which was Catholic) in the late 16th century. The Republic consisted of seven separate northern provinces, each with a great deal of independence but under the umbrella of a federal government called the States-General. Protestant merchants who had lived in Catholic cities, such as Antwerp, moved north to escape persecution, taking with them their capital and trading links. Also, many Flemish artisans who were skilled in textile production (primarily weaving wool, silk, and linen) emigrated to the northern

cities of Haarlem, Leiden, and Amsterdam, boosting the Dutch Republic's economy further.

As the 17th century progressed, the Republic really began to prosper. Various factors came together to make this small region successful. Most importantly, the nation had a strong tradition of seafaring, giving it a huge advantage over many other countries. In addition, its citizens had a strong work ethic—largely due to the Protestant belief that

worldly work was a duty and a route to salvation—so productivity was high. There was also a growing population (especially of the urban middle classes) and an expanding major city—Amsterdam—which proved an ideal center for trade. All of these contributing factors resulted in the Dutch economy moving increasingly toward shipping, trading, and finance.

Exploration and trade

As a coastal nation, the Dutch Republic produced notable sailors and explorers, so long-distance trading was a natural consequence of the country's maritime history. In addition, advancements in ship-building technology in the Republic enabled the Dutch merchant fleet to expand rapidly; by 1670, the Dutch had more merchant ships than the rest of Europe put together. **»**

Agricultural revolution

The expanding population of the Dutch Republic in the 17th century encouraged farmers to make agriculture much more productive. In large part, this was achieved through continued land reclamation—a process that was already well underway by the late Middle Ages. The Dutch also changed the way they used their land. Instead of growing grain one year and letting the land lie fallow the next, farmers began planting certain nitrogen-producing crops (such as peas,

turnips, and clover, which they could use as animal feed), in order to improve the soil ready for the next corn crop. Growing more fodder meant that farmers could keep larger herds, thereby increasing production of meat and milk as well as manure, which could be used as fertilizer. This greater productivity helped to sustain a growing population, although some wheat still had to be imported to make up the shortfall. It also freed up larger segments of the population to work in trade or finance rather than agriculture.

The expanding merchant class saw large potential profits in the spice trade with Asia and, as in other maritime cultures such as Spain and Portugal, navigators sought new sea routes to the east. The Dutch traveled all over the globe and set up colonies, including one in North America: New Amsterdam, which they officially settled in 1624 and was renamed New York when the British took over. In 1596, the Dutch explorer Willem Barentsz tried to find a northern passage to Asia and in the process discovered Svalbard (Spitsbergen), which later became a destination for Dutch whalers.

Most importantly for their prosperity, from 1595 the Dutch began to make regular journeys to Southeast Asia to trade in spices, particularly pepper, nutmeg, cloves, and cinnamon. They established colonies in the region and founded the city of Batavia, later renamed Jakarta. With this permanent base,

Batavia was the headquarters of the Dutch East India Company in Asia. The port city was founded by the Dutch in 1619, after razing the existing city of Jayakarta to the ground.

the Dutch had the ability to trade long-term, producing a massive boost to their economy.

A need for investment

While the wealth generated by exploration and trade was injected back into the Dutch economy, at the same time investment was required to cover the considerable costs of overseas expeditions. A trading voyage to Asia in the 17th century was a very risky venture—the potential profits were high, but storms at sea, pirates, warfare, or an accident could lead to the loss of a ship, crew, or cargo and wipe out all the profits. It therefore made sense for many people to invest in each voyage and spread the risk, rather than one entity shouldering all the costs and responsibilities. Private trading companies were set up, each investing a small amount in a larger whole, and all being well they would then receive a commensurate share of the profits.

Birth of the Exchange

In 1602, these trading companies merged to form the Dutch East India Company, and shares in the

> If one were to lead a stranger through the streets of Amsterdam and ask him where he was, he would answer "among speculators."
> **Joseph Penso de la Vega**
> *Confusion of Confusions* (1688)

enterprise were allocated at the new Stock Exchange in Amsterdam. It was established at the outset that the owners could buy and sell these shares, and very quickly other companies were listing their own shares on the Stock Exchange in order to raise money. The ease of buying and selling shares meant that the Stock Exchange became very busy indeed, fueling the growth of capitalism in this part of Europe; increased investment resulted in more industry, which then led to further investment and the generation of greater wealth.

A history of trading

The Amsterdam Stock Exchange did not develop in a vacuum. Buying and selling securities— tradable financial assets such as shares—already had a long history in Europe. By the 14th century, possibly earlier, merchants in rich Italian trading cities, such as Venice and Genoa, had traded in securities. However, the prevailing conditions in the Netherlands in the 17th century meant that the market was especially buoyant. Since the 16th century, there had

been a strong financial market in Amsterdam, where there was a tradition of trading in commodities and speculation in everything from whale oil to tulips. The idea of buying and selling shares therefore appealed in this entrepreneurial society, especially as there was a good prospect of healthy profits from the Asian trade. In addition, the unique way in which the exchange traded—opening for limited hours only—encouraged rapid buying and selling and produced a very fluid market.

Boosts to the economy

The opening of the Amsterdam Stock Exchange was followed in 1609 by the foundation of the Bank of Amsterdam—the forerunner of modern national banks. The bank provided a secure place to keep money and bullion, and it assured that local currency kept its value. It thus helped to make the Dutch Republic more financially secure, underpinning the vigorous and often risky trading activity that went on in this burgeoning market.

In 1623, the market had a further boost when the Dutch East India Company negotiated a new charter, paying investors a regular dividend and permitting those who wanted to leave the company to sell their shares on the Stock Exchange. This action further increased trade on the Stock Market, which was also pioneering other lucrative activities such as futures trading.

The insurance business was also thriving in Amsterdam during this time—particularly marine insurance, which had been created in the 16th century to protect ship owners and investors against the risks of long-distance voyages. When the Stock Exchange opened, a special area was set aside for the buying and selling of insurance.

Dutch explorers discover **new sea routes** and the **Dutch merchant fleet expands**.

↓

Trading voyages to the spice-producing countries of Asia yield **high profits** but pose a **high risk**.

↓

The **Dutch East India Company** is set up to share the **financial risk of voyages** between **multiple investors**.

↓

The Amsterdam Stock Exchange is formed to allow shares in the East India Company to be traded.

↓

Rapid **buying and selling** creates a fluid financial market, encouraging speculators to **take more risk**.

A flourishing culture

The very buoyant financial activity prevalent in Amsterdam in the 17th century encouraged the expanding middle classes to buy consumer goods, including fine furniture and oil paintings, further fueling the economy of this already successful region. A particularly strong art market developed, allowing major painters—such as Vermeer and Rembrandt, as well as numerous lesser followers—to flourish. Many artists were specialists, satisfying a growing demand for portraits, landscapes, seascapes, and still lifes, although great artists like Rembrandt excelled in all genres and art forms, including painting, drawing, and printmaking.

The increasing wealth also led to the expansion of towns, with new town halls, warehouses, and merchants' homes springing up. Numerous brick houses, owned by the middle classes, survive in cities such as Amsterdam and Delft, many of them set on the banks of the canals that were built during this period—a time of economic boom that combined elegance and artistic flair with success in trade. ∎

AFTER VICTORY, TIGHTEN THE CORDS OF YOUR HELMET
THE BATTLE OF SEKIGAHARA (1600)

IN CONTEXT

FOCUS
The Edo Period

BEFORE
1467 The Warring States Period begins, with the emperor losing power to conflicting factions led by daimyos and shoguns.

1585 Toyotomi Hideyoshi is given the title of Imperial Regent by the emperor.

AFTER
1603 Tokugawa Ieyasu is appointed shogun.

1610–1614 Missionaries are expelled from Japan and Christian activity is banned.

1616 Tokugawa Ieyasu dies.

1854 After years of being closed to the West, Japan opens its ports to American shipping and trade.

1868 The Tokugawa shogunate finally ends with the restoration of imperial power under Emperor Meiji.

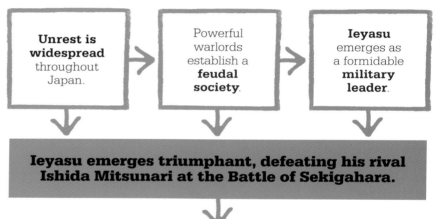

Unrest is widespread throughout Japan.

Powerful warlords establish a **feudal society**.

Ieyasu emerges as a formidable **military leader**.

Ieyasu emerges triumphant, defeating his rival Ishida Mitsunari at the Battle of Sekigahara.

Ieyasu becomes **shogun** and political power is **unified** under the **Tokugawa shogunate**.

On October 21, 1600 there was a momentous battle in Sekigahara, central Japan, between two warring factions—the Eastern and Western armies—who were both fighting for control of the country. The Eastern Army, under the leadership of the lord Tokugawa Ieyasu, won a decisive victory. Three years later, the Emperor of Japan awarded Ieyasu the title of shogun, granting him the power to rule the country on the emperor's behalf.

Ieyasu brought stability and peace to Japan and transferred the nation's capital to Edo (now Tokyo), creating a new focus for Japanese culture as well as a central power base.

Factional struggles
Since 1192, the Emperor of Japan had been little more than a figurehead. He delegated power to the shogun: a hereditary, high-ranking military commander who ruled with absolute authority. However,

See also: Minamoto Yoritomo becomes Shogun 98–99 ▪ The opening of the Amsterdam Stock Exchange 180–83 ▪ The Meiji Restoration 252–53 ▪ The Second Opium War 254–55

Tokugawa Ieyasu

The samurai leader Tokugawa Ieyasu (1542–1616) was the son of a minor Japanese warlord from Mikawa in central Japan. As a young man, he received a military training before becoming an ally of more powerful warlords, such as Oda Nobunaga (1534–82)—one of the most brutal leaders of the turbulent Warring States Period in Japan—and his successor, Toyotomi Hideyoshi (1536–98). Working with Nobunaga and Hideyoshi, Ieyasu not only built up large personal landholdings, but he also learned the key values of loyalty and military power that enabled Hideyoshi to bring a brief period of unity to Japan. When Hideyoshi died, Ieyasu rose to the fore. As shogun, he was able to impose stability on his country, but he formally abdicated after only two years in favor of his son, Hidetada, to secure a smooth succession and establish a pattern of shoguns passing on their office, helping to ensure that the Tokugawa shogunate was long lasting. Although Hidetada had officially become shogun, Ieyasu remained the effective ruler of Japan until his death.

by the 1460s the local feudal lords (*daimyos*) were so powerful that few shoguns had control over them, as they and their armies of samurai warriors fought to win the right to appoint the shogun's successor. By the time of the Battle of Sekigahara, Japan had endured bitter factional struggles between its ruling classes for over a century.

Ieyasu's victory at the battle put an end to this Warring States Period. His steady rule, followed by that of the Tokugawa shoguns who succeeded him, ushered in a 250-year period of stability.

The Tokugawa shoguns
In many respects, the Tokugawa shoguns modeled themselves on earlier rulers—particularly Toyotomi Hideyoshi. Although he was not sufficiently high-born to become a shogun, Hideyoshi (who ruled under the lesser title of imperial regent) had brought unity to Japan in the 1580s by imposing a military, feudal style of rule whereby he wielded great power through the daimyos and their samurai warriors. The

Tokugawa shoguns decided to govern in the same way, with the daimyos keeping order in their local areas. As an extra precaution, Ieyasu made the daimyos spend alternate years in Edo to ensure they would not build up local power bases; he also suppressed rivals ruthlessly.

The shoguns encouraged an ethic of loyalty and developed an elite bureaucracy. They improved Japan's road network, promoted education, and standardized the currency. The shogunate also tried to reduce foreign influence in Japan by expelling foreigners and limiting contact with the outside world. Exceptions were made for strictly controlled trade with the Chinese, Koreans, and the Dutch East India Company; all other Europeans were distrusted, as the shoguns believed that they had plans to convert the Japanese to Christianity and gain political power. Furthermore, the Japanese people were forbidden to travel and build ocean-going ships. This policy of isolation virtually cut off Japan from Western influence until the mid-19th century.

The "floating world"
The capital Edo became the center of a thriving urban culture during the Tokugawa shogunate. Japanese literary forms, such as the *haiku* (a short poem consisting of three lines and 17 syllables) flourished, as did the distinctive theatrical forms of *kabuki* (which combines theater and dance) and the *bunraku* puppetry theater. It was also a time of major achievements in the visual arts, particularly landscape painting and woodblock printing.

The capital's elite became increasingly hedonistic, with their lifestyle frequently described as the "floating world" (*ukiyo*). Originally, Buddhists had used the term *ukiyo* to mean "sorrowful world," reflecting their opinion that life on earth was transitory and expressing a desire to reach a more permanent place, free from suffering and all earthly desires. However, in the Edo Period the homonym *ukiyo* ("floating") was used to describe the joyful aspect of the ephemeral material world, reflecting the pleasure-seeking mood of the day. ▪

USE BARBARIANS TO CONTROL BARBARIANS

THE REVOLT OF THE THREE FEUDATORIES (1673–1681)

IN CONTEXT

FOCUS
China's Three Emperors

BEFORE
1636 The Manchu establish the Qing dynasty in their homeland of Manchuria.

1644 The Qing dynasty conquers northern China.

AFTER
1683 The Qing destroy all Ming resistance and establish their rule across China.

1689 Emperor Kangxi's peace settlement with Russia, the Treaty of Nerchinsk, checks Russia's eastward expansion.

1750 The Summer Palace—a masterpiece of Chinese landscape design—is built.

1751 Tibet becomes a Chinese protectorate.

1755–60 Emperor Qianlong removes Turk and Mongol threats to northeastern China.

1792 Invasion of Nepal by the Qing.

Qianlong employed the Italian Jesuit Giuseppe Castiglione as court painter, and his imperial portraits fused elements of Chinese scroll painting with Western realism and perspective.

In 1644, the Manchu—a semi-nomadic people who had built a large state to the northeast of China's Great Wall—seized Beijing from the crumbling Ming regime and established their own dynasty, the Qing, as the rulers of northern China. Seventeen years later, after fierce fighting on an epic scale, the Qing had overcome the determined resistance of Ming loyalists, and extended their power across mainland China. However, their dynasty was still not secure—in 1673, Kangxi, the second emperor, was forced to confront a major uprising, which became known as the Revolt of the Three Feudatories.

The Three Feudatories were vast areas of south China that had been granted as semi-independent fiefdoms to three turncoat Ming generals who had assisted the Qing in their conquest of China. Over time, the fiefdoms became increasingly autonomous, but when Kangxi declared that they would not be hereditary, the generals rebelled. The ensuing struggle was hugely costly in terms of loss of life and economic disruption, and for a while, it seemed that one general, Wu Sangui, would topple the Qing. However, he was finally defeated by Kangxi's supporters, and in 1683, the Qing eliminated the last stronghold of Ming support on Taiwan, which they then occupied.

With the Qing now undisputed rulers of China, Kangxi embarked on military campaigns that added parts of Siberia and Mongolia to the Chinese empire, and extended its control over Tibet. Under his exceptional leadership, and that

See also: Marco Polo reaches Shangdu 104–05 ▪ Hongwu founds the Ming dynasty 120–27 ▪ The Second Opium War 254–55 ▪ The Long March 304–05

The Revolt of the Three Feudatories fails, marking the end of resistance to Manchu power.

→ The first three Qing emperors **legitimize their foreign rule** by adopting Chinese ways.

→ In the stability that follows, China **triples in size** and the economy **expands rapidly.**

In the 18th century, China becomes the **biggest manufacturing power** in the world.

By the end of the 19th century, the Qing are a power in name only, as the **pressures of European imperial expansion** and **growing internal dissent** fatally weaken the regime.

Qing society

The era of the Three Emperors was conservative in many ways: Han Chinese men were required to wear the Manchu hairstyle, in which the front and sides of the head were shaved, and the remaining hair plaited into a braid; society was rigidly hierarchical, and there were strict conventions regarding the conduct of women, laws against homosexuality, and censorship. Yet the country's economy grew substantially in the early part of the Qing period, thanks to a strong demand in the West for luxury products such as silk, porcelain, and tea.

However, by the beginning of the 19th century, the regime's repressive treatment of the Han Chinese people, together with famine and widespread addiction to opium—which had been brought into China by European traders—had sent the country into decline. These factors sowed the seeds of rebellions, trading disputes, and wars with European trading partners in the mid-19th century. ▪

of his two immediate successors, China enjoyed a golden age of peace, economic prosperity, and political stability that lasted until the late 18th century.

A global superpower

During his 61-year reign, Kangxi won the cooperation and loyalty of his native Han Chinese subjects—who had once viewed the Manchu as barbarians—by preserving and honoring China's cultural heritage. He also continued the preceeding dynasty's form of government, and allowed Ming officials to retain their provincial posts alongside Manchu appointees, although the latter supervised most of the work.

Qing China became immensely powerful during the reigns of the next two emperors—Yongzheng (1722–35), who also kept a tight control on government and the bureaucracy and increased state revenues by reforming the tax system, and Qianlong (1735–96),

under whom the empire's borders reached their greatest extent and the population boomed. Qianlong was an avid patron of the arts who wrote poetry and sponsored literary projects that enhanced his people's reputation—although at the same time, he banned or destroyed books that were judged to be anti-Qing.

The Jesuits in China

In 1540, Ignatius of Loyola, a Catholic theologian from Spain, founded the Society of Jesus—the Jesuits—with the aim of spreading the faith through the teachings of Jesus. The Catholic Church sent Jesuit missionaries to China during the Ming and early Qing periods, and initially they were welcomed. Kangxi was curious about the Jesuits' knowledge of science (especially mathematics and astronomy) and technology (particularly the manufacture of weapons and

pumps). He appointed Jesuits to the imperial board of astronomy, and it was a Jesuit who made the first accurate map of Beijing.

Kangxi gave Catholics freedom of worship in China, and the Jesuits allowed Chinese converts to continue their rites of ancestor worship (they saw these as commemorations of the dead rather than true acts of veneration). However, when a visiting Vatican envoy ruled against the ancestral rites, and the pope followed suit, Kangxi expelled Jesuit missionaries who opposed the practice.

I HAVE IN THIS TREATISE CULTIVATED MATHEMATICS SO FAR AS IT REGARDS PHILOSOPHY

NEWTON PUBLISHES *PRINCIPIA* (1687)

IN CONTEXT

FOCUS
Scientific revolution

BEFORE
1543 Copernicus publishes his heliocentric version of the universe.

1609 German Johannes Kepler describes the planets' elliptical orbits and speeds.

1620 Francis Bacon publishes the *Novum Organum*.

1638 Italian Galileo Galilei's *Discourses* lay the foundation of the science of mechanics.

1660 The Royal Society is founded in England.

AFTER
1690 Dutchman Christiaan Huygens publishes his theory of the wave motion of light, *Traité de la lumière*.

1905 Albert Einstein's Special Theory of Relativity shows that Newton's Laws of Motion are only approximately correct.

English scientist Isaac Newton published the first edition of his *Mathematical Principles of Natural Philosophy*, or *Principia*, in 1687. The book examines the way objects behave in motion, describes gravity, and explains the movements of planets and satellites. Although it built on the work of earlier scientists such as Galileo, Huygens, and Kepler, the work was revolutionary. By illustrating how the same force—gravity—is responsible for movements both on Earth and in the heavens, it united two scientific realms that had previously been thought separate.

A lasting influence

Newton's use of mathematics-based theory to explain phenomena was part of a wider scientific revolution. English essayist Francis Bacon insisted that scientists test their observations using reasoned argument, and French philosopher René Descartes championed the use of mathematics and logic to address scientific questions. By emphasizing the importance of human reason, such philosophers broke free from the notion that explanations of the physical world depended on Christian faith and church doctrine. This paved the way for the intellectual movement called the Enlightenment, and even for the work of later scientists such as Albert Einstein, who modified and refined Newton's theories. ∎

[Newton] spread the light of mathematics on a science which... had remained in the darkness of conjectures and hypotheses.
Alexis Clairaut
French mathematician and astronomer (1747)

See also: The founding of Baghdad 86–93 ▪ Brunelleschi designs the dome of Florence Cathedral 152–55 ▪ Diderot publishes the *Encyclopédie* 192–95 ▪ Darwin publishes *On the Origin of Species* 236–37

AS FAR AS I THINK IT POSSIBLE FOR MAN TO GO
THE VOYAGES OF CAPTAIN COOK (1768–1779)

IN CONTEXT

FOCUS
Pacific and Australasian exploration

BEFORE
1642–1644 Dutchman Abel Tasman becomes the first European to reach New Zealand and Tasmania.

1768–1771 James Cook makes his first voyage to Australia and New Zealand.

1772–1775 Cook sails close to Antarctica, and around the southern Pacific.

1776–1779 Cook's third voyage takes him to Hawaii, where he is killed in a fight with local people.

AFTER
1788 The first convicts from Britain arrive at the Port Jackson (Sydney Harbour) penal colony.

1802 British navigator Matthew Flinders circumnavigates Australia.

I n 1768, British navigator James Cook sailed to Tahiti to make scientific observations of the Transit of Venus across the Sun, a rare event that could be seen only from the southern hemisphere. Having recorded the event, Cook sailed on in search of the rumored "unknown land of the South." He mapped the New Zealand coast, and then traveled northwest, discovering the eastern coast of Australia in the process. Claiming the land for Britain, he named it New South Wales. Working closely with botanists Joseph Banks and Daniel Solander, he also produced unique records of the indigenous peoples, flora, and fauna.

An enduring link
Cook's voyages were part of a wider tradition of European exploration of the Pacific by navigators such as Dutchman Abel Tasman, after whom Tasmania is named. Cook forged the enduring connection between Australasia and Europe, beginning a process that continued with colonization, the transportation

We were regaled with the pleasing sight of the Mountains of New Zealand— after an absence from Land of 17 weeks and 3 days... how changed the scene!
Richard Pickersgill,
**Third lieutenant on the *Resolution*
(1773)**

of British convicts into exile, and the founding of cities such as Sydney and Melbourne.

In his later voyages, Cook used the chronometer, newly developed by Englishman John Harrison. It facilitated accurate timekeeping at sea, and so the calculation of precise longitude, which was invaluable to Cook in charting his discoveries. ∎

See also: Marco Polo reaches Shangdu 104–05 ▪ Christopher Columbus reaches America 142–47 ▪ The Treaty of Tordesillas 148–51 ▪ The voyage of the *Mayflower* 172–73

I AM THE STATE

LOUIS XIV BEGINS PERSONAL RULE OF FRANCE (1661)

On the death of his chief minister Cardinal Mazzarin, the 23-year-old Louis XIV of France declared that he would now rule alone, as an absolute monarch. During his 72-year reign (1643–1715), Louis dominated his subjects, cultivating the image of a "Sun King" around whom the country orbited. Louis saw his power as God-given, and himself as the embodiment of the state, with the nobility, the middle classes, and peasants dependent on him for justice and protection.

To maintain this position, Louis controlled the historically unruly aristocracy. He compelled them to attend his court, where he dispensed privileges and positions via a system of etiquette. He filled the treasury's depleted coffers by appointing members of the upper-middle classes to collect taxes in the provinces. Taxation was extensive and the burden fell mainly on the peasantry. Louis's finance minister, Jean-Baptiste Colbert, whose overhaul of France's trade and industry helped to make the country Europe's leading power, increased the efficiency of the revenue system.

Expanding France

Louis's tax income paid for his court at the dazzling Palace of Versailles, an old hunting lodge extended into a royal palace, and the venue for extravagant entertainments. From 1682 it became the permanent base of the royal court, and the seat of government. Louis also waged a series of costly dynastic wars to make some territorial gains along France's frontiers, leading the other European nations to form coalitions against him.

Peace was finally achieved at the Treaty of Utrecht in 1713, but brought few gains for France. The country was plunged into debt and opinion turned against the Crown. In spite of this, Louis established a pattern of absolutism in France that lasted, in a more enlightened form, for most of the 18th century until attempts to reform the system resulted in the overthrow of the monarchy in 1792 during the French Revolution. ∎

See also: The execution of Charles I 174–75 ▪ Diderot publishes the *Encyclopédie* 192–95 ▪ The storming of the Bastille 208–13 ▪ The Battle of Waterloo 214–15

DON'T FORGET YOUR GREAT GUNS, THE MOST RESPECTABLE ARGUMENTS OF THE RIGHTS OF KINGS
THE BATTLE OF QUEBEC (1759)

IN CONTEXT

FOCUS
Seven Years' War

BEFORE
1754 Fighting between France and Britain in North America, the so-called French and Indian War, begins.

1756 Frederick II of Prussia begins the Seven Years War by invading Saxony to prevent Russia from creating a base there.

1757 Prussia inflicts a significant defeat on superior French and Austrian forces at Rossbach.

1759 Russia wipes out two-thirds of the Prussian army at Kunersdorf.

AFTER
1760 French forces at Montreal surrender to the British.

1763 The Seven Years' War comes to an end with the treaties of Paris and Hubertusburg.

O n September 13, 1759, 24 British men scaled the cliffs below Quebec, opening the way for British forces commanded by General James Wolfe to capture the city. The crucial battle ended French dominance in Canada and was a key event in the Seven Years' War (1756–1763).

The war involved most of the chief European nations in a struggle for territory and power. It centered on two main clashes: one maritime and colonial, involving land battles in North America and India between Britain and Bourbon France; the other a European land war that chiefly pitted France, Austria, and Russia against Prussia. Overseas colonies also became involved, making this the first true global conflict.

Competing powers
Britain achieved notable victories over France. A French invasion attempt on Britain was thwarted by Britain's superior navy, and Britain scored colonial victories over France in West Africa, the Caribbean, and

Without supplies
no army is brave.
Frederick the Great, 1747

North America where there were major successes in Canada. Britain forced France to cede all of their territory east of the Mississippi River, effectively ending the threat France posed to Britain's North American colonies.

There were similar victories in India. The British general Robert Clive, wrongfooted the French by defeating the Nawab of Bengal at Plassey in 1757 and acquiring his territory for Britain, paving the way for the British domination of India. The end of the Seven Years' War left Britain the leading colonial power. ∎

See also: Christopher Columbus reaches America 142–47 ▪ The Defenestration of Prague 164–69 ▪ The voyage of the *Mayflower* 172–73 ▪ The Battle of Waterloo 214–15 ▪ The Battle of Passchendaele 270–75

ASSEMBLE ALL THE KNOWLEDGE SCATTERED ON THE EARTH

DIDEROT PUBLISHES THE *ENCYCLOPÉDIE* (1751)

IN CONTEXT

FOCUS
The Enlightenment

BEFORE
1517 The Reformation begins, challenging the authority of the Catholic Church.

1610 Galileo Galilei publishes *Sidereus Nuncius* (*Starry Messenger*), containing his observations of the heavens.

1687 In *Principia,* Newton outlines a concept of the universe based on natural, rationally understandable laws.

AFTER
1767 American thinker and diplomat Benjamin Franklin visits Paris, and transmits Enlightenment ideas to the US.

1791 English writer Mary Wollstonecraft adds feminism to Enlightenment ideas in the pioneering *A Vindication of the Rights of Women*.

In the mid 18th century, the French philosopher Denis Diderot invited some of his country's leading intellectuals—literary men, scientists, scholars, and philosophers to write articles for a huge "Classified Dictionary of Sciences, Arts, and Trades," for which he was both editor-in-chief and contributor. The first volumes of his *Encyclopédie* appeared in 1751, and the full work was completed 21 years later, made up of 17 volumes of text and another 11 volumes of illustrations.

The *Encyclopédie* was not the first large encyclopaedia to be published, but it was the first to feature content by named authors,

See also: Newton publishes *Principia* 188 ▪ The signing of the Declaration of Independence 204–07 ▪
The storming of the Bastille 208–13 ▪ Stephenson's *Rocket* enters service 220–25 ▪ The Slave Trade Abolition Act 226–27

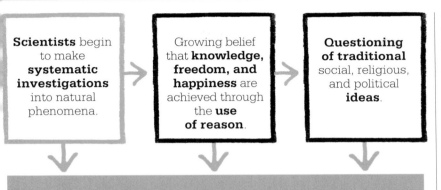

Scientists begin to make **systematic investigations** into natural phenomena.

Growing belief that **knowledge, freedom, and happiness** are achieved through the **use of reason**.

Questioning of traditional social, religious, and political **ideas**.

The Enlightenment movement begins, spearheaded by the publication of the *Encyclopédie*.

and to give close attention to the trades and crafts. Its most striking feature, however, was its critical approach to contemporary ideas and institutions: its authors were champions of scientific thought and secular values. They sought to apply reason and logic to explain the phenomena of the natural world, and humankind's existence, rather than religious or political dogma. As such, the work challenged both the Catholic Church and the French monarchy, which derived their authority from traditional ideas such as a divinely ordained, unchanging order.

A revolution in thought
The mission of the *Encyclopédie* was to catalog the collective knowledge of the Western world in the spirit of the Enlightenment. This was a multifaceted intellectual movement that took root around 1715, although its origins lay in work done by the pioneers of modern scientific and philosophical thought in the previous century. The work's multidisciplinary articles, which

numbered around 72,000, distilled the ideas and theories of France's key Enlightenment thinkers— including the writers and philosophers Voltaire, Jean-Jacques Rousseau, and Montesquieu.

The articles were extremely wide-ranging, but centered on three main areas: the need to base society not on faith and the doctrines of the Catholic Church but on rational thought; the importance of observations and experiments in science; and the search for a way of organizing states and governments around natural law and justice.

Diderot organized the *Encyclopédie*'s articles into three main categories: memory (subjects connected to history); reason (philosophy); and imagination (poetry). Controversially, there was no special category for God or the divine—religion, like magic and superstition, was treated as part of philosophy. This approach was groundbreaking, and contentious. Religion had been at the very heart of life and thought in Europe for

centuries: the *Encyclopédie,* and the Enlightenment itself, denied it this key position.

In spite of repeated efforts by the authorities to censor some of its articles, and to intimidate and threaten its editors, the *Encyclopédie* became the most influential and widely consulted work of the period. The ideas that it transmitted inspired the revolutions that exploded in France and the US at the end of the 18th century.

Science and reason
The Enlightenment movement was characterized by a focus on the power of human reason and skepticism of accepted knowledge. This marked a break from earlier generations in which beliefs about the world derived from religious teachings and the doctrines of the Church. These governed everything from the laws of marriage to the way people understood the movement of the planets and the creation of the universe. For Enlightenment thinkers, however, the evidence of a person's senses and the use of one's reason was far more important than their blind »

Dare to know! Have courage to use your own reason!
Immanuel Kant
"What is Enlightenment?" (1784)

Voltaire

François-Marie Arouet, who chose to be known publicly by the name Voltaire, was one of the greatest writers and social activists of the Enlightenment, renowned for his wit and intelligence. He was born in Paris in 1694, and spent much of his long life there, although he traveled widely and spoke several languages. He was a hugely prolific writer, producing works in almost every literary genre: novels, plays, poems, essays, historical studies, and philosophical books as well as countless pamphlets.

Voltaire was an outspoken supporter of social reform, including the defense of civil liberties and freedom of religion and speech; he also denounced the hypocrisy of the political and religious establishment. This led to the censorship of some of his work, and also to short spells of imprisonment and periods of exile in England—after which he converted his experiences into an influential book, *Philosophical Letters on the English*—and Geneva, Switzerland, where he wrote his most famous work, the philosophical novella *Candide*.

In all the ages of the world, priests have been the enemies of liberty.
David Hume

adherence to a faith. They argued that "truths" about the tangible world, which had been set down in antiquity by Aristotle and others, and upheld by the Church, should be tested through experimentation and observation, checked, and then discussed in a rational way.

This radical mode of thinking had its origins in the scientific revolution of the 17th century. Scientists and philosophers including Francis Bacon, Johannes Kepler, Isaac Newton, and Galileo Galilei had transformed the study of nature and the physical universe, making it more observational. They conducted careful experiments and subjected their results to mathematical analysis; in the process they drastically updated and expanded the fields of physics, chemistry, biology, and astronomy.

Enlightenment scientists took this investigation of reality further, making possible, for example, Swedish botanist Carl Linnaeus's development of a proper, rational biological classification in the early 18th century. The inquiring, reason-based approach of the Enlightenment also triggered dramatic technological advances. In the 1760s, the Scottish physician Joseph Black discovered carbon dioxide, while in 1769, Scotsman James Watt made improvements to the steam engine that increased its efficiency, thereby enabling the improvement of factories. The *Encyclopédie* helped to publicize these, and other, achievements of 18th-century scientists, as well as those of their precursors.

The work also found an audience in the learned societies, academies, and universities that flourished in the Enlightenment period. Although many teachers and scholars at Europe's older, Church-dominated universities remained deaf to the new scientific way of thought, more progressive ones helped to teach and promote it.

Equality and freedom

The scientific revolution and the Enlightenment also encouraged the belief that reason could reveal natural laws in human affairs. Instead of drawing fact from faith, Enlightenment thinkers believed that politics should be separated from religion, that neither should curtail the rights of the individual, and that people should be free to express their opinions, worship in their own way, and read what they want to. This political doctrine, which is often labeled liberalism, had roots in the work of 17th-century philosophers such as Englishman

Scepticism is the first step towards truth.
Denis Diderot
Philosophical Thoughts (1746)

> To renounce liberty is to renounce being a man.
> **Jean-Jacques Rousseau**
> *The Social Contract (1762)*

John Locke—the father of liberalism. Locke asserted that there are certain intrinsic human rights that are not dependent on law or custom—in other words, they exist quite separately from what the Church or monarch might decree. These rights could be expressed in different ways, but included the right to life, the right to liberty, and the freedom to own what one has produced. These ideas were central to Enlightenment thinkers, following Locke, who felt that such natural rights should form the basis of any system of government.

Liberal ideas also found expression in the work of Enlightenment writers. For example, Voltaire, in books such as the *Philosophical Dictionary,* highlighted the injustices and abuses of the Catholic Church, and espoused values such as tolerance, freedom of the press, and the promotion of reason over doctrine and religious revelation. In his *Spirit of the Laws*, Montesquieu advocated the separation of governmental powers (legislature, executive, judiciary) and pressed for an end to slavery. In *The Social Contract*, Jean-Jacques Rousseau rejected the power of the monarch in favor of that of the people, who,

he said, must balance rights with duties, and should be able to decide the laws that govern their lives. The contributors to the *Encyclopédie* also promoted liberal values in economics. They were critical of fairs—where goods were sold by visiting dealers at the expense of local traders, who often had to close their businesses for the duration—and favored markets, which allowed local traders to meet the needs of the local population.

Ideas such as these spread across Europe. Conversations and debates on philosophical, political, and scientific subjects took place in the coffee-houses that had sprung up in English, French, German, and Dutch cities a century earlier. These coffee-houses now served as information-sharing hubs where men from all walks of life, including writers, politicians, philosophers, and scientists, could congregate to exchange views.

Into the light
In Europe, the Enlightenment movement, and the *Encyclopédie* itself, which helped promote its ideals, had a profound impact on social, political, and intellectual life. Its proponents believed that they were sweeping away an oppressive medieval worldview and ushering in a new era that they hoped would be characterized by freedom of thought, open-mindedness, and tolerance.

The Enlightenment's questioning, rational approach, and urgent demand for liberty, paved the way for the granting of new civil rights. The movement affected the policies of monarchical rulers, such as the freeing of serfs in the Holy Roman Empire in the 1780s. Monarchs who accepted Enlightenment values took on the movement's name, titling

themselves Enlightened Despots. Enlightenment thought also provided the intellectual fuel for the French Revolution of 1789–99—begun by citizens inspired by Enlightenment notions of individual freedom and equality—and the campaign to abolish the Atlantic slave trade in the 19th century.

Liberalism and other aspects of Enlightenment political philosophy began to influence leaders in many parts of the world when they came to draw up legal systems and to establish rights for their citizens—most notably in the fledgling United States, whose Constitution (1787) adopted Montesquieu's idea of the separation of power into branches of government.

More generally, the movement promoted the pursuit of knowledge for its own sake and recognized that one person's quest for understanding could benefit the entire human race. ∎

In 1783, France's Montgolfier brothers gave the first demonstration of their new invention, the hot-air balloon, bringing science to the forefront of public attention in a spectacular way.

I BUILT ST. PETERSBURG AS A WINDOW TO LET IN THE LIGHT OF EUROPE
THE FOUNDING OF ST. PETERSBURG (1703)

IN CONTEXT

FOCUS
The rise of Russia

BEFORE
1584 The emperor Ivan the Terrible dies. The following succession of rulers bring greater unity to Russia.

1696 Peter the Great assumes sole rule of Russia.

AFTER
1709 Russia wins a decisive victory over Sweden at the Battle of Poltava.

1718 Peter's son Alexis, opposed to his father's reforms, dies under torture.

1721 Russia and Sweden sign the Treaty of Nystad, pledging mutual defence.

1725 Peter the Great dies, ushering in an era of less competent emperors.

1762 Catherine the Great becomes empress and continues Peter's work of reform and expansion.

R ussian ruler Peter the Great founded St. Petersburg, on the estuary of the River Neva, on May 27, 1703. This new city, fortress, and port by the Baltic gave Russia direct sea access to Europe, opening new opportunities for both trade and military conquest. In 1712, Peter made his new city Russia's capital, stripping the title from the ancient seat of Moscow.

An admirer of Western palaces, Peter employed European architects to design the government buildings, palaces, university, and houses in the fashionable baroque style. He also pressed 30,000 peasants each year into construction gangs, along with Russian convict laborers

St. Petersburg offered a new vision for the country. Its strategic location facilitated trade, its ethos encouraged education, and its architecture provided a showcase for Russian achievement.

and Swedish prisoners of war. The regimen was strict, and living conditions stark: more than 100,000 workers died, but those who survived could earn their freedom.

The lavish design and vast scale of Peter's architecture showed not only his appreciation of European culture, but also his determination to be an exalted, absolute ruler in the manner of Western despots such as Louis XIV. Peter used his power to make significant changes in

See also: Louis XIV begins personal rule of France 190 ▪ Diderot publishes the *Encyclopédie* 192–95 ▪ The storming of the Bastille 208–13 ▪ Russia emancipates the serfs 243 ▪ The October Revolution 276–79

Russia. He founded the Russian navy and reformed the army, which until then had relied on bands of men led by untrained village elders. He reorganized the army along European lines and developed new iron and munitions industries to equip it. By 1725, Russia had a professional army of 130,000 men.

A new and modern culture

Peter transformed his court, making his courtiers adopt French-style dress instead of traditional robes, and ordering them to cut off their long beards. He founded colleges, forced the nobility to educate their children, and promoted people to high rank according to their merit rather than their birth, as had previously been the case.

The emperor was also known for his harsh treatment of rebels, his aggressive foreign policy, and particularly for his successful war against Sweden, which gave him control of the Baltic Sea. This style of rule was continued under later monarchs, notably Catherine II, also "the Great," who extended

> Peter I visits **Western Europe**, absorbing **ideas and influences**.

> Contemporary theories of rulership provide a **model of enlightened despotism**.

> Baroque western palaces and **cities demonstrate their rulers' power**.

> **Peter founds St. Petersburg as the capital of a Westernized Russian empire.**

the modernizing trend Peter had begun. Influenced by the ideas of the European Enlightenment, she promoted education and the arts, sponsored translations of foreign literary works, and wrote books herself. She too increased Russia's imperial strength, gaining military victories over the Ottoman Empire.

The rulers were also influenced by the example of Prussia, a north-German state that expanded in

the 18th century due to an efficient bureaucracy, a powerful army, and strong leadership under kings such as Frederick II. Between Prussia and Russia lay Poland, whose territories these two powers and Austria carved up and took over in a series of partitions. This left Russia with influence over an area stretching from Eastern Europe to Siberia that it still largely retains today. ▪

Peter the Great

Peter (1672–1725) became ruler of Russia in 1682, at first jointly with his half-brother Ivan as co-tsar and their mother as regent, and then as sole monarch. Well-educated and constantly curious, Peter travelled to the Netherlands and England to learn about Western life, government, and architecture. He also studied disciplines such as shipbuilding and woodworking, and practised many with distinction. His rule was greatly influenced by these travels and by Western advisers, leading him to carry out military reforms and adopt a dictatorial

style of rule. The position and grand architecture of his new city illustrated how his focus had been directed towards Western European culture and power.

Although Peter forged lasting diplomatic ties with Western Europe, he failed in his attempt to form a European alliance against the Ottomans. He was more successful in his war against Sweden, his reforms, and in establishing himself as emperor of a vast empire and monarchy that survived until the revolution in 1917.

FURTHER EVENTS

THE FOUNDING OF THE SAFAVID DYNASTY, PERSIA
(1501)

The Safavid dynasty rose to power under Shah Ismail I, a leader of the Twelver School of Shia Islam, which believes in 12 imams as successors to the prophet Muhammad. In a series of military campaigns lasting until 1509, Shah Ismail conquered Persia (now Iran) and areas of Iraq, in the name of Shia Islam. His son, Shah Tahmasp (r. 1524–76), defended these lands against the neighboring Ottoman empire, whose rulers followed the opposing Sunni School of Islam. The Safavid dynasty established strong Shia rule in Persia, created an efficient government and bureaucracy, and lasted until 1736.

CHARLES V BECOMES HOLY ROMAN EMPEROR
(1519)

One of the most powerful European monarchs as Hapsburg king of Spain and ruler of Burgundy and the Netherlands, Charles V was elected Holy Roman Emperor in 1519, bringing much of central Europe and northern Italy under his rule. This gave him unprecedented power but also brought challenges from the empire's neighbors—France on one flank and the Ottomans on the other—and from Protestants within his territories. When Charles abdicated, the Spanish crown passed to his son Philip and the title of emperor to his son Ferdinand.

HENRY VIII BREAKS WITH ROME
(1534)

English king Henry VIII faced a dynastic crisis: he needed a male heir to ensure the succession, but he and his wife, Catherine of Aragon, were unable to produce one. Henry wanted to divorce Catherine, but the Pope refused him permission to do so. In response, Henry cut off ties with Rome and declared himself head of the church in England. Although under Henry the English church remained largely Catholic in doctrine and practice, the king's move paved the way for England's later acceptance of Protestantism. In addition, Henry dissolved the monasteries, which brought him a new source of land and wealth, and removed a key link with the Roman Catholic church.

CARTIER EXPLORES CANADA
(1534–42)

French navigator Jacques Cartier explored the northern coast of Canada and Newfoundland, sailing up the St. Lawrence River as far as what later became Montreal. Although he did not establish a colony there, Cartier sparked the French interest in Canada, and his explorations were crucial when French travelers began to found settlements there and make claims on the land in the 17th century. Canada has had a significant French heritage ever since.

THE START OF THE DUTCH REVOLT
(1568)

In 1568, the Protestant northern provinces of the Netherlands rebelled against their Catholic ruler, Philip II of Spain, and declared their independence, beginning an 80-year period of war before their Republic was recognized. Philip had imposed his Catholic beliefs uncompromisingly on his Dutch subjects, so many Protestants from the southern Netherlands, which remained loyal to the crown, moved north. This influx helped the Republic, which soon grew into a financially and culturally stable nation thanks to sea trade, scientific progress, and impressive artistic achievements.

THE ST. BARTHOLOMEW'S DAY MASSACRE
(1572)

In 16th-century France, violent clashes, and, from 1562, civil war, erupted between Catholics and Protestants. One of the worst episodes took place in 1572, when the Protestant claimant to the French throne, Henry of Navarre, was married in Paris and several thousand Protestants were massacred. After Henry became king of France, he issued the Edict of Nantes in 1598, ordering religious tolerance. However, the edict was revoked in 1685 by Louis XIV, who ruthlessly oppressed France's Protestant population;

under his reign many Protestants were imprisoned and many others fled from the country.

THE SPANISH ARMADA
(1588)

In 1588, the Catholic monarch Philip II of Spain attempted to conquer Protestant England by sending a fleet of 130 ships to invade the country. After the English succeeded in destroying part of the fleet using fire ships, a defeat at Gravelines sent the remains of the Spanish Armada retreating northward toward Scotland, where many more ships were wrecked by storms. Only 86 vessels made it back to Spain. The defeat was a blow to Spain, ending this campaign to capture England for Catholicism and confirming England's status as a secure Protestant nation under Elizabeth I.

THE JAPANESE INVASIONS OF KOREA ARE DEFEATED
(1592–98)

The Japanese samurai leader Toyotomi Hideyoshi launched attempts to conquer Korea in 1592 and 1597, part of a larger campaign that was intended to culminate with an invasion of China. Both times, Japan made major advances, but the Koreans, with the support of Chinese forces, managed to fight back. However, they were unable to expel the Japanese completely, which led to a stalemate on land, although Korea's Admiral Yi inflicted frequent naval defeats on Japan. Beaten at sea and confined to a few fortresses on land, Japan abandoned its attempts to invade.

Korea remained independent until 1910, after which there was a 35-year period of Japanese rule.

THE SIEGE OF DROGHEDA
(1649)

England's parliamentarian leader Oliver Cromwell launched his campaign to conquer Ireland in 1649, after Irish Catholics took control of the country from its English administrators in 1641. Once Cromwell had taken Dublin, Drogheda became a base for Irish Catholic leaders. Cromwell laid siege to the town, massacring the people inside its walls when they refused to surrender. Most of the garrison of about 2,500 men, plus many civilians, were killed. Although the killings did not break the military code of the time, their ruthlessness and the sheer number of victims were both unprecedented, and they embittered future relations between the English and the Irish Catholics.

THE DUTCH ESTABLISH A COLONY AT CAPE TOWN
(1650)

Although Portuguese explorers were the first Europeans to discover the Cape of Good Hope in the 15th century, it was the Dutch who founded Cape Town. In 1652, a group from the Dutch East India Company under Jan van Riebeeck established a colony there, creating a stopping point for Dutch ships on their way to and from Asia. The settlement became the center of a large community of people of Dutch origin, who dominated trade and agriculture in the region, evolved

their own language—Afrikaans— and played a central role in the history of South Africa.

THE OTTOMAN SIEGE OF VIENNA
(1683)

By 1683, the Turkish Ottoman Empire was at its greatest extent and included large areas of North Africa, the Middle East, and Eastern Europe. Austria was on the empire's western border, and the Turks had already attempted to conquer Vienna. In 1683 they besieged the city for the last time: the forces of the Habsburg-ruled Holy Roman Empire and of Poland came to defend Vienna, and the Ottomans were vanquished. From this point on, the Ottomans' power declined. No longer a threat to Christian Europe, they steadily lost their European territories.

THE BATTLE OF CULLODEN
(1746)

At the Battle of Culloden, Scotland, an army being led by the Duke of Cumberland, son of the Hanoverian king George II, defeated a smaller force of Jacobites (including many from the Highland clans) under Prince Charles Edward Stuart. The Stuart prince had hoped to restore his bloodline to the British throne, but Culloden effectively put an end to his campaign. It also led to the disarming of the Scottish Highlands, where Jacobite support was strongest, the dismantling of the clan system there, and a ruthless suppression of Highland culture that included bans on wearing Highland dress and speaking Gaelic.

CHANGIN
SOCIETIE
1776–1914

GS

The Declaration of Independence is signed. It asserts basic **human rights** and creates a new nation: the **United States of America**.

The Slave Trade Abolition Act is passed in Britain, **outlawing trading in slaves**; however, slavery itself is not banned until 1833.

Simon Bolívar establishes Gran Colombia, a new **South American** republic **independent** of Spanish rule; it lasts until 1830.

Revolts occur all over Europe as demands for **liberalism**, **socialism**, and **national self-determination** grow; all are suppressed by force.

1776   **1807**   **1819**   **1848**

1789   **1815**   **1830**   **1856**

The storming of the Bastille signals the start of the French Revolution, in which the **monarchy is overthrown** and a **republic is established**.

Napoleon is defeated at the Battle of **Waterloo** by the British, Dutch, and Prussians, ending 23 years of war in Europe.

George Stephenson's *Rocket* **steam engine** powers the world's first **commercial rail** service, which links Liverpool and Manchester.

The Second Opium War is launched by **Western powers** to force China to give access to **Chinese ports for trade**.

From the late 18th century, history took on a perhaps delusory air of "progress." Change accelerated and seemed to have a clear direction. The world population exceeded 1 billion in 1804 and was approaching 2 billion by 1914. This growth was made possible by tremendous increases in economic output. Agriculture became more efficient and large areas of new land were put to productive use. The exploitation of new sources of energy—especially steam power—the application of new technology, and organized industrial production in factories revolutionized the manufacture of commodities. Railways made it possible for humans to travel faster than a horse for the first time and cities expanded—for instance, the population in London increased from 1 million in 1800 to 7 million in 1910. Improvements in public health and medicine increased life expectancy in the most advanced countries.

Human rights and equality

Despite these advancements, it is debatable whether progress was detectable in the quality of life. At the start of this period, political revolutions in America and France enunciated principles of human rights and equal citizenship that radically challenged the existing order of society. By the early 20th century, liberals and democrats in Europe and North America could look with some complacency upon successes such as the widespread expansion of voting rights, the abolition of slavery, and freedom of speech. However, women remained mostly excluded from voting, and there was no economic equality. Extremes of wealth and poverty polarized the world's wealthiest and most advanced societies, and conditions of life for industrial workers were often very miserable. Artists and intellectuals of the Romantic movement criticized the impact of mechanized industry on people and the environment, while socialist movements looked forward to further revolutions that would end the exploitation of man by man and create egalitarian societies.

Western imperialism

The most obvious losers in the new world order created by industrial capitalism were the inhabitants of countries at the periphery of the global economy. The industrializing countries of the West, needing places to invest their excess capital,

Charles Darwin publishes *On the Origin of the Species*, in which he introduces his controversial **theory of evolution**.

During the **American Civil War**, US president Abraham Lincoln gives the Gettysburg Address, one of the **greatest speeches** in history.

The Suez Canal opens, **linking** the Red and Mediterranean seas, and dramatically **reducing sailing times** between **Europe** and the **East**.

A coalition of various reform groups, known collectively as the **Young Turks**, overthrow the **authoritarian Ottoman sultan** and attempt to rule.

1859   **1863**   **1869**   **1908**

1860   **1868**   **1892**   **1913**

Giuseppe **Garibaldi** leads 1,000 volunteers to **overthrow** the French Bourbons in **southern Italy** and **Sicily**; Italy was unified one year later.

The Tokugawa shogunate is **ousted** and the Emperor Meiji becomes **ruler of Japan**; the nation emerges as a major **imperial power**.

Ellis Island opens in New York Harbor to process **arrivals of immigrants** to the **United States**; most become US citizens. The island closed in 1954.

Emily Davison steps out under King George V's horse at the Derby and is killed, raising the profile of **women's suffrage** worldwide.

---

raw materials for their factories, and markets for their new products, found them in Asia, Africa, and Latin America. They also sought land for their expanding populations to settle in thinly populated zones such as the North American plains and Australia. Peoples who stood in their way were swept aside. The Europeans started expanding the areas under their direct rule or control. The British takeover of the Indian subcontinent, more or less complete by the mid-19th century, was the most spectacular example of imperialism in action, and Sub-Saharan Africa was divided among the European powers as if the local population did not exist.

The world's response to Western imperialism was mixed. Resistance was widespread in the form of wars and uprisings against European dominance. On the other hand, the growing superiority of the West in technology, science, military power, and social organization led several non-European governments to try to modernize based on the Western model. In the Muslim world, Egypt, Turkey, and Iran attempted, with only partial success, to pursue a modernizing agenda. In East Asia, Japan successfully transformed itself into an efficient modern state, becoming an imperialist power in its own right. China, by contrast, experienced turmoil and invasion, and imperial rule collapsed in the early 20th century.

Rising nationalism

Most Europeans and people of European descent gloried in a sense of their own racial and cultural superiority to the rest of the world, but Europe remained a deeply divided continent. Militant nationalism, unleashed by the French Revolution, was a threat to stability. By 1815, the Napoleonic Wars had generated battles of unprecedented scale. After the wars of the mid-19th century that created a unified Italy and Germany, the great powers maintained large conscript armies and formed mutually hostile alliance systems. These armies were equipped with high-explosive shells and rapid-fire weapons.

European military power, which was supported by highly organized state systems and economies, was certainly one of the key elements in European world domination. There would be disaster when European states turned this power against one another. ■

WE HOLD THESE TRUTHS TO BE SELF-EVIDENT, THAT ALL MEN ARE CREATED EQUAL

THE SIGNING OF THE DECLARATION OF INDEPENDENCE (1776)

IN CONTEXT

FOCUS
The American Revolution

BEFORE
1773 Boston Tea Party protests tax on tea imports.

1775 Armed clashes take place between patriot militia and British forces.

AFTER
1777 British defeat at Saratoga persuades France to support the American rebels.

1781 The British surrender at Yorktown, Virginia.

1783 Britain recognizes American independence.

1787 Drafting of the Constitution begins.

1789 George Washington is elected as the first president of the United States.

1790 The US Constitution is ratified.

There has been no more daring assertion of statehood than that proclaimed by the Declaration of Independence, adopted by the Second Continental Congress on July 4, 1776 and signed by all 56 delegates present. What would become the United States consisted of 13 British colonies, steadily established since the 17th century, and scattered along the east coast of North America. They were not just geographically remote from their mother country; most were also geographically remote from each other. Their economies were fragile, and they had no coherent political identity—citizens of

See also: The Battle of Quebec 191 ▪ The storming of the Bastille 208–13 ▪ The Slave Trade Abolition Act 226–27 ▪ The 1848 revolutions 228–29 ▪ The Gettysburg Address 244–47 ▪ The California Gold Rush 248–49

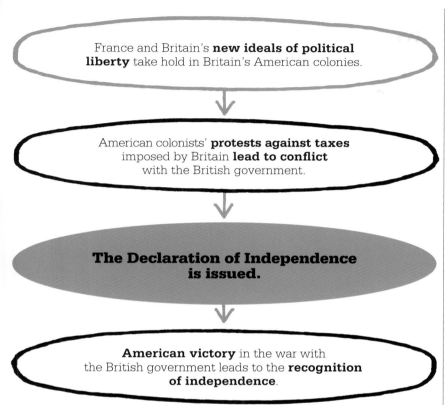

France and Britain's **new ideals of political liberty** take hold in Britain's American colonies.

↓

American colonists' **protests against taxes** imposed by Britain **lead to conflict** with the British government.

↓

The Declaration of Independence is issued.

↓

American victory in the war with the British government leads to the **recognition of independence**.

Virginia considered themselves to be Virginians, for example, not Americans—beyond an increasingly strained loyalty to the British crown.

However, the colonies were also remarkably self-aware and acutely conscious of Enlightenment notions of political liberty, and they were concerned that their freedom would come under threat as a result of British rule. Unable to assert their own natural rights, and subjected to what they considered unreasonably imposed taxes, the colonists questioned why a distant parliament and a distant king should impose their will on them. Impelled by a series of exceptional leaders, in 1776 they not only rejected British authority, but they set about establishing an entirely new kind of state in which government would derive from "the consent of the governed." This explosively novel idea would lead to the creation of a new and enduring republican government.

Support for a formal assertion of American independence was far from universal in the colonies, however. Five states in particular— New York, New Jersey, Maryland, Delaware, and Pennsylvania— feared it would damage trade and, if unsuccessful, provoke harsh reprisals from Britain. In the same way, as many as 500,000 of a population of 2.5 million remained loyal to the British crown to the end of the conflict, many subsequently settling in Canada.

The conflict takes shape

It would take a drawn-out and bitterly fought war to make independence into reality. Britain was determined to assert what it saw as its legitimate rule, while the hastily assembled forces of the nascent United States were no less determined to assert what they saw as their right to independence. The two modest armies—Britain's, because of the difficulties of sending forces en masse to America; the colonists', because they consistently lacked the means to raise and equip any substantial fighting force— confronted each other in a series of minor engagements over six years.

At their peak, the American forces numbered scarcely 40,000 and had almost no navy at all. Britain deployed about the same number of soldiers but in addition had a vastly greater number of ships. In 1778, however, France declared support for the colonists and sent 5,000 troops and a substantial fleet. Facing certain »

These United Colonies are, and of Right ought to be Free and Independent States.
Richard Henry Lee
Proposed resolution at the Second Continental Congress (June 1776)

defeat, in October 1781, the British surrendered at Yorktown, Virginia. The war would not formally end for another year, but in every important respect, the colonists—and their French allies—had dealt a huge blow to their British masters.

The French involvement in the creation of this new nation owed everything to a desire to reverse the humiliations of the Seven Years' War. But the debts incurred would, ironically, be among the many causes of the bankruptcy of the French crown that led to the French Revolution in 1789. There was a profound irony, too, in absolutist France seeking to win Americans the freedoms that it was unwilling to accord its own citizens.

Revolutionary ideals

At the heart of the American Revolution was the new political philosophy encapsulated by the Declaration of Independence.

It was the work of a distinctly patrician Virginian, a haughty, wealthy slave-owner named Thomas Jefferson. He was one of a committee of five charged with writing the Declaration, yet the two drafts it went through in June 1776 were almost entirely his own. It is hard to overstate the importance of the Declaration of Independence. It made, for the time, an astonishing claim: "that all men are created equal." It further claimed "that governments are instituted among men deriving their just powers from the consent of the governed."

These were actively seditious sentiments that neither George III of England nor Louis XVI of France could have any sympathy with. They nonetheless formed the bedrock of what would become the United States and, indeed, liberal political systems across the Western world. These political creeds, derived from the work of

In *Declaration of Independence*, by John Trumbull, the drafting committee is shown presenting its work to Congress. Thomas Jefferson can be seen standing in a red waistcoat.

British and French Enlightenment thinkers, led to the creation of the first modern state and, in doing so, changed the world.

The destiny of America

Jefferson remains an enigma. He loathed monarchy yet loved pre-Revolutionary France, where he was the United States' first ambassador, delighting in its civilized elegance. He claimed to despise high office yet served two terms as President of the United States. And, as president, in 1803 he drove through the Louisiana Purchase, which saw a vast area west of the Mississippi transferred at a bargain price from France, its nominal ruler, to the United States.

The god who gave us
life, gave us liberty
at the same time.
Thomas Jefferson

He understood that the destiny of the US lay in its colonization of the vast lands to the west, he assented to the notion that its indigenous inhabitants should be driven off, and he owned slaves. "Blacks," he asserted, "are inferior to the whites in the endowments both of body and mind." Whereas George Washington, also a patrician Virginian, freed his slaves, Jefferson opted not to.

None of this, though, can diminish Jefferson's significance in articulating notions of liberty that resonate today. And even though he felt slavery was wrong, his personal belief was that emancipation would be bad for both slaves and white Americans—unless they were returned to Africa.

A new constitution

Although Jefferson can readily be considered the guiding spirit behind the Declaration of Independence, he played no formal role in the drawing up of the next great document that shaped the nation: its Constitution. The United States was legally able to assert its independence from Britain in 1783. But for the next four years, it existed in an increasingly unstable political vacuum, its fate decided by an ever-more divided Confederation Congress, meeting variously in Pennsylvania, New York, and New Jersey.

There were serious reasons to believe the new nation might fail, torn apart by those arguing for the primacy of the rights of the individual states over the central government, and those in favor of a strong central government or even the creation of an American monarchy. In the spring of 1787, a Constitutional Convention took place in Philadelphia. The written, formalized Constitution proposed would not be provisionally ratified until June the following year, and then only after prolonged disputes. The result was an assertion of a new form of government. It was both a bill of rights and a blueprint for an ideal government, whose three branches—executive, legislative, and judiciary—would keep each other in check. It would have a profound influence on that issued in Revolutionary France in 1791 and remains a model of its kind.

"Unfinished business"

The founding fathers were rightly optimistic about the United States' potential, but they had failed to resolve one crucial question. Jefferson's first draft of the Declaration of Independence called slavery "an execrable commerce" and "a cruel war against human nature itself." However, to placate the slave states of the south and the slave traders of the north, these radical statements were later dropped. Almost 90 years later, it would take a civil war and 620,000 dead to end the practice and complete what Abraham Lincoln saw as the "unfinished business" of the Declaration of Independence and the Constitution. ∎

George Washington

Born in 1732, George Washington served the British crown with distinction during the Seven Years' War (1756–63) against France. He represented Virginia in the House of Burgesses and in the Continental Congresses of 1774 and 1775. With the outbreak of the Revolutionary War, he was the unanimous choice to lead the Continental Army, which he did with imagination and great fortitude, especially in the very difficult early years of the conflict: his "skeleton of an army," under-equipped and close to starving, was forced to endure an exceptionally harsh winter in 1777–78 at Valley Forge in Pennsylvania. From 1783, Washington sought to establish a constitutional government for the new nation. The nation's first president, he served two terms, retiring in 1797 in the face of increasing disputes between Jefferson's Democratic Republicans and the Federalists, who were led by the quick-tempered Alexander Hamilton. Washington died in 1799 and was buried at his Virginia plantation, Mount Vernon, overlooking the Potomac River.

SIRE, IT'S A REVOLUTION

THE STORMING OF THE BASTILLE (1789)

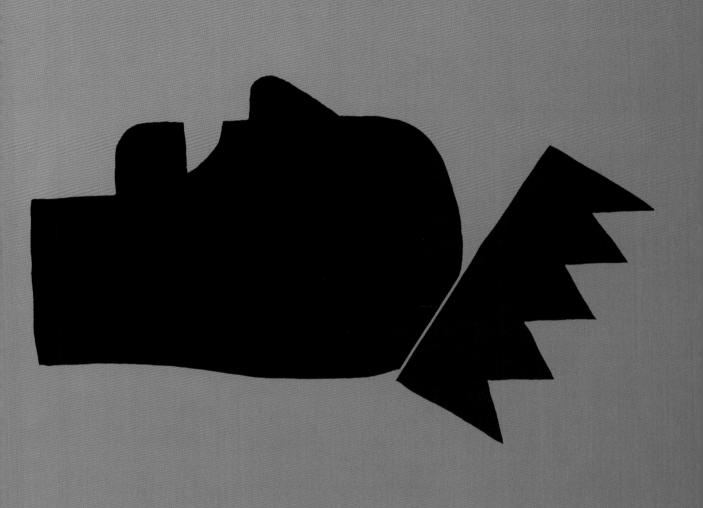

IN CONTEXT

FOCUS
The French Revolution

BEFORE
May 1789 Louis XVI
summons the States General.
In June, the commons forms
the National Assembly, taking
effective power in the name
of the people.

AFTER
April 1792 The Legislative
Assembly declares war on
Austria and Prussia. The first
French Republic is declared.

January 1793 Louis XVI
is executed.

March 1794 The Terror is at
its peak. In July, Robespierre,
its prime exponent, is executed.

October 1795 Napoleon
forcibly restores order to a
turbulent Paris.

November 1799 Napoleon
effectively becomes the ruler
of France.

On July 14, 1789, an enraged Parisian mob, seeking weapons to defend their city from a rumored royal attack, stormed the crumbling fortress known as the Bastille and murdered its governor and guards. This violent defiance of royal power has become the symbol of the French Revolution, a movement that not only engulfed France but also reverberated around the world. The ideas articulated in the revolution spelled the beginning of the end for Europe's absolute monarchies and inspired their eventual replacement by more democratic governments.

The French Revolution originally set out to sweep away aristocratic privilege and establish a new state based on the Enlightenment principles of *liberté*, *égalité*, and *fraternité*. But although it was introduced by a surge of optimism, the revolution soon degenerated into a violence that played out over several years and that would be brought to an end only by the dictatorship of Napoleon Bonaparte. It remains a story of confusion and chaos, of a collision between a privileged old order, the *ancien*

The French Revolution was the greatest step forward in the history of mankind since the coming of Christ.
Victor Hugo
Les Misérables (1862)

régime, and a new world that struggled, often violently, to create a coherent new order.

A country in disarray

The French king, Louis XVI, well meaning but indecisive, was hardly the man to confront any crisis, let alone one as grave as that facing France in 1789. In the previous century, his great-great-great-grandfather Louis XIV, the Sun King, had established France as an absolute monarchy, with all power concentrated in the king's hands, and his palace at

---

Enlightenment **thought** establishes a belief in a **new political order** based on **liberty**.

→

A **political crisis** arises in France, and the **overthrow of the old order** seems suddenly possible.

→

The Bastille prison is attacked by a violent mob.

↓

The underpinning belief in *liberté*, *égalité*, *fraternité* changes not just France, but the world.

←

A sustained period of **instability, rioting, civil war, and state-sanctioned executions** follows.

←

Attempts are made to construct a **new society:** the monarchy is abolished and a **republic declared**.

See also: Louis XIV begins personal rule of France The Battle of Quebec 191 ▪ Diderot publishes the *Encyclopédie* 192–95 ▪ The signing of the Declaration of Independence 204–07 ▪ The Battle of Waterloo 214–15 ▪ The 1848 revolutions 228–29

The storming of the Bastille symbolized the start of the French Revolution. The prison held only seven prisoners in July 1789, but its fall had great importance.

royal family to Paris, ransacking the palace for good measure. In what would become an unnerving foretaste of the violence to come, the severed heads of the guards at Versailles were paraded on stakes as Louis and his family were escorted to the capital.

It had been comparatively easy to overthrow the existing royal government, but it would prove infinitely harder to establish a new government. It was presumed that a kind of constitutional monarchy would be the most obvious solution. In the event, France found itself wrenched between those arguing for this more or less moderate option, and those in favor of a much more radical republican alternative.

The First Republic
Although in most important respects Louis's reign seemed by now to be doomed, the king had not entirely »

Versailles as the most sophisticated court in Europe and a bastion of aristocratic privilege.

Louis XVI thus ruled over a country where nobles refused to surrender any privileges, and taxes were paid almost exclusively by an oppressed peasantry: France was effectively bankrupt. In the late 18th century, France's population was expanding rapidly, but unlike England, France had not had an agricultural revolution and remained particularly vulnerable to any failure of the harvest, as happened in 1787 and, again, in 1788. These desperate summers were followed in 1788–89 by a bitterly harsh winter, leading to mass starvation.

The king's response
The financial crisis critical, Louis was desperate to raise further funds while preserving his authority, so he summoned what was called the States General, a

semi-parliamentary body that had last assembled in 1614. It consisted of clergy, the first estate; nobles, the second; and the commons (essentially a kind of bourgeoisie, lawyers predominating), the third. The States General met at Versailles on May 5, 1789. Almost instantly, the nobles and clerics tried to assert that their votes should be worth more than those of the commons. In response, on June 17, the commons declared itself a National Assembly, vesting power in itself instead of the crown. In August, with peasant uprisings across rural France, the Assembly abolished feudal taxes and aristocratic privileges and issued what it called the Declaration of the Rights of Man, a statement asserting fundamental freedoms.

In October 1789, events were suddenly accelerated when a vast crowd, outraged by the lack of bread in Paris, descended on Versailles and forcibly removed the

Terror is nothing more than speedy, severe, and inflexible justice; it is thus an emanation of virtue.
Maximilien Robespierre, February 1794

abandoned hope of reasserting his authority. Large numbers of French aristocrats—emigrés—had already fled France, fearing the revolution had made it unsafe. In trying to persuade other European regimes—Austria above all, whose emperor was the brother of the French queen Marie-Antoinette—they stirred up opposition to the revolution, but their principal impact was to reinforce a determination in France to see the revolution succeed.

In June 1791, Louis attempted to escape, but was intercepted near the border with the Low Countries and brought back to Paris to the jeers of the increasingly violent, politicized common folk, the sans-culottes, their name a reference to their striped, baggy trousers. There was an increasingly hostile stand-off between political factions in Paris, such as the Girondins and more extremist Jacobins, which attracted the support of the sans-culottes, and the French government.

An external threat

Whatever the obvious instability, progress toward a new social order was being made. In September 1791, a constitutional monarchy was proclaimed. Similarly, the church's privileged position was forcibly ended, though this, too, provoked lasting upheaval and violence. Equally critically, the freedom of the press was asserted.

At the same time, revolutionary France faced an external threat from Austria and Prussia, both determined to reassert the primacy of hereditary monarchy and to forestall revolutionary tendencies in their own countries. In April 1792, France declared war on both, a war that would continue, in different guises, for 23 years. By August, the combined Austrian and Prussian forces were within 100 miles (160km) of Paris.

A kind of hysteria gripped the city. A mob stormed the Tuileries, where the royal family was held, slaughtering its Swiss Guards. The following month, a further round of killings, the September Massacres, was unleashed against anyone suspected of royalist sympathies. September 1792 also marked the establishment of the directly elected National Convention and of the First French Republic. Almost

So, legislators, place Terror on the order of the day!... The blade of the law should hover over all the guilty.
Committee of General Security, September 1793

its first act was to put Louis XVI on trial as a traitor. In January 1793, he was executed, an early victim of the guillotine, championed as a humane and egalitarian means of death.

The sense of crisis continued to grow. In April 1793, the Committee of Public Safety was created to safeguard the revolution. For a year or more, under the chairmanship of a provincial lawyer, Maximilien Robespierre, the most influential of the now-dominant Jacobins, it would effectively become the government of France. Its impact on France, however short-lived, was devastating. This was the Terror. Counter-revolutionary movements across the country were ruthlessly suppressed, most obviously in the Vendée region of the southwest, where up to 300,000 died. Churches proved particularly rich targets. The Terror's victims were less likely to be remaining aristocrats and more anyone Robespierre suspected of impure thoughts, including almost all of his political opponents.

Louis XVI was executed in 1793. Using the guillotine as the only means of execution for all people—royals and paupers alike—was meant to reinforce the revolutionary principle of equality.

The French Revolution set out with the idea of building a new state that would take the Enlightenment principles of liberty, equality, and brotherhood as its foundation.

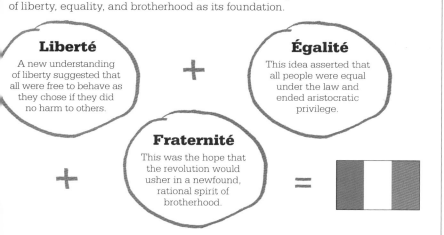

Liberté
A new understanding of liberty suggested that all were free to behave as they chose if they did no harm to others.

+

Égalité
This idea asserted that all people were equal under the law and ended aristocratic privilege.

+

Fraternité
This was the hope that the revolution would usher in a newfound, rational spirit of brotherhood.

=

Maximilien Robespierre

Robespierre (1758–94), a lawyer and a member of the third estate in 1789, was the chief architect of the Terror that gripped France between September 1793 and July 1794. He was a consistent champion of the dispossessed, as well as a remarkable orator, capable of astonishingly intense speeches that electrified his supporters and opponents alike. He was also a fierce opponent of the Revolutionary Wars, believing that a strengthened army risked becoming a source of counter-revolutionary fervor. Initially, at least, he was also opposed to the death penalty. His change of mind was startlingly absolute. When persuaded that terror was the most effective means of preserving the revolution, he embraced it implacably, arguing that it was the natural handmaiden of the virtue he felt should drive the revolution. He remains the original, chilling model for all those who have since championed state violence in the interests of a supposed greater good.

Robespierre's single-minded pursuit of revolutionary purity reached an improbable climax with his creation in 1794 of a new religion, the Cult of the Supreme Being. It was intended as a focus of, and spur to, patriotic and revolutionary virtues, the superstition of the Catholic Church replaced by a belief dedicated to reason celebrating the natural laws of the universe. The megalomania it suggested contributed significantly to his sudden downfall, and at the end of July 1794, Robespierre was put to the guillotine.

Order restored
With the end of the killings—more particularly with the establishment of yet another government, the Directory, at the end of 1795—order of a sort was restored. Tellingly, it was achieved in part by the Directory's willingness to use force against the Paris mob, ordered by Napoleon Bonaparte, then a young general in the revolutionary army.

Furthermore, France's armies, boosted by mass conscription, were reversing early defeats, apparently poised to carry the revolution into new territories. Emboldened, France reinforced its assertion of what it claimed were its "natural frontiers" on the Rhine, which in reality meant an audacious extension of French rule into Germany. By 1797, it had inflicted crushing defeats over Austria in the Low Countries and in northern Italy. France was ready to re-assert what it saw as its natural primacy in Europe.

Historical significance
Whatever the importance of the French Revolution, it remains the subject of continuing and intense historical debate. Its notional goals were clear: the ending of repressive monarchy and entrenched privilege; the establishment of representative government; and the championing of universal rights. But the reality was confused and often violent.

Furthermore, by 1804 Napoleon had effectively swapped one form of absolutism for his own, albeit one vastly more effective than any France had known since Louis XIV. Yet the revolution's consequences reverberated well into the 20th century. It remains a pivotal moment in the belief that freedom should underpin the civilized world. ■

I MUST MAKE OF ALL THE PEOPLES OF EUROPE ONE PEOPLE, AND OF PARIS THE CAPITAL OF THE WORLD
THE BATTLE OF WATERLOO (1815)

IN CONTEXT

FOCUS
Revolutionary and Napoleonic Wars

BEFORE
1792 The Revolutionary Wars against Republican France begin.

1799 Napoleon seizes power in a military coup.

1804 Napoleon names himself Emperor of the French.

1805 The British are victorious against France and Spain in the naval Battle of Trafalgar.

1807 France invades Portugal.

1809 Austria is defeated in Napoleon's last major victory.

1814 A series of defeats lead to Napoleon's abdication.

AFTER
1815 Napoleon is exiled for the final time, and the Bourbon monarchy is restored.

1830 The Bourbon monarchy is overthrown.

France institutes **mass conscription**, creating armies of an unprecedented size.

Napoleon becomes emperor and vows to **restore France's dominant role** in Europe.

Sweeping conquests create the largest **European empire** since Charlemagne's time.

The invasion of Russia leaves **Napoleon overextended** and French manpower depleted.

The rate of Napoleonic conquest becomes **unsustainable**.

Napoleon is finally defeated at Waterloo.

N apoleon's defeat at the Battle of Waterloo, south of Brussels, on June 18, 1815 marked his final overthrow as Emperor of the French, ending 23 years of European warfare. It was an epic encounter, fought on rain-soaked ground, in which 118,000 British, Dutch, and Prussian forces finally prevailed over a French army of 73,000, hastily assembled by Napoleon.

France's Revolutionary Wars, which began in 1792, had been launched to extend Revolutionary principles to neighboring states and to defend France against its enemies. Under Napoleon,

See also: Louis XIV begins personal rule of France 190 190 ▪ The Battle of Quebec 191 ▪ Diderot publishes the *Encyclopédie* 192–95 ▪ The signing of the Declaration of Independence 204–07 ▪ The storming of the Bastille 208–13 ▪ The 1848 revolutions 228–29

Napoleon Bonaparte

Born in Ajaccio, on the island of Corsica, to a family with claims to minor Italian nobility, Napoleon Bonaparte (1769–1821) was commissioned in the French army in 1785 and was an enthusiastic supporter of the Revolution. In 1796, at the age of 26, he was appointed to command the Army of Italy, winning a series of impressive victories. Two years later, Napoleon led an unsuccessful French invasion of Egypt.

Increasingly convinced of his destiny, by 1800, having staged a coup d'état, he dominated France

as he would subsequently dominate Europe. He was as brilliant and tireless an administrator as he was a soldier. His most enduring reform was the introduction, in 1804, of the Napoleonic Code, which is still the basis of French law. Forced to resign in 1814, Napoleon was exiled to the island of Elba in the Mediterranean, from where he escaped before his final defeat at Waterloo. In 1815, he was despatched to St. Helena in the South Atlantic, where he died six years later.

they became, in effect, wars of conquest, despite being waged in the name of the Revolution.

A continent reshaped

During the Revolutionary Wars, France had established sister republics in northern Italy and the Low Countries; under Napoleon, many of these were reformed into kingdoms, whose monarchs came from the Emperor's family. States across Germany were carved up, at the expense of Prussia, to become a French puppet state, while the 800-year-old Holy Roman Empire was abolished. From 1807, much of Poland was controlled by the French as the Grand Duchy of Warsaw. These were states recast on French lines: clerical power was reduced, serfdom abolished, and aristocratic privilege ended. But such reforms provoked inevitable resentments.

Napoleon's conquests were the result not just of military genius but also of greatly enlarged French armies. Conscription, introduced in 1793, swelled the French army from 160,000 men to 1.5 million.

Only Britain, protected by the English Channel, remained undefeated, its position as the world's leading maritime power underscored by victory at Trafalgar, off southern Spain, in 1805. But maritime muscle alone was not enough to beat Napoleon. Britain's most significant role was financing the endlessly shifting alliances confronting the French.

In response, Napoleon imposed the Continental System, which forbade trade between continental Europe and Britain. However, Portugal and Russia continued to trade with Britain, prompting Napoleonic invasions in 1807 and 1812 respectively.

Resistance to Napoleonic rule was mounting; the Spanish began a brutal guerrilla war that drained French resources and came to be referred to by Napoleon as the "Spanish ulcer."

The final defeat

Napoleon had bred a sense of French invincibility, and this made his eventual defeat all the more

traumatic for the nation. Of the 450,000 men he led against Russia in 1812, barely 40,000 survived.

Napoleon had overreached himself. At Leipzig, Germany, in 1813, outnumbered three to one by forces from Austria, Prussia, Russia, and Sweden, he suffered his first major defeat. By Waterloo, his forces had recovered slightly, and the ratio was only two to one, but Napoleon's military genius failed to redress the balance and his imperial ambition ended in the Waterloo mud. ▪

All Frenchmen are in permanent requisition for the services of the armies.
Declaration of Conscription, 1793

LET US LAY THE CORNERSTONE OF AMERICAN FREEDOM WITHOUT FEAR. TO HESITATE IS TO PERISH

BOLÍVAR ESTABLISHES GRAN COLOMBIA (1819)

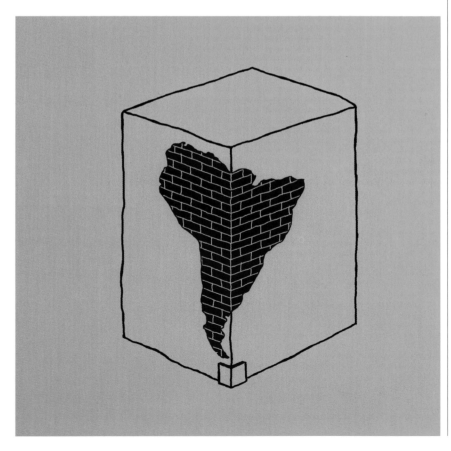

The establishment in 1819 of the Republic of Colombia, or Gran Colombia as it came to be known, by Simón Bolívar, the self-styled *Libertador*, marked a pivotal moment in the emergence of an independent Latin America.

By 1825, the continent had successfully ended almost 300 years of Spanish and Portuguese rule. In Brazil, which won its independence in 1822, the process was relatively easy and also largely bloodless. Elsewhere, it was complex, long, drawn out, and violent. It was a reflection of societies comprising a teeming range of classes and races—ruling Europeans, native

See also: The signing of the Declaration of Independence 204–07 ▪ The storming of the Bastille 208–13 ▪ The Slave Trade Abolition Act 226–27 ▪ The Mexican Revolution 265 ▪

Notions of **political liberation** spread across the territories colonized by the Spanish and Portuguese in **South America**.

↓

These ideas **destabilize the Spanish administration** on the continent.

↓

Bitter wars against colonial powers in South America lead to the creation of a newly independent Gran Colombia.

↓

The new state faces **divisions and instabilities**. Hopes of unity are undermined by political infighting.

↓

South America, racially and socially **dislocated**, consistently **struggles to assert itself** economically and politically.

Simón Bolívar

Born in Caracas, Venezuela, in 1783, Simón Bolívar came from one of the oldest and wealthiest noble families in the city. His education was completed in Europe, where he absorbed the republican ideals of the American and French revolutions. The idea of independence for Hispanic America accordingly took root in his imagination.

His revolutionary career began with an abortive uprising in Caracas in 1810. In 1814, the charismatic Bolívar declared himself "liberator" and head of state of the new republic of Venezuela. In 1817, he staged a daring invasion of Colombia and went on to complete the conquests of Ecuador and Peru in 1824. Bolívar's dream was to unite all of South America—except Argentina, Brazil, and Chile— in a single great republic. However, his dictatorial tendencies and the brutality of his armies eventually led to dissent and the fracturing of Gran Colombia in 1830, the year of his death.

Indians, black people, and those of mixed race—that were never likely to produce coherent political wholes. Plagued by bitter disputes, the short-lived republic of Gran Colombia would break up in 1830.

Brazilian independence

Though partly influenced by the liberal doctrines of the American and French revolutions, the drive to independence in South America was seldom the product of a desire for social justice or representative government. Aside from two abortive Mexican revolutions in 1810 and 1813, it was a struggle for supremacy between ruling elites, none of which had much interest in the kind of social change—society recast on liberal principles—that underpinned the French Revolution. That said, this drive was also significantly affected by the Napoleonic Wars. Napoleon's invasion of Portugal in 1807 forced the Portuguese king, João VI, and his court to flee to safety in its Brazilian colony. João remained there even after the fall of Napoleon in 1815, returning to Portugal only in 1821. However, his son and heir, Pedro, stayed in Brazil.

As in Spain's Latin American colonies, Brazil was dominated by a land-owning elite, a great many of whom, over many generations, had been born in South America. »

Pedro I of Brazil, whose coronation is illustrated in this painting by Jean-Baptiste Debret, was the son of the king of Portugal. He had been left in Brazil to rule as regent.

Miguel Hidalgo, appalled at the obvious inequalities of Mexico, led a popular revolution that ended the following year in its brutal suppression and Hidalgo's execution. Another popular uprising led by a second Catholic priest, José Morelos, between 1813 and 1815 was similarly put down. When, in 1821, Mexico did gain independence, it was by force against more or less token Spanish resistance, and under the leadership of Augustín de Iturbide, a Mexican general who proclaimed himself emperor of Mexico the following year. His rule lasted less than a year. By 1838, Mexico had lost all its Central American territories, and by 1848, it had lost all its North American territories.

Gran Colombia
Events in Spanish South America—which included the triple Vice-Royalties of New Granada, Peru, and Rio de la Plata—followed a very different course. The key figure here was Simón Bolívar. Born in modern-

They came to resent the fact that ultimate authority was exercised by a distant monarchy, and saw no reason why they should submit to it.

There are clear parallels with the American Revolution. But while in North America it was the fundamental liberties of free-born men that were disputed, in Brazil the issue was narrower—it was simply a question of who would govern.

In 1822, to protect the interests of the native-born elite, Pedro declared Brazil an independent constitutional monarchy and himself its emperor. This was a revolution only in the sense that it produced Brazilian independence in the interests of those already ruling it. One of the more obvious consequences was that, with no change to the social or economic order, slavery remained legal in Brazil until 1888, later than anywhere else in the Western world.

Governing Spain's colonies
In Spain's colonies, the drive for independence stemmed partly from the desire of the native-born ruling class—the creoles—to assert their interests, not least in the face

of Spain's restrictive control of South American trade and punitive taxation policies, both to the disadvantage of the colonies. In the short term, however, it was a reaction to Napoleon's invasion of Spain in 1808 and his deposition of the Spanish king, Ferdinand VII, in favor of Napoleon's brother, Joseph. In effect, Spain's colonies no longer had a legitimate ruler of their own, so it was their plain duty to become rulers themselves, at least until the monarchy could be restored.

While South American liberals saw Joseph as the harbinger of a new, more just social order in place of the absolutism of Ferdinand VII, monarchists in the colonies saw any such liberalizing tendencies as inherently destabilizing. The seeds of internal conflict were being sown.

Social revolution in Mexico
At the time, Mexico, which was known as the Vice-Royalty of New Spain, encompassed an immense area that extended almost from present-day Wyoming to Panama and that included most of Texas. Events there took a different turn. In 1810, a priest,

> For my blood, my honour, my God, I swear to give Brazil freedom.
> **Prince Pedro**
> **Future Emperor Pedro I of Brazil (1822)**

May slavery be banished forever together with the distinction between castes.
José Morelos
Leader of the failed
Mexican Revolt of 1813–15

day Venezuela, he was a creole, aristocratic, and highly educated. He had visited Europe several times and was an enthusiastic supporter of modern nation-building on the model established by the French Revolution. He believed, in particular, that the diverse peoples and interests of South America could be brought together by the assertion of a shared South American identity, expressed by the creation of a vast new South American state. This was to be Gran Colombia, which embraced an immense area of northern South America, essentially the modern states of Ecuador, Colombia, Venezuela, and Panama.

Bolívar's vision of an independent South America consistently fell foul of a series of political realities. His military successes—for example, in 1824, the routing of the remaining Spanish strongholds in Peru, when his armies attacked from the north and the south in a pincer

movement in the Central Andes—proved impossible to translate into enduring and stable states.

Bolívar was an idealist and a passionate opponent of slavery. He considered that so disparate a land and a people could only be ruled by a strong central government. Seeing himself as its natural leader, he proposed himself as the lifelong president of Gran Colombia. This provoked predictably bitter opposition.

Gran Colombia breaks up

By 1830—the year Bolívar died, aged 47, of tuberculosis—Gran Colombia had already broken up. Arguably, it was the result of the kind of nationalism already surfacing in Europe, with the independence of Greece and, the following year, of Belgium. More particularly, it was due to a failure to agree on the future of Gran Colombia. There were disputes over whether its government was to be liberal, conservative, or authoritarian. Venezuela, in particular, was subjected to bitter wars throughout the 19th century that cost the lives of an estimated 1 million people.

This lack of direction resulted in instability and a social inequality that would persist for a century or more. It would also produce a series of authoritarian military leaders acting in the interests of the landowners. An inevitable consequence was a persistently oppressed underclass, urban and agricultural, black and white. The *hacienda*—vast acres inefficiently worked by armies of peasants in the interests of a complacently cruel, land-owning elite—dominated.

In 1910, Mexico descended into another revolution. This was partly a result of being wrenched between ineffectual liberal regimes that sought to alleviate the obvious suffering of the poor but did little to address fundamental economic weaknesses and self-serving authoritarian regimes that cared more for bombast than real reform.

Bolívar's visions of a recast, independent South America could never contend with the reality of an unequal society that shared no common belief in its own destiny and that was consistently the victim of competing, mostly violent efforts to assert special interests. ∎

The Battle of Ayacucho (1824) saw the defeat of the Spanish army at the hands of the South American liberation troops. It marked the end of Spanish rule in Peru and in South America.

LIFE WITHOUT INDUSTRY IS GUILT

STEPHENSON'S *ROCKET* ENTERS SERVICE (1830)

IN CONTEXT

FOCUS
The Industrial Revolution

BEFORE
1776 Adam Smith's *The Wealth of Nations* is published.

1781 Watts's first rotating steam engine is invented; the world's first iron bridge is built at Coalbrookdale, England.

1805 The Grand Junction Canal, between Birmingham and London, is completed.

1825 The world's first commercial steam-powered railway, linking Stockton and Darlington, opens.

AFTER
1855 The Bessemer furnace is introduced.

1869 The first transcontinental railroad is completed in the US.

1885 The first practical petrol-driven internal-combustion engine is installed in a motor vehicle, in Germany.

A **scientific revolution** in the West creates a sense that the world can be **better understood** and **better exploited**.

The development of steam-powered machinery encourages the growth of **factory-based mass production**.

Stephenson's *Rocket* heralds a new form of faster, more reliable transport.

The West imposes itself across the rest of the globe, creating **interlocked global markets**.

Industrial societies' **dependence on fossil fuels** leads to a strain on the **natural environment**.

O n September 15, 1830, the world's first commercial passenger rail service to be powered by a steam engine—George Stephenson's *Rocket*—was opened. This was the Liverpool and Manchester Railway, which was 35 miles (56km) long and served by locomotives, also designed by Stephenson, that were capable of reaching speeds approaching 30 mph (48km/h).

Stephenson's *Rocket* symbolized what remains the key development in world history over the past 250 years: the transformation from an agricultural society that relied on windmills, watermills, horses, and other beasts of burden, to an industrial one, in which steam engines were capable of generating reliable power on a scale that was previously unimaginable.

The background

The industrialization process that started in Britain around the mid- to late 18th century was initiated by the scientific revolution in Europe in the late 17th century. Of similar importance were financial changes pioneered in the Netherlands, then imported to Britain: more readily available credit helped boost entrepreneurial activities. It had never been easier for members of the increasingly wealthy middle class, looking for ways to invest their money, to support new inventions and technologies.

A third factor was an agricultural revolution, which began in the Netherlands and Britain, where farmers realized that crop rotation made it unnecessary to leave land fallow every third year. In both of these countries, land reclamation increased the acreage available

See also: The opening of the Amsterdam Stock Exchange 180–83 ▪ Newton publishes *Principia* 188 ▪ Diderot publishes the *Encyclopédie* 192–95 ▪ The construction of the Suez Canal 230–35 ▪ Darwin publishes *On the Origin of Species* 236–37 ▪ The opening of the Eiffel Tower 256–57

for farming. Crop yields were thus boosted, just as selective breeding produced larger and more profitable domesticated animals—sources of food and wool alike. With any likelihood of famine now receding, the population of Britain grew, between 1750 and 1800, from 6.5 million to over 9 million. This, in turn, meant new markets and an expanded workforce.

Finally, in Britain, an improved transport network allowed goods, produced on an ever-larger scale, to be transported faster and more reliably. Between 1760 and 1800, as many as 4,250 miles (6,840km) of canals were built in England.

Thinkers sought to understand the impulses behind these societal changes. The publication in 1776 of *The Wealth of Nations* by the Scottish philosopher Adam Smith underpinned what was becoming known as political economy, and the central role of the profit motive and of competition in increasing efficiencies and lowering prices.

This economic transformation also contributed to and was, in turn, boosted by the emergence of global markets—a consequence of burgeoning European colonial empires, which offered greater access to raw materials and also provided markets for finished goods. A better-mapped world, and improvements in ship types and position-finding at sea, also facilitated global trade.

Steam power

The overriding force behind the economic transformation, though, was the development of the steam engine. In an astonishingly short

A hundred years ago business was limited in area, now it is world-wide.
Frank McVey
Modern Industrialism (1903)

time, it would revolutionize Britain, making it the world's first industrial power, and ultimately transform the world. Even so, it might never have had its dramatic global impact had Britain not had huge reserves of the fuel needed to make it work: coal. The replacement of wood as the prime source of fuel was critical to industrial development. In exactly the same way, the development of coke (processed coal that burns at much higher temperatures than coal) at the beginning of the 18th century would make the production of iron—the indispensable core material of the new technologies— faster and simpler.

Steam engines of varying degrees of reliability had been developed as early as 1712, when Thomas Newcomen built an "atmospheric engine." But it was only with James Watts's first rotating steam engine in 1781 that the extraordinary potential of machine »

Stephenson's *Rocket* was the steam engine on the world's first passenger railway, which linked Liverpool and Manchester. This photograph shows it outside the Patent Office in London.

power became clear. The earliest steam engines had been used mainly as pumps. Watts's rotating engine, on the other hand, could power machinery. The engineering company he and Matthew Boulton established in Birmingham in 1775 produced over 500 steam engines.

When Watts's patents expired in 1800, others started producing their own steam engines. The textile industries in the northwest benefited in particular from the increased availability of steam power, and large-scale, almost entirely mechanized, factory production soon replaced small, home-based manufacturing. By 1835, there were more than 120,000 power looms in textile mills. No longer dependent on rivers as power sources, factories could be built anywhere, and they came to be concentrated in towns in the north and Midlands of England that rapidly grew into major industrial centers as the century progressed.

Social changes

Huge numbers of workers were drawn to these new cities, which became synonymous with poor living and working conditions for the workforce, many of whom were children. This influx led to the creation of an urban underclass. It took a long time before the workers saw any improvement in their lives, and the realization that they should share in the rewards of this social and economic transformation, rather than simply be exploited as mere drudges, came very slowly. In the meantime, however, the increasingly wealthy factory owners emerged as a significant political voice.

The wider world

As late as 1860, Britain was, by some way, still the world's leading industrial and mercantile power, but other Western nations were quick to see how they too could benefit. In continental Europe, industrialization was initially uncertain, inhibited by the kind of political instability Britain had managed to avoid, such as the revolutions of 1848. Later, the pace of its development would rival Britain's. In 1840, Germany and France each had around 300 miles (480km) of railway lines; in 1870, both had 10,000 miles (16,000km). Similarly, pig-iron output from each rose from about 125,000 tons in 1840 to 1 million in 1870.

The Bessemer process, devised by the English engineer of the same name to convert iron into steel, improved the efficiency of all industries—from transport to the military.

However, the most startling developments came in the United States, where there were around 3,300 miles (5,300km) of railway in 1840, almost all in the northeast. By 1860, this had increased to 32,000 miles (51,500km), and by 1900 it had soared to 193,000 miles

Isambard Kingdom Brunel

No figure better encapsulates the determination, ambition, and vision that drove the first phase of the Industrial Revolution in Britain than the prodigiously hard-working Isambard Kingdom Brunel (1806–59). He was responsible for an extraordinary series of firsts: the world's longest bridge (the Clifton Suspension Bridge), the world's longest tunnel (Box Tunnel in Wiltshire), and the world's largest ship (the *Great Eastern*). In 1827, still only 21, he was appointed chief engineer of the Thames Tunnel. In 1833, he became engineer to the newly

formed Great Western Railway, which by 1841 linked London directly with Bristol, whose docks he had rebuilt from 1832. Believing it should be possible to travel directly from London to New York, Brunel also designed the world's first practical ocean-going steamship, the *Great Western*. He followed this with the screw-driven iron-built *Great Britain*. Despite his great vision, delays and cost overruns dogged many of Brunel's projects, but his works include some of the grandest feats of engineering the world had yet seen.

(310,600km) of rail track. The production of pig iron rose similarly: in 1810, it was a little less than 100,000 tons a year; in 1850, it was approaching 700,000 tons; in 1900, it was over 13 million.

The role of steel

By about 1870, in both Europe and the United States, a second wave of industrialization began, in which oil, chemicals, electricity, and steel became increasingly important. The production of steel had been transformed after 1855, when English engineer Henry Bessemer devised a way to make the metal lighter, stronger, and more versatile; from that point forward, steel would prove the linchpin for industry. In 1870, total world steel production was 540,000 tons, but within 25 years it had risen to 14 million tons, and railways, armament production, and the shipbuilding industry all benefited from its ready availability.

While Germany was beginning to threaten Britain's industrial preeminent position in Europe, quadrupling its industrial output between 1870 and 1914, the United States was rapidly becoming the world's largest industrial power. In 1880, Britain was still producing more steel than the United States, but by 1900 the United States was producing more steel than Britain and Germany together.

At the same time, steam-powered ships were also being introduced. Sailing times, no longer dependent on the vagaries of the wind, became more controllable, and journey times were shortened. The ships were significantly larger, too. While the largest wooden ships rarely exceeded 200ft (60m) in length, the *Great Eastern*, launched in 1858, was 689ft (210m) long.

Total world steamship tonnage in 1870 was 1.4 million. By 1910, it had reached 19 million.

Winners and losers

The benefits of industrialization were unevenly spread. Southern Europe was slow to react to it, and Russia also struggled to catch up. The Chinese and Indian empires proved unwilling or unable to industrialize, Latin America did so only intermittently, and Africa was dominated by technologically superior powers. By contrast, after 1868, Japan's single-minded pursuit of industrialization made it a world power.

Industrialization also made possible a new kind of warfare, one capable of bringing death on a scale never seen before. An enduring irony of industrialization is that the nations that benefited most from it turned it against themselves in two world wars, deploying weapons of extraordinarily destructive power.

The Industrial Revolution laid the foundations for the modern world. Fueled by an enormous sense of new possibilities, in some places it raised living standards across all sections of society in ways unimaginable in earlier ages. However, in the wealthy West, it also produced a sense that material superiority was equivalent to a kind of moral superiority, one that not merely made it possible for the West to dominate the world, but demanded that it do so. ∎

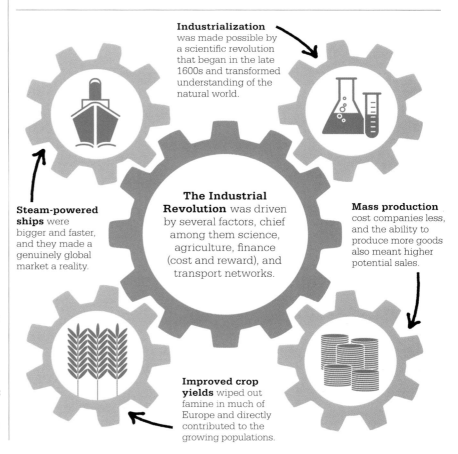

Industrialization was made possible by a scientific revolution that began in the late 1600s and transformed understanding of the natural world.

The Industrial Revolution was driven by several factors, chief among them science, agriculture, finance (cost and reward), and transport networks.

Steam-powered ships were bigger and faster, and they made a genuinely global market a reality.

Mass production cost companies less, and the ability to produce more goods also meant higher potential sales.

Improved crop yields wiped out famine in much of Europe and directly contributed to the growing populations.

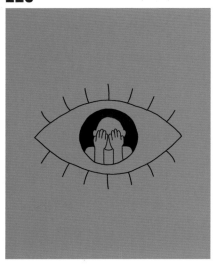

YOU MAY CHOOSE TO LOOK THE OTHER WAY, BUT YOU CAN NEVER AGAIN SAY YOU DID NOT KNOW

THE SLAVE TRADE ABOLITION ACT (1807)

IN CONTEXT

FOCUS
Abolitionism

BEFORE
1787 The Society for the Abolition of the Slave Trade is founded in London.

1791 Slaves revolt in the French Caribbean island of Haiti (St. Domingue). Independence is successfully declared in 1804.

AFTER
1823 The Anti-Slavery Society is founded. It campaigns to abolish slavery throughout the British Empire.

1833 Slavery is outlawed throughout the British Empire.

1848 Slavery is abolished in France's colonies.

1865 The Thirteenth Amendment outlaws slavery in the United States.

1888 Brazil abolishes slavery, the last country in the Americas to do so.

Radical notions of freedom in Britain combine with the religious belief that **slavery** is an **abomination**.

→

Merchants and **plantation owners** resist calls for an end to slavery.

↓

After several parliamentary defeats, the Slave Trade Abolition Act is passed by an overwhelming majority.

↓

Britain campaigns vigorously to persuade other nations to **oppose the shipping of slaves**.

→

Slavery is abolished in the British Empire in 1833. It does not finally end in the US until 1865.

The passing in 1807 of the Act Prohibiting Importation of Slaves in the United States and the Abolition of the Slave Trade Act in Britain marked a radical shift in Western thinking. Even as late as the 1780s, the trade in slaves was still regarded as a "natural" economic activity. Both the newly created United States, "conceived in liberty," and the European colonies in the Caribbean depended on slave labor that was relatively easily obtained in West Africa. Portuguese-ruled Brazil was even more dependent on slaves. Yet Britain in particular found itself in an uncomfortably anomalous

See also: The formation of the Royal African Company 176–79 ▪ The signing of the Declaration of Independence 204–07 ▪ The storming of the Bastille 208–13 ▪ The Siege of Lucknow 242 ▪ Russia emancipates the serfs 243 ▪ The Gettysburg Address 244–47 ▪ The Second Opium War 254–55

William Wilberforce, portrayed here by Karl Anton Hickel, was a fervent Christian and the British politician who campaigned most vociferously against the slave trade.

position. Not only had slavery never been legal there—a point critically reinforced in 1772 in what was called the Somersett case, which ruled that any slave was free once on British soil—but Britons prided themselves on their robust defense of such fundamental freedoms.

Even so, Britain was also, by some margin, the West's leading slave-trading nation. It was this contradiction that offended both religious and Enlightenment political sensibilities alike.

Global changes

To a number of high-minded and unusually active campaigners such as William Wilberforce and Thomas Clarkson, the abolition of slavery became an imperative. A remarkably effective campaign was launched that, despite entrenched opposition, rapidly won wide public and parliamentary support. For much of the 19th century, the Royal Navy would be at the forefront of the campaign to intercept those still engaged in slave trading.

While Britain took the lead, the movement had important supporters elsewhere. The revolutionary French National Convention outlawed slavery in 1794 (though this would partially be overturned by Napoleon in 1802). Brazil aside, where slavery would not be banned until 1888, all the

The state of slavery is repugnant to the principles of the British constitution and of the Christian religion.
Thomas Fowell Buxton
British politician (1823)

newly independent states that emerged in Latin America after 1810 likewise outlawed slavery.

It wasn't until 1833 that slavery itself, as opposed to the trade, was made illegal in the British Empire. Whatever the efforts of a new set of campaigners, not least Elizabeth Heyrick, the motive was not entirely humanitarian. The Haitian slave revolt, which began in 1791 and led to the emergence of an independent Haiti in 1804, had left the West uncomfortably aware that any such uprisings might prove difficult to suppress. A slave revolt in British-ruled Jamaica in 1831 reinforced the point: in the longer term, freeing slaves might prove less trouble than enslaving them.

The United States, forward-looking and expansive, remained the great troubling sore. The more abolitionists in its industrializing northern states denounced slavery, the more its southern states, their agrarian economies dependent on slave labor, were determined to retain it. It would take a four-year civil war and 670,000 dead to settle the question. ▪

The Haitian Revolt

Few uprisings illustrate the contradictions of the revolutions that swept across the late 18th-century Western world better than that in Haiti (1791–1804). This French Caribbean colony, known as St. Domingue, owed its enormous prosperity to slave labor. The revolt, led by a freed slave, Toussaint L'Ouverture, was inspired by the American and French revolutions. Yet neither country supported it: The US was concerned it might inspire similar revolts in its slave states; France, despite its pledge to abolish slavery, was wary of the damage to its trade. Spain, which ruled the eastern half of the island, also opposed it, as did Britain, fearing it would spread to its own colonies. Even the South American colonies seeking independence refused to back it, fearful of its impact among their substantial slave populations. Yet the occasional combined resources of all these states were unable to quell the uprising. This was the only slave revolt to result in the emergence of an independent state.

SOCIETY WAS CUT IN TWO
THE 1848 REVOLUTIONS

IN CONTEXT

FOCUS
Labor movements, socialism, and revolution

BEFORE
1814–15 The Congress of Vienna restores the French monarchy.

1830 Charles X of France is overthrown. Greece obtains its independence from the Ottoman Empire.

1834 An uprising of French silk-weavers is suppressed.

AFTER
1852 The Second French Republic, established in 1848, is dissolved. Louis-Napoleon is proclaimed Napoleon III.

1861 Victor Emmanuel II is declared king of a united Italy.

1870–71 The Franco-Prussian War ends with the unification of Germany under Prussia. The Paris Commune is overthrown, and a Third Republic declared.

On February 24, 1848, Louis-Philippe of France, the "Citizen King," abdicated as Paris erupted in protest at the government's refusals to initiate reforms—demanded by the middle and working classes alike—to introduce political liberalization and to end inequalities. In his place, a Second Republic was declared. In June, fearful that one authoritarian government had been exchanged for another, the Parisian working classes rose again, but the uprising was savagely put down. In December, Louis-Napoleon Bonaparte—nephew of Napoleon, who had died in

This painting by Horace Vernet shows the barricades at Rue Soufflot, Paris. In June 1848, fighting erupted between the liberal republican government and Parisian workers seeking social reform.

1821—was elected president. In 1851, he staged a coup, and the following year he was proclaimed as Emperor Napoleon III.

France was plagued by political instability throughout the 19th century. The 1848 revolution came after a similar upheaval in 1830, and it would be followed by an even more violent uprising 23 years later, in 1871.

The spark for the revolution of 1848 was a famine during the previous two winters. This provoked widespread unrest among the dispossessed urban poor, along with demands from a burgeoning bourgeoisie for liberal political reforms. The ardor of the revolution sparked similar revolts across continental Europe, most obviously in the German Confederation, in multi-ethnic Austria, and in Italy. Every single revolt was subdued, in most instances by force.

The rise of socialism
Before and after the final defeat of Napoleon in 1815, and concerned about citizens rising up elsewhere, Europe's statesmen met in Vienna to try to create a political order that would stifle such an occurrence.

See also: The signing of the Declaration of Independence 204–07 ▪ The storming of the Bastille 208–13 ▪
The Expedition of the Thousand 238–41 ▪ Russia emancipates the serfs 243 ▪ The Gettysburg Address 244–47 ▪
France returns to a republican government 265

Workers of the world,
unite! You have nothing
to lose but your chains!
*The Communist
Manifesto*

Their goal was the preservation of aristocratic ruling elites, sustaining the old order, and holding frontiers.

This desire, however, was to be countered by a new political reality informed by a number of factors, including the desire to ensure that the liberties championed by the French Revolution were upheld. This new reality was also the result of what came to be called nationalism: the right of peoples, however they were defined, to determine their own futures as independent nations.

Just as important was the emergence of a new political creed—socialism—that sought to end the inequalities accelerated by the Industrial Revolution and led to impoverished workers being exploited by factory owners.

The old order is restored
In the feverish atmosphere of 1848, however, these aims would prove irreconcilable. As chaos threatened, the liberally minded middle classes sided much more naturally with existing political elites in restoring order than with the radicals seeking to rebuild societies and create new nations.

The ultimate beneficiaries of the revolutions were the monarchies in Italy and Germany, which would exploit a kind of popular nationalism to unify their countries. But at the same time, as economic shifts brought social change in their wake, the gradual emergence of trade unions—at least in Western European liberal democracies—led to improving standards of living for the previously dispossessed. ▪

The Communist Manifesto

The Communist Manifesto was published in London in 1848, the same year as the revolutions that engulfed Europe. Although its impact on those upheavals was negligible, its resonance in years to come on social thought almost everywhere would be overwhelming. The pamphlet was the work of two Germans: Friedrich Engels, son of a textile manufacturer, and the similarly privileged Jewish academic Karl Marx. In 1847, both men had joined a semi-subversive

French group, the League of the Just, which later re-emerged, in London, as the Communist League. Engels subsequently financed Marx's seminal work, *Das Kapital*, its first volume published, again in London, in 1867. It was a detailed attempt to demonstrate how what Marx called capitalism contained the seeds of its own downfall, and the inevitability of the proletarian revolution that would create a classless society free of exploitation or want.

The **Congress of Vienna** attempts to **stifle nationalism** and the threat of **future revolt**.

The **promise of liberalism** proves impossible to extinguish. Demands for **national self-determination** grow.

France, in particular, after the restoration of the monarchy, sees **violent uprisings**.

The French Revolution of 1848 spawns rebellions in Germany, Austria, and Italy. All are suppressed by force.

Conservative elites exploit **nationalism** to drive the **unifications** of Italy and Germany.

THIS ENTERPRISE WILL RETURN IMMENSE REWARDS

THE CONSTRUCTION OF THE SUEZ CANAL (1859–1869)

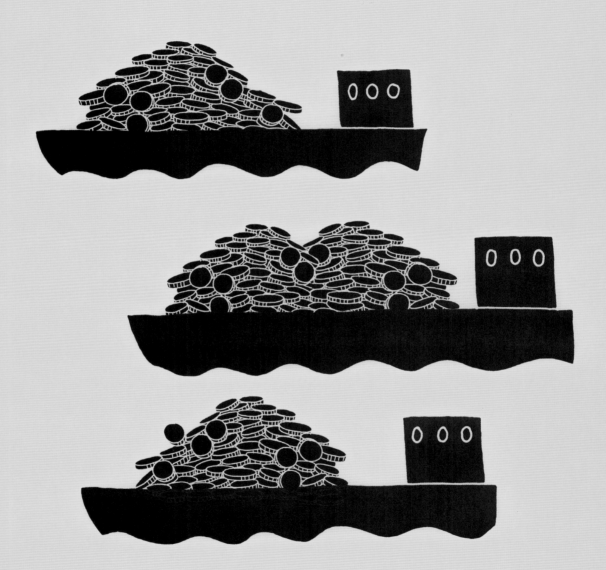

IN CONTEXT

FOCUS
Imperial economies

BEFORE
1838 The first Atlantic crossing under steam power alone is made.

1858 The first transatlantic telegraph cable is laid.

AFTER
1869 The Suez Canal opens, slashing sailing times between Europe and the East.

1878 The Gold Standard is adopted in Europe; the US follows suit in 1900.

1891 The Trans-Siberian railway is begun. It is completed in 1905.

1899–1902 Britain aims to secure control of South Africa in the Second Boer War.

1914 The Panama Canal, linking the Atlantic and Pacific oceans, opens.

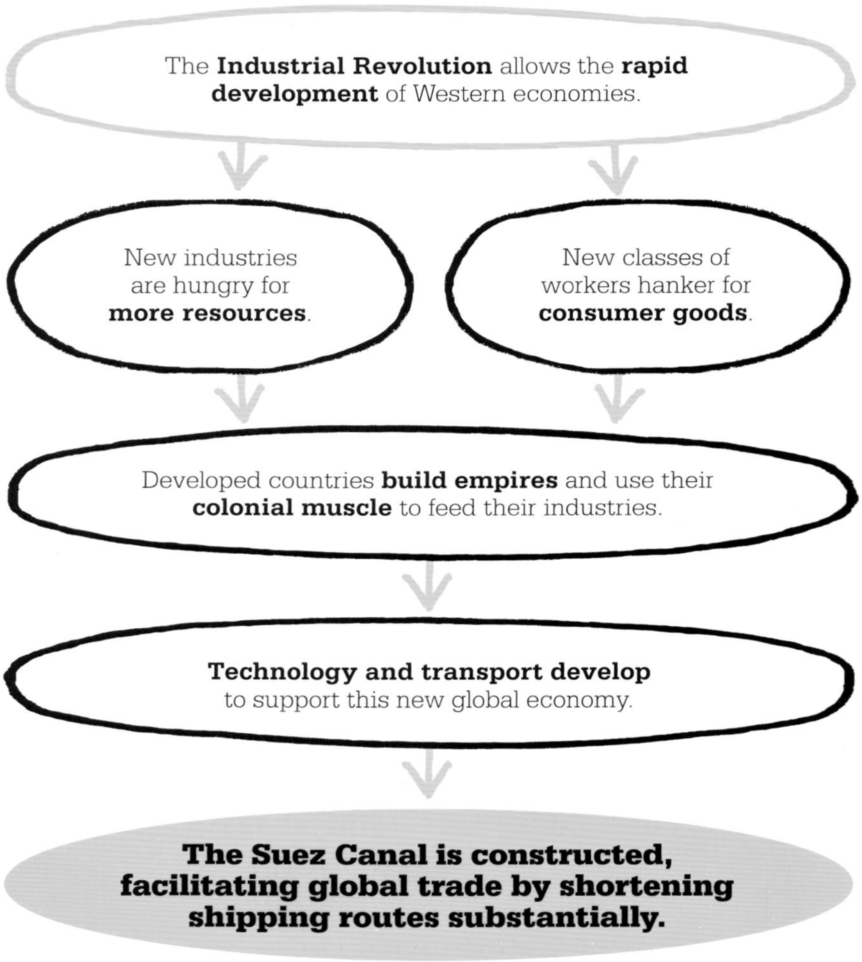

The **Industrial Revolution** allows the **rapid development** of Western economies.

New industries are hungry for **more resources**.

New classes of workers hanker for **consumer goods**.

Developed countries **build empires** and use their **colonial muscle** to feed their industries.

Technology and transport develop to support this new global economy.

The Suez Canal is constructed, facilitating global trade by shortening shipping routes substantially.

The ceremonial opening, on November 17, 1869, of the Suez Canal, linking the Mediterranean and the Red seas, was an emphatic declaration of European—specifically, French – technological and financial means. It was also a significant illustration of a rapidly emerging and increasingly interdependent global economy, featuring goods from all parts of the world being traded on an ever-larger scale. This was a process dominated by Europe's colonial powers and the United States, overwhelmingly its principal beneficiaries. It was simultaneously a further boost to Europe's imperial ambitions.

The Suez Canal reduced the sailing route between London and Bombay by 41 percent, and the route between London and Hong Kong by 26 percent. The impact on trade was plain to see. However, reduced sailing times also greatly simplified the defense of India and its crucial markets, Britain's key imperial goal. By the end of the 19th century, trade in the Indian Ocean, protected by no fewer than 21 Royal Navy bases, had become almost a British monopoly, a point further underlined when Britain gained control of the Suez Canal in 1888 after having invaded and occupied Egypt six years earlier. This "gunboat diplomacy" proved a remarkably effective means of protecting British interests.

The Panama Canal
The Suez Canal was just one of a number of similar massive engineering undertakings in the interests of imperial trade. An even more challenging project was the construction, begun in

See also: Marco Polo reaches Shangdu 104–105 ▪ The opening of the Amsterdam Stock Exchange 180–83 ▪ Stephenson's *Rocket* enters service 220–25 ▪ The California Gold Rush 248–49 ▪ The Meiji Restoration 252–53 ▪ The opening of the Eiffel Tower 256–57

The Suez Canal opened in 1869 and dramatically cut sailing times between Europe and Asia. This provided a massive boost for trade, which, in turn, spurred technological advances.

Roosevelt's successor, William Taft, pursued a more legalistic variant of the policy—Dollar Diplomacy – by which American commercial interests, chiefly in Latin America and East Asia, were to be secured by the full backing of the US government, and huge overseas investments encouraged.

Trains and telegraphs

At the same time, major new railways were constructed in both the US and Europe. The east and west coasts of the US were first linked by rail in 1869, with the opening of the 1,907-mile (3,070-km) Central Pacific Railroad. By 1905, there were eight more transcontinental rail lines across the United States and one in Canada.

The building of the Trans-Siberian Railway in Russia, between 1891 and 1905, was undertaken in »

1881, of the Panama Canal in Central America, linking the Atlantic and Pacific oceans. It too was a French initiative, but one dogged by controversy and a consistently hostile climate that cost the lives of 22,000 laborers. The United States eventually completed the Panama Canal in August 1914, stepping in when the French finally admitted defeat. It was the largest and most expensive engineering project in the world. It, too, dramatically reduced sailing times, shortening the Liverpool to San Francisco route by 42 percent, and the New York to San Francisco route by 60 percent.

US involvement

The fact that the United States took over the construction of the Panama Canal underlines a crucial shift in US attitudes: they were committing not just to expanding trade but also to advancing US overseas interests. This had begun in 1898, when the US itself became a colonial power, taking over the Philippines from Spain.

The process began to accelerate under the presidency of Theodore Roosevelt (1901–09). He actively advocated US military involvement, above all, in Latin America, to ensure stability as a means of advancing American interests. One consequence of this was his strengthening of the US Navy, the Great White Fleet.

May the Atlantic telegraph, under the blessing of Heaven, prove to be a bond of perpetual peace and friendship between the kindred nations.
President Buchanan
Telegram to Queen Victoria (1858)

The scheme in question is the cutting of a canal through the Isthmus of Suez.
Ferdinand de Lesseps
French diplomat on his proposals for the Suez Canal (1852)

the same spirit. A remarkable 4,600 miles (7,400km) long and spanning seven time zones, it remains the longest continuous railway in the world. It played a key role not just in the settlement of Russia's vast Siberian territories, but in Russia's encroachment on parts of northern China, too.

The impact of the telegraph was just as significant, allowing messages to be communicated along electrical lines. Samuel Morse devised the system in the United States in the 1830s, and the first telegraph line was inaugurated in May 1844. Within a decade, there were 20,000 miles (32,200km) of telegraph cable in the US.

The first telegraph cable across the Atlantic, laid in 1858, worked for only two weeks. But by 1866, a new cable had been installed, capable of transmitting 120 words per minute. By 1870, a telegraph link had been established between London and Bombay; this was then extended to Australia in 1872 and New Zealand in 1876. By 1902, the United States was linked to Hawaii. This was the first near-instant international communications system.

The RMS *Mauretania*, built at Wallsend, Tyne and Wear, UK, was the largest and fastest ship in the world. In 1909, it set a record, sailing across the Atlantic in less than five days.

The *Great Eastern*

The ship responsible for laying the transatlantic cable in 1866 was the *Great Eastern*, designed by the most visionary engineer of the first phase of the Industrial Revolution, Isambard Kingdom Brunel. Designed to carry 4,000 passengers from England to Australia non-stop (and to return to England without refueling), the ship was overly ambitious in concept and a commercial failure.

However, it was indicative of a trend toward larger, faster, and safer ships. Unlike the *Great Eastern*, which was built of iron, later, steel-built, propeller-driven ships would prove more versatile. Their introduction coincided with the development of more powerful and efficient steam engines.

Steamships and trade

The decline of the sailing ship further transformed imperial trade. One notable result was the introduction of a series of ever-larger passenger ships. The transatlantic route saw the most obvious developments. In 1874, the British steamer *Britannic*, capable of generating 5,500 horsepower, set a new east–west Atlantic record of just under eight days. In 1909, the *Mauretania*, which generated 70,000 horsepower and carried over 2,000 passengers, set a new record of four days and 10 hours, cruising at an average speed of 26 knots, or 30mph (48km/h).

New types of merchant ships—mainly refrigerated vessels—were also being built. Such developments show how technology helped drive trade, making it possible to reach global markets. The cattle and sheep farms in South America (especially Argentina), Australia, and New Zealand were growing

Although gold and silver are not by nature money, money is by nature gold and silver.
Karl Marx
Das Kapital

in size in line with their own populations. At the same time, the number of people in Europe was also increasing—for example, Britain swelled from 28 million to 35 million between 1850 and 1880. Feeding and clothing the populations were important priorities. Wool could be easily transported, but lamb and beef could not be shipped because it would rot en route—until 1877, when 80 tons of frozen beef were shipped from Argentina to France on board the world's first refrigerated ship. By 1881, regular shipments of frozen meat were traveling between Australia and Britain. The first shipment of lamb from New Zealand was made the following year. There was a vast increase in the export of meat from all three countries—New Zealand, for example, exported 2.3 million frozen sheep in 1895, 3.3 million in 1900, and 5.8 million in 1910.

The demand for cotton—above all in the great textile mills of the northwest of England, which by 1850 were producing up to 50 percent of the world's cloth—led to an enormous surge in cotton growing. In the southern states of the US, raw cotton production increased from

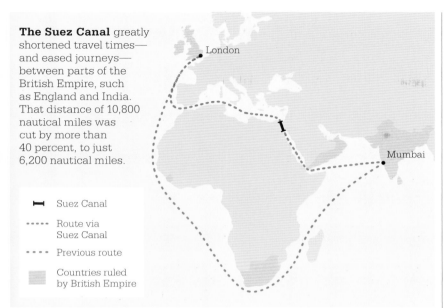

The Suez Canal greatly shortened travel times—and eased journeys—between parts of the British Empire, such as England and India. That distance of 10,800 nautical miles was cut by more than 40 percent, to just 6,200 nautical miles.

London

Mumbai

- Suez Canal
- Route via Suez Canal
- Previous route
- Countries ruled by British Empire

100,000 bales in 1800 to 4 million in 1860. During the American Civil War, the southern, Confederate states restricted exports of cotton in an attempt to force European intervention in the war. However, the ploy failed, since Britain merely increased its imports of raw cotton from India. After weaving the cotton, it then exported it back to India at substantial profits.

Global finance

This complex trading network could not have grown without developments in banking and financing. Throughout the late 19th century, new banks were established, their capital used to support enterprises across the world. At the same time, London emerged as the world's financial capital. By the end of

the 19th century, the British pound sterling, its value pegged at 113 grains of gold, was the currency against which all others were measured.

Western overseas investments dramatically increased. By 1914, the United States had overseas assets worth $3.5 billion, Germany $6 billion, France $8 billion, and Britain almost $20 billion. Between them, North America and northern Europe's share of world income in 1860 was about $4.3 billion a year, 35 percent of the world's total. In 1914, it was $18.5 billion, 60 percent of the world's total.

Patterns of imperialism varied over the 19th century. In the British Empire, for example, clear and increasing distinctions were drawn between those colonies—in Africa and Asia, above all—whose native populations were governed by Europeans, and those—such as Canada, South Africa, Australia, and New Zealand—deemed capable of self-government. By 1907, all four had been granted dominion status. It was not a privilege extended to a single British African colony or to India. ∎

Working conditions in South Africa's gold mines were harsh, and the work force—mainly young black men—was exploited and underpaid.

The Great Mineral Rush

The search for new sources of minerals, both precious and industrial, reached new heights toward the close of the 19th century. Discoveries of diamonds and gold in the US, Canada, Australia, and—most significantly of all—South Africa sparked a frenzy of development. Diamonds were discovered in South Africa's Orange Free State in 1867, and gold in the Transvaal in 1886. Both were independent Boer republics, established by the descendants of the original

Dutch settlers of what had become the British Cape Colony. Their heightened economic importance reinforced Britain's determination to annex them, which they could do only after the bitter Boer War (1899–1902), which stretched Britain's military resources to their limits. The exploitation, both before and after the conflict, of the mineral resources of what in 1910 became the Union of South Africa by armies of underpaid black workers would later prove to be critical in the institutionalizing of Apartheid.

ENDLESS FORMS MOST BEAUTIFUL AND MOST WONDERFUL HAVE BEEN, AND ARE BEING, EVOLVED

DARWIN PUBLISHES *ON THE ORIGIN OF SPECIES* (1859)

IN CONTEXT

FOCUS
Scientific advance

BEFORE
1831–36 The voyage of the HMS *Beagle* takes the young naturalist Charles Darwin around the world.

AFTER
1860 Thomas Huxley defends Darwin from an attack by the established Anglican church.

1863 Gregor Mendel demonstrates how genetics influence all plant life.

1871 Darwin's *The Descent of Man* advances the view of sexual selection, whereby the most successful members of a species are naturally attracted to perpetuate the species.

1953 Discovery of DNA demonstrates how traits are passed on genetically.

Geologists begin to understand that **the Earth has existed** for previously unimaginable **eons of time**.

It became clear to scientists that the Earth had undergone a series of **immense changes and extinctions**.

Charles Darwin publishes *On the Origin of Species*.

Darwin's book explains the **diversity of animal species** and posits that all life on Earth is **related to a common ancestor**.

Modern science decisively **reinforces the evidence and conclusions** presented in **Darwin's landmark text**.

Perhaps the most important scientist of the 19th century, Charles Darwin originally intended to follow his father into medicine and was subsequently sent to Cambridge to train as an Anglican cleric. Endlessly curious, he was interested in almost any scientific question.

The publication of his book *On the Origin of Species* (1859) introduced a new scientific understanding of what gradually came to be known as evolution. In the book, Darwin asked a fundamental question. The world teems with plant and animal life: where and what had it come from? How had it been created?

See also: The voyages of Captain Cook 189 ▪ Diderot publishes the *Encyclopédie* 192–95 ▪ Stephenson's *Rocket* enters service 220–25

Darwin was far from the first to propose that a process of change over vast periods had produced this diversity, but he was the first to suggest an explanatory theme, which he called "natural selection".

Natural selection

At the heart of Darwin's idea was his contention that all animal life was derived from a single, common ancestor—that the ancestors of all mammals, humans included, for example, were fish. And in a natural world that was never less than relentlessly violent, only those most able to adapt would survive, in the process evolving into new species.

These views were largely formed by the around-the-world voyage he made as the naturalist on the British survey vessel HMS *Beagle* between 1831 and 1836, most of it spent in South America. It would take him 10 years to work up his voluminous notes and to go through all the samples he collected on his voyage.

Darwin's book inevitably generated controversy, outraging Christian views that the world had

The finches on the Galapagos Islands were key to Darwin's work. The 13 species he found there all had different types of beaks, which had evolved to deal with the food available to the birds.

been created intact and unchanging by a benevolent deity. Yet however heated the initial debate, quite rapidly there was widespread acceptance of Darwin's views and a realization that he had made

a decisive contribution to the understanding of the world. In the process, the status of science generally was immensely boosted.

The primacy of science

Despite everything, it was possible for Darwinism to be warped. What came to be called "the survival of the fittest" would later prove to be influential in justifying imperialism, racism, and eugenics.

On the Origin of Species was published at a time when a growing understanding of the natural world and rapid technological progress meant scientific study had a greater practical worth than ever before. Darwin was one of the last amateur gentleman scientists in a discipline that was becoming professionalized as society came to view science more highly. Partly as a result of Darwin's work, but also because of these changing attitudes, science began to have a more central place in public life. By the end of Darwin's life, continual progress in scientific knowledge had become an almost standard expectation. ▪

Charles Darwin

Charles Darwin (1809–82) was only the fifth choice for the position of naturalist on the voyage of the HMS *Beagle* in 1831. However fortuitous his selection, it would transform his life. Although he was constantly seasick during his time aboard the craft, Darwin proved himself an assiduous observer of the world around him. He would take as much amazed delight in the jungles of Brazil as he would in the pampas of Argentina or in the arid wastes of the Galapagos Islands. Upon returning to England, he settled into a life

of persistent hard work—the model high-Victorian scientist, aided by considerable private means and a notably happy family life, despite the deaths of three of his ten children. Although his own health may effectively have been severely damaged by the time he spent on the *Beagle*, his output remained prodigious, as did his level of intrigue at almost any subject in the natural world. In the absence of the exotic, he was as fascinated by pigeons as by parasites, barnacles as much as earthworms.

LET US ARM. LET US FIGHT FOR OUR BROTHERS

THE EXPEDITION OF THE THOUSAND (1860)

IN CONTEXT

FOCUS
Nationalism

BEFORE
1830 Greece obtains its independence from the Ottomans.

1848 Nationalist revolutions sweeping across Central Europe and Italy are crushed.

1859 Austria is driven out of Lombardy, which is then annexed by Piedmont.

AFTER
1861 The Italian kingdom is established.

1866 Austria is forced to cede Venetia in northeast Italy to the new Italian kingdom.

1870 The Papal States are incorporated within Italy.

1871 Germany is united under Prussian control. Rome is declared the capital of Italy.

On May 11, 1860, the Italian patriot and guerrilla fighter Giuseppe Garibaldi landed in Sicily, then part of the Bourbon-ruled Kingdom of the Two Sicilies, in southern Italy, leading a force of volunteers drawn from across Italy and just over 1,000 strong, hence their name, *I Mille* (The Thousand). Their goal was to overthrow the Bourbons, but there was much uncertainty as to what government would replace the ruling family.

Like the other great 19th-century stalwart of Italian liberty, Giuseppe Mazzini, who in 1849 had briefly established a Roman republic, Garibaldi was committed to ending royal, clerical, and aristocratic

See also: The storming of the Bastille 208–13 ▪ The 1848 revolutions 228–29 ▪ Russia emancipates the serfs 243 ▪ The opening of the Eiffel Tower 256–57 ▪ The Young Turk Revolution 260–61 ▪ France returns to a republican government 265 ▪ The October Revolution 276–79 ▪ The Treaty of Versailles 280

privilege. He was also driven by the goal of ending Austrian rule in the north of the country and by the idea of a united Italy. The desire to form new political entities based on common national elements such as geography and history came to be known as nationalism.

Reaching a compromise

In 1859, much of Italy had already been united under the state of Piedmont-Sardinia, in the northwest of Italy, a process directed by its shrewd and pragmatic prime minister Camillo Cavour and critically boosted by French military assistance in expelling the Austrians.

For Cavour, unification meant the creation not of a republican Italy, but of a centralized state under a constitutional monarchy. He believed this was the only way that

Ideas of **national self-determination**, inspired by the **French Revolution**, begin to proliferate across Europe.

The **Greek War of Independence** epitomizes the struggles necessary to free nations from **foreign domination**.

The **failed revolutions of 1848** illustrate the **ruling elites' resistance** to notions of national independence.

Garibaldi lands in Sicily and overthrows the Kingdom of the Two Sicilies, but Italy remains a constitutional monarchy.

German unification under Prussia reinforces **conservative nationalism** at the expense of **republican liberties**.

Giuseppe Garibaldi, in the red shirt that symbolized his makeshift army, managed to overthrow Bourbon rule in the Kingdom of the Two Sicilies but had to compromise on governmental form.

Italy could realize its potential— above all, to press ahead with industrialization and compete with the leading powers of Europe.

The Redshirts' forces, swelled by locals flocking to join them, soon overcame the ineptly led armies of the Kingdom of the Two Sicilies.

When it came to deciding upon a government for the newly united Italy—Venice and Rome excepted, though both would subsequently be incorporated, in 1866 and 1870 respectively—Garibaldi recognized the inevitability of Piedmontese domination. In November 1860, with Garibaldi at his side, Victor Emmanuel II of Sardinia entered Naples. In March 1861, he was crowned king of Italy.

Divided goals

The difference between Garibaldi's and Cavour's goals illustrates the contradictions at the heart of nationalism in 19th-century Europe. Prompted by the notions of liberty and equal rights promised by the French Revolution, nationalism developed an idealistic view of a more just society. National groups oppressed by alien rule believed they should be able to assert their independence as a natural right. »

Nationalism was furthermore characterized by a romantic view of the right of peoples to lay claim to their historic destinies and rule themselves: independence. In place of loyalty to an established ruling dynasty, new loyalties to national groups defined by language, culture, history, and self-identity were formulated. The idea of the nation-state became increasingly common, and likewise a belief in the right to national self-determination.

The failure of the revolutions of 1848 in central Europe and Italy, intended to advance these very goals, made plain the resolve of Europe's ruling elites to oppose such initiatives and to preserve the Europe created by the Congress of Vienna in 1814–15, after the defeat of Napoleon—a Europe of monarchs, multinational empires, and pre-French Revolution frontiers.

Metternich's failures

The new Europe was far from stable, and the principal architect of the Congress of Vienna, the Austrian Prince Metternich, would later admit: "I have spent my life in shoring up rotten buildings." By 1830, Belgium had revolted against the Kingdom of the Netherlands, of

> A people destined to achieve great things for the welfare of humanity must one day or other be constituted a nation.
> **Giuseppe Mazzini, 1861**

which it was a province; the next year, it secured its independence with British support. Similar nationalist uprisings followed in Poland in 1831 and in 1846, both savagely repressed by Russia.

German nationalism

Rising nationalism had momentous consequences, especially in the various states across Germany. The country's unification under the chancellorship of Otto von Bismarck of Prussia in 1871 and the declaration of a German empire jolted Europe into a new era. For Bismarck, much as they had been for Cavour, the benefits of unification were clear. It would be the means by which a common German nationality could be expressed, allowing the country to fill the need to underline an overarching German character that the philosopher Georg Hegel had identified. It would also break the dominance of Habsburg Austria over the German-speaking world—in particular, to lever the southern Catholic German states, Bavaria above all, away from Austrian influence.

In the interests of building this great German state, Bismarck pressed into service a kind of conservative nationalism. The goal was not social or democratic reform to establish a more just or liberal state; it was the creation of a country to challenge the world. German nationalism under Bismarck translated into a determined adoption of industrialization and the creation of ever larger and more efficient armed forces.

And it was military means that Bismarck single-mindedly deployed to create this new Germany. He mounted three major campaigns. The first, against Denmark in 1864, saw Prussia subsuming the southern Danish territories

The Ottoman army's brutality in suppressing Greek revolts—as seen in Eugène Delacroix's painting *The Massacre at Chios*—led to increased support for the Greek cause.

of Schleswig and Holstein, with Austrian support. In 1866, Prussian troops routed Austria itself; finally, in 1870–71, an army from across Germany comprehensively and humiliatingly defeated France, toppling Napoleon III's government, and starving Paris into submission. These military victories underlined a seemingly irresistible German destiny whose logical consequence was a unified German empire under the Prussian king, now emperor, Wilhelm I.

Nationalist aspirations

Nowhere were the conflicting impulses of nationalism more tangled than in the Habsburg Austrian Empire, an immense patchwork of ethnic groups across Central Europe under the nominal rule of Vienna. In 1867, following Austria's defeat by Prussia the previous year, Hungary was able to secure almost complete independence from Austria. The "dual-monarchy" that resulted— the Austrian Empire, now the Austro-Hungarian Empire—not

The proclamation of Wilhelm I as Emperor of Germany took place in Versailles in 1871. It was heralded by a series of military campaigns, including one against France.

only greatly boosted an assertive sense of Hungarian self-identity but also secured Hungary significant territorial concessions from Vienna, notably in Transylvania and Croatia. Yet whatever the continuing tensions between Austria and Hungary, the two warily preferred to remain united precisely for fear of further nationalist agitations from their own splintered ethnic populations. The Hungarians, for example, were notably reluctant to concede the kind of political rights they demanded for themselves to their substantial Slovak, Romanian, and Serb populations. At the same time, waning Ottoman control of the Balkans also encouraged nationalist aspirations—Serbia, for example, had emerged as a more or less independent state as early as 1817. Wallachia and Moldavia, essentially modern Romania, could lay similar claims to independence by 1829. The Greeks, portraying themselves as the legatees of ancient Greek

civilization, a role that won them support from liberals across Europe, had secured their independence by 1830 after a nine-year war.

Both Austria and Russia competed to fill the void left by the Ottomans. Austria's provocative occupation of Bosnia in 1878, which it peremptorily annexed in 1908, would create tensions that led directly to the outbreak of World War I in 1914. The Balkan Wars of 1912–13—in effect a bitter squabble for supremacy between Serbia, Bulgaria, and Greece—were further evidence of the destabilizing effect of nationalist-driven state building.

The consequences

The notion that social justice could be secured by peoples pursuing the right to self-determination would rarely be realized in the 1800s—Vienna would continue to rule over its multi-ethnic empire until its defeat at the end of World War I in 1918, for example. Likewise, the people of Poland were denied any means of exercising such nationalistic rights to self-determination. And the Jews of Europe remained persistently oppressed, whatever the promise of Zionism from the 1890s to create a Jewish nation in the Holy Land. ■

Otto von Bismarck

Minister-president of Prussia from 1862 and chancellor of Germany 1871–90, Otto von Bismarck (1815–98), also known as the Iron Chancellor, towered over continental Europe after engineering the unification of Germany. Bismarck's main goals were to ensure Prussian leadership of the German world at the expense of Austria and to contain the threat of renewed French hostility. A supreme opportunist, despite starting three wars, in 1864, 1866, and 1870, Bismarck thereafter worked tirelessly to maintain the balance of power in Europe, a task in which, juggling competing interests, he was remarkably successful. He committed Germany to a huge program of industrialization, oversaw the further growth of the German armed forces, and launched a program of colonization. Despite being socially conservative, Bismarck also introduced the world's first welfare system, though his motive was as much to outflank his socialist opponents as to protect the interests of German workers.

THESE SAD SCENES OF DEATH AND SORROW, WHEN ARE THEY TO COME TO AN END?

THE SIEGE OF LUCKNOW (1857)

IN CONTEXT

FOCUS
British rule in India

BEFORE
1824 The British conquest of Burma is launched; it is largely completed by 1886.

1876 Queen Victoria is declared Empress of India.

May 1857 The first revolt by Indian sepoy troops against British rule occurs at Meerut.

AFTER
1858 The rule of the East India Company in India is formally ended. Control of India passes directly to the British crown.

1869 The Suez Canal is opened, dramatically reducing sailing times to and from India.

1885 The Indian National Congress is founded—the first pan-Indian political movement. It later forms the core of a nationalist movement.

The Siege of Lucknow, which took place between May and November 1857, led to scenes that were duplicated across much of north-central India during the Indian Mutiny of 1857–58: of British enclaves enduring great suffering at the hands of previously loyal local troops. When the British began to restore order, the retribution was no less severe. The violence from both sides shocked the public and led to immediate calls for reform.

The mutiny began when the Indian troops—sepoys—became convinced that the cartridges of their new rifles had been greased with cow and pig fat, offensive to Hindus and Muslims alike. But its roots lay in the dislocation that many in India felt at British control—the uprooting of traditional rulers, the apparent threat to local religions, and the aggressive assertion of alien rule.

Britain's initial response after the mutiny was intended to reassure India of Britain's peaceful intentions toward it, but in reality it underlined the fact that India was now entirely subservient to Britain, both economically and politically.

As the number of European-educated Indian elites grew, they would challenge Britain's rights over the subcontinent. Britain would continue to assert its imperial destiny, but increasingly had to confront the improbability that it could. If there was an enduring truth, it was that British rule in India was never as robust as it seemed. ∎

We hold ourselves bound to the natives of our Indian territories by the same obligations of duty which bind us to our other subjects.
Queen Victoria

See also: The Battle of Quebec 191 ▪ The construction of the Suez Canal 230–35 ▪ The Second Opium War 254–55 ▪ The Berlin Conference 258–59 ▪ The Sikh Empire is founded 264

BETTER TO ABOLISH SERFDOM FROM ABOVE, THAN TO WAIT FOR IT TO ABOLISH ITSELF FROM BELOW
RUSSIA EMANCIPATES THE SERFS (1861)

IN CONTEXT

FOCUS
Tsarist Russia

BEFORE
1825 The Decembrist revolt against tsarist rule is suppressed.

1853–55 Russia's defeat by Britain and France in the Crimea highlights its military weaknesses.

AFTER
1881 Tsar Alexander II is assassinated by the People's Will, an underground terrorist movement.

1891 Work on the Trans-Siberian railway begins, leading to massive new settlements in Siberia.

1894 The last tsar, Nicholas II, allows finance minister Sergei Witte to launch further industrialization.

1905 Russian expansion in East Asia is halted by a humiliating defeat by Japan.

Alexander II's emancipation of Russia's 20 million serfs (unfree laborers) in 1861 was not a humanitarian act. Its goal was a further attempt to modernize a Russia that, regardless of potential, was being left behind by the industrializing nations of the West. To take what it saw as its rightful place in the world, Russia adopted wide-ranging reforms across political, social, economic, and military areas.

The effects of these reforms were mixed at best. Emancipation did very little to improve the serfs' well-being or agricultural productivity, and Alexander refused to consider any real constitutional reform: he remained an autocrat to the last, convinced of his divine right to rule as an absolute monarch. However, his reforms had raised hopes that a degree of political liberalization might follow.

A police state
His assassination in 1881 provoked a predictably reactionary backlash. His successor, Alexander III, showed greater willingness to embrace industrial reform but also created a kind of police state: introducing strict censorship, suppressing protest, and making trade unions illegal.

Nonetheless, tsarist Russia was emerging into the industrialized world. The country could lay claim to substantial, if not always efficient, military means. Politically, however, its unwillingness to reform would ultimately ensure its complete destruction in a Soviet revolution. ∎

We must give the country such industrial perfection as has been reached by the United States of America.
Sergei Witte
Russian minister

See also: The founding of St. Petersburg 196–97 ▪ The 1848 revolutions 228–29 ▪ The construction of the Suez Canal 230–35 ▪ The Crimean War 265 ▪ The October Revolution 276–79

GOVERNMENT OF THE PEOPLE, BY THE PEOPLE, FOR THE PEOPLE, SHALL NOT PERISH FROM THE EARTH

THE GETTYSBURG ADDRESS (1863)

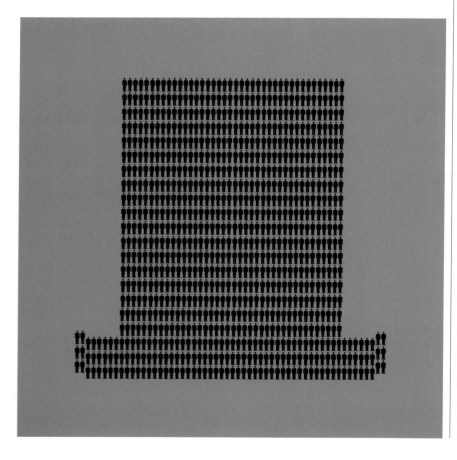

On November 19, 1863, little more than halfway through the American Civil War, in Gettysburg, Pennsylvania, US President Abraham Lincoln gave what came to be known as the Gettysburg Address. In it, he characterized the Civil War as a struggle both for national unity and to guarantee equality for all people.

Lincoln was talking at the dedication of the Soldiers' National Cemetery, which commemorated the 7,058 soldiers killed at the Battle of Gettysburg, an encounter fought between July 1–3, the same year that had left 27,224 more wounded. Gettysburg had been the bloodiest battle of the American Civil War, as

See also: The signing of the Declaration of Independence 204–07 ▪ The storming of the Bastille 208–13 ▪ The 1848 revolutions 228–29 ▪ The California Gold Rush 248–49 ▪ The opening of Ellis Island 250–51

The Battle of Gettysburg took place in 1863. After three days of fighting and the death of more than 7,000 soldiers, the Confederate army was forced to retreat.

well as a turning point in which the outnumbered, outgunned, yet improbably successful Southern army led by Robert E. Lee, the Army of Northern Virginia, suffered its first major defeat.

The causes of the war
The American Civil War was not simply a war about slavery; it was a war about whether so divisive an issue could be allowed to break up the United States. The United States, as Lincoln said, was a nation "conceived in liberty and dedicated to the proposition that all men are created equal," yet its Southern states had a population of almost 4 million black slaves. Under the constitution of the United States, they were legally owned property. For the abolitionists of the rapidly industrializing north—always a minority, but still exceedingly vocal—slavery was morally repugnant and an outrage against their Christian sensibilities.

However, slavery was not just the backbone of the agricultural prosperity of the Southern states; for slave-owning Southerners, it was a right. For them, "liberty" had an additional meaning: the liberty to possess slaves.

The disagreement underlined the question over States' Rights— the extent to which the rights of individual states trumped the authority of the federal, or central, government in Washington. This question repeatedly resurfaced as territories in the west were settled and sought admission to the Union: would they be slave or "free" states?

The 1820 Missouri Compromise stated that slavery would be allowed only in new states south of a line extending westward from the southern border of Missouri. It was later agreed that the settlers of new states should decide for themselves whether theirs would be free or slave states—a decision that was reinforced by the »

Abraham Lincoln

When he arrived in Washington in February 1861 for his presidential inauguration, Abraham Lincoln (1809–65) was widely dismissed in political circles as an ignorant, socially awkward backwoodsman. By the time of his assassination just four years later, he had come to dominate America. Lincoln had not just won the Civil War, but he had also established himself as a kind of irresistible political oracle.

Born in a log cabin in Kentucky, Lincoln qualified as a lawyer by his late 20s. He became an increasingly articulate champion of what would emerge as the anti-slavery Republican party. Despite having no military experience, Lincoln proved an increasingly shrewd judge of how the Civil War should best be fought, actively arguing in favor of General Grant. He never lost sight of his wider aims: the maintenance of American liberties and the essential dignity of humanity. He pushed on with the war with unflinching determination, yet he understood precisely what loss of life on the scale of the Civil War meant.

Kansas-Nebraska Act of 1854. Since both Kansas and Nebraska were to the north of Missouri's southern border, what followed, in Kansas above all, was a sudden inrush of pro- and anti-slavery settlers, each desperate to prevail. The two sides clashed repeatedly and violently.

The South breaks away

This conflict led to the founding of a new anti-slavery party, the Republicans, on whose ticket Abraham Lincoln, with practically no support from any slave state, was voted into office in November 1860. Lincoln's victory prompted the almost immediate decision by South Carolina to secede, to leave the Union. By February, a further six Southern states had broken away, and the seven declared themselves a new nation: the Confederate States of America. By May, when Richmond in Virginia was made the capital of the new country, four more states had joined them. However, five slave states, the so-called Border States, opted to remain within the original Union.

The Confederacy argued that the Constitution had been freely adopted, and as such any state could legitimately break away from the Union if it felt oppressed.

I cannot raise my hand against my birthplace, my home, my children.
Robert E. Lee
On his resignation (April 1861)

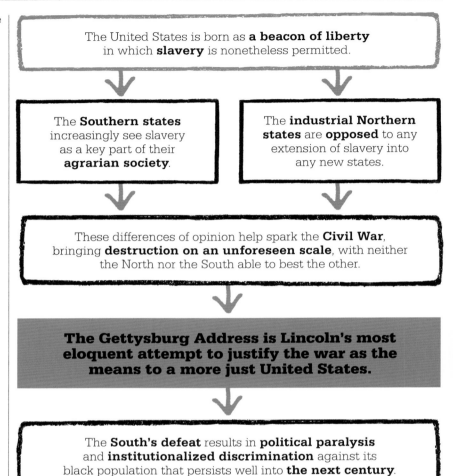

The United States is born as **a beacon of liberty** in which **slavery** is nonetheless permitted.

The **Southern states** increasingly see slavery as a key part of their **agrarian society**.

The **industrial Northern states** are **opposed** to any extension of slavery into any new states.

These differences of opinion help spark the **Civil War**, bringing **destruction on an unforeseen scale**, with neither the North nor the South able to best the other.

The Gettysburg Address is Lincoln's most eloquent attempt to justify the war as the means to a more just United States.

The **South's defeat** results in **political paralysis** and **institutionalized discrimination** against its black population that persists well into **the next century**.

As free-born men, the citizens of the South had an "inalienable" right to shape their own destinies, just as the founding fathers had when they rejected the tyranny of British rule. In the minds of many Southerners, the US government was guilty of precisely the same kind of tyranny in seeking to limit these freedoms.

It was a deeply held position. As Alexander Stephens, vice-president of the Confederacy, asserted, the cornerstone of this new state "rest[ed] upon the great truth that the negro is not equal to the white man; that slavery… is his natural and normal condition…".

As a supreme political operator, Lincoln realized the need to proceed with caution. Initially at any rate, his position was that he sought only to restrict the expansion of slavery while preserving the Union. On the second point, Lincoln was immovable; he felt the authority of the federal government overrode that of individual states.

The United States, the only fully democratic country on Earth, had been created as what Lincoln called "a great promise to the world," so ensuring its survival was an absolute moral duty. By the time of his Emancipation Proclamation in January 1863, Lincoln felt politically secure enough to order the freeing of all Southern slaves. But in the short term, the Civil War was fought to keep this "great promise" intact.

Eventual Northern victory

The outcome of the American Civil War was dictated ultimately by the human and material discrepancies between the North and the South. There were 21 Union states with a population of 20 million, and 11 Southern states with a population of 9 million, 4 million of whom were slaves, and therefore not allowed to bear arms. Despite the fact that by 1864, 44 percent of males in the North between the ages of 18 and 60 were in military service, versus 90 percent in the Southern states, the North was still able to enlist 2.2 million men over the whole war, compared to the South's 800,000.

The North was three times richer than the South. It had 2.4 miles (3.8km) of railroad to every 1 mile (1.6km) in the South. Its factories manufactured 10 times more goods. It produced 20 times more iron than the South, 38 times as much coal, and 32 times as many firearms. The only area in which the South exceeded the North was in cotton production, at 24 to 1.

In the face of this superiority, the fact that the South was not only able to resist the Union forces for four years but also to come close to victory in 1862 and 1863 was a reflection of the Southern soldiers' profound belief in their cause. It was also the result of its plainly superior generals—the Virginian Robert E. Lee above all. By contrast, at least until the emergence of Ulysses S. Grant and William Sherman as the two leading commanders of the Union forces, the North had been able to muster only a succession of timid and inept generals who frittered away the advantages they so abundantly possessed.

Reinvigorated by Grant and Sherman, the North prevailed. The razing of Atlanta in September 1864 was followed by Sherman's "march to the sea" at Savannah, Georgia. Completed in December, it left a 60-mile- (96.5-km-) wide swathe of destruction, deliberately targeting

> Grant stood by me when I was crazy, and I stood by him when he was drunk, and now we stand by each other.
> **William Sherman**

civilian property. "War is cruelty," Sherman asserted. "The crueler it is, the sooner it will be over."

A new freedom

The US Civil War was the world's first major industrial war, the first to make widespread use of railroads, and the first widely reported in a new kind of popular press. There was concentrated death on a scale never seen before: around 670,000 dead, 50,000 of them civilians, in little more than four years.

For Abraham Lincoln, the war represented what in the Gettysburg Address he called "unfinished business." The Constitution had left unresolved the question of how slavery could exist in a nation "conceived in liberty." Despite the destruction and the huge death toll, the war brought a chance at "a new birth of freedom." The end of slavery, confirmed by the Thirteenth Amendment in 1865, represented an opportunity for the US to be recast as a genuinely free land for all its citizens, black and white. ∎

This Thomas Nast illustration shows life for black Americans before and after emancipation. Abraham Lincoln is also portrayed.

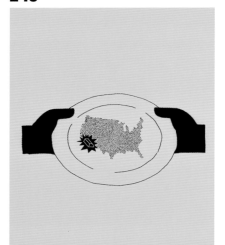

OUR MANIFEST DESTINY IS TO OVERSPREAD THE CONTINENT

THE CALIFORNIA GOLD RUSH (1848–1855)

Knowledge of the lands in the **American West** encourages an interest in **settlement** there.

↓

The gold rush in California sparks a global frenzy to share in the new riches, accelerating the settlement of the West Coast.

↓

Telegraph and railway lines **improve links** between the east and west coasts.

American Indians are forced away from their **ancestral lands**.

↓

Better communications spur the development of **industry** in the United States.

The belief by sometime journalist John L. O'Sullivan that the "manifest destiny" of the United States lay in its expansion to the west was boosted immensely by the discovery of gold in a river in northern California in January 1848. Even allowing for the inevitable difficulties of communication and travel at the time, the find sparked a frenzied reaction. Over the next five years, as many as 300,000 "49ers"—a reference to the year in which the influx began in earnest—were drawn to what, in 1850, would become the 31st state of the US. The immediate consequence was both a wild lawlessness in the pursuit of instant riches and the confirmation of America's Pacific coast as a promised land. The

See also: The signing of the Declaration of Independence 204–07 ▪ The Gettysburg Address 244–47 ▪
The opening of Ellis Island 250–51 ▪ The Trail of Tears 264

John Gast's painting *American Progress* (1872) depicts the concept of manifest destiny. In it, the figure of Columbia, representing the USA, lays telegraph wires and leads settlers west.

population of San Francisco, the main point of entry, was hardly 200 in 1846. By 1852, it was more than 30,000, and by 1870, it was 150,000.

A handful, mostly early arrivals, made fortunes, while some made modest profits, and most made nothing at all. The California gold rush seemed a national obsession. In reality, it was no more than an extreme instance of the determination to colonize North America by the United States that was underway long before the gold had been found. By 1803, Vermont, Kentucky, Tennessee, and Ohio had become states. In addition to the annexation of Texas in 1845, a further 13 states had been added by 1848. In the same year, New Mexico and California were also seized from Mexico.

New technologies
To reach California, the 49ers endured journeys of astonishing hardship, traveling by wagon across the immensities of the Great Plains or by ship around Cape Horn, and sometimes across the Isthmus of Panama. Six months was the expected minimum for these enormous undertakings.

However, an extraordinary resolution to link these vast new territories was being forged, harnessing new technology in pursuit of nation-building on an epic scale. In 1861, the first telegraph line was established between the east and the west coasts. In 1869, the first transcontinental railroad line was completed, slashing journey times: by 1876, it was possible to travel from New York to California in three and a half days.

Settlers and victims
Immigration was the fuel that powered these transformations, the new lands demanding a vast influx of new settlers. In 1803, the US population stood at 4 million. By 1861, it was 31 million; and by the turn of the century, 76 million. There were inevitable costs to such rapid growth, and American Indians paid the highest price. Driven off their tribal lands with relentless brutality, their numbers dropped from perhaps as many as 4.5 million to 500,000. Herded into reservations, their traditional way of life destroyed, they were helpless in the face of this seemingly irresistible expansion. ▪

The Battle of the Little Bighorn

Gold was behind the most famous confrontation between the new settlers of the west and the native populations: the Battle of the Little Bighorn on June 25, 1876. The US government authorized the settlement of the Black Hills of South Dakota after gold was discovered there. By doing so, however, it broke a treaty with the Sioux of the Great Plains. When, in return, substantial numbers of Sioux and Cheyenne refused to move to reservations, federal cavalry were sent to Montana to round them up. Among them was a troop of 600 under Lieutenant Colonel George Custer. Some 200 of these men, led by Custer himself, discovered the Indian encampment in the Little Bighorn Valley. In just one hour, the combined American Indian warriors, led by Sitting Bull, killed Custer's entire force. Their deaths renewed the government's determination to force the Sioux and Cheyenne into reservations at any cost.

AMERICA IS GOD'S CRUCIBLE, THE GREAT MELTING POT

THE OPENING OF ELLIS ISLAND (1892)

IN CONTEXT

FOCUS
Mass migration and population growth

BEFORE
1840s The Irish potato famine leads to mass emigration.

1848 The failure of liberal revolutions sparks large-scale German emigration.

c.1870 Major emigration of Jews from Russia begins as they flee persecution.

1882 Restrictions are placed on the entry of Chinese into the United States.

1880s Mass emigration from Italy begins.

AFTER
1900 The population of Europe reaches 408 million; the United States, 76 million.

1907 The largest number of immigrants in a single year enters the United States: more than 1 million.

1954 Ellis Island closes.

Industrialization, urban growth, and lower infant mortality **boost European populations**.

Political and religious freedoms and **economic opportunities** in young countries such as the US attract **millions of immigrants**.

Steamships make **ocean voyages** to distant lands **safer, faster, and cheaper**.

An immigration station on Ellis Island opens to process arrivals to the United States.

By the mid-19th century, the world was experiencing an unprecedented boom in population, particularly in Europe. This increase would continue into the 20th century and beyond. It was partly due to improvements in health, backed by more ready access to food as a consequence of improved agricultural methods. It was also a result of industrialization and urban growth, as well as the affluence and improved living standards that both produced. Political stability played a role, too. After the defeat of Napoleon in 1815, Europe enjoyed almost 100 years of largely unbroken peace.

See also: The signing of the Declaration of Independence 204–07 ▪ Bolívar establishes Gran Colombia 216–19 ▪ The 1848 revolutions 228–29 ▪ Russia emancipates the serfs 243 ▪ The California Gold Rush 248–49 ▪ The Irish famine 264

Nature also had a hand in the increase in migration. The Irish potato famine of the 1840s, resulting from a failed crop, may have been the last major European famine, but it brought suffering on a startling scale: up to 1 million people died. It provoked among survivors a vast wave of emigration of over 1 million, almost all to the US. The population of Ireland in 1841 was 6.5 million; by 1871, it had dipped to 4 million.

Urban underclass

Industrialization produced a similar paradox. Whatever the civic pride and bombast of the immense new urban centers of the Industrial Revolution, especially in Britain, a new urban underclass was being created, desperately impoverished and living in extreme squalor.

For the citizens of continental Europe, the lure of new lands in which to be free and prosper would prove irresistible. Substantial numbers of Germans, Czechs, and Hungarians left central Europe after the suppression of the nationalist

I had always hoped that this land might become a safe and agreeable asylum to the virtuous and persecuted part of mankind, to whatever nation they might belong.
George Washington

revolts of 1848. From 1870, huge numbers of Russian and Polish Jews—1.5 million in 1901–10 alone—similarly emigrated, fleeing anti-Semitic pogroms.

The numbers involved in this huge transfer of populations were remarkable. From the mid-19th century to 1924, 18 million people emigrated from Britain, 9.5 million from Italy, mostly from its deprived south, 8 million from Russia, 5 million from Austria-Hungary, and 4.5 million from Germany. Between 1820 and 1920, the US attracted 33.6 million immigrants, where they often found themselves in poor living conditions in rapidly expanding cities such as Chicago and New York, aiding the growth of American industry with their cheap labor. Over the same period, 3.6 million Europeans settled in South America, and 2 million in Australia and New Zealand.

Unwelcome guests

This process of relocation was not exclusively European. Indians settled in South Africa, Chinese migrants spread across the East Indies, and Japanese migrants settled in California; many found themselves unwelcome.

There were also victims of enforced emigration. Unknown numbers of black African slaves were still being shipped around the world.

By 1910, more than one in seven of the US population had been born outside of the United States. ▪

In its first 30 years, Ellis Island saw 80 percent of United States' immigrants passing through— almost 12 million people.

Ellis Island

Opened on January 1, 1892, Ellis Island, along with the Statue of Liberty, became a symbol of the vast stream of immigrants that poured into the United States. This immigration center processed perhaps 12 million people, and it is claimed that as much as 40 percent of the immigrant population of the United States has at least one relative who was fed through this immense bureaucratic machine. Built on nothing more than a nondescript sandy island, close to the New Jersey shore in New York Harbor, Ellis Island had at its heart a vast, echoing hall. Here, shuffling forward, the newly arrived immigrants, speaking a dazzling array of languages, would be processed. They were examined medically before being subjected to a series of simple questions to establish their eligibility. The great majority would then become accepted as citizens of the United States, with scarcely 2 percent turned away. Ellis Island finally closed its doors on November 12, 1954.

ENRICH THE COUNTRY, STRENGTHEN THE MILITARY
THE MEIJI RESTORATION (1868)

IN CONTEXT

FOCUS
Modernizing Japan

BEFORE
1853 A US naval force arrives in Japan and demands trading links.

1854–55 The US, Britain, the Netherlands, and Russia force trading treaties on Japan.

1866 The rulers of the Choshu and Satsuma regions form a secret alliance against the ruling Tokugawa shogunate.

1867 The Tokugawa shogunate ends.

AFTER
1868–69 Defenders of the shogunate are defeated.

1871 Feudalism is abolished, and Japan launches a far-reaching program of reform.

1894–95 The Sino-Japanese War underlines expansionist Japanese aims in the region.

1904–05 The Russo-Japanese War ends in Japanese victory.

Aggressive Western demands for trading rights in Japan highlight the **weakness of its ruling elite**.

⬇

Leading feudal barons reassert the authority of the boy emperor Meiji and oust the shogunate.

⬇ ⬇

The barons see the adoption of Western political and social methods as the best way to **strengthen Japan**.

Military might is seen as an essential way of fulfilling Japanese ambitions.

⬇ ⬇

Modernization and Westernization encapsulate the Meiji period, and Japan emerges as an **imperial power**.

The overthrow in 1868 of the Tokugawa shogunate, rulers of Japan for 250 years, was led by feudal barons from the southern provinces of Choshu and Satsuma and was the direct consequence of its weakness in the face of aggressive demands by the United States, Britain, Russia, and the Netherlands to establish trading links. In place of the shoguns, the pliant 14-year-old Meiji emperor would "exercise supreme authority." The goal of the barons was not to take over and maintain Japan as it had existed under the shogunate—rigidly hierarchical and deliberately isolated from the wider world.

See also: The Battle of Sekigahara 184–85 ▪ Stephenson's *Rocket* enters service 220–25 ▪ The construction of the Suez Canal 230–35 ▪ The Second Opium War 254–55 ▪ Nazi invasion of Poland 286–93 ▪ The Long March 304–05

This image of Yokohama in 1874 depicts the modernity of Meiji-era Japan in the form of steam-powered trains and ships, which also served to open up the country to trade.

The spur to modernization had largely been the fear that Japan, like China, would become another Western colonial pawn. In fact, the opposite occurred.

Rather, it was that they felt Japan's clear destiny could only be established by the adoption not just of Western technological means but of Western political and financial systems, too.

Japan transformed

There followed a transformation of a kind no society had seen before or has seen since. Modeling itself on the West, in 30 years Japan became one of the most dynamic industrial powers in the world and the leading military power in East Asia.

Almost no aspect of Japanese society was left untouched by this whirlwind of change. In 1871, Japan abolished feudalism and established the yen as its currency. By 1872, the first railway was under construction; within 15 years, there was 1,000 miles (1,600km) of track. In 1873, Japan introduced conscription, along with Western weapons and uniforms. The same year saw an overhaul of the education system, and in 1877 Japan's first university was established in Tokyo. Japan also introduced a new legal code in 1882, and a new constitution seven years later. As industry boomed, so exports grew. Cities similarly expanded, as did the population, swelling from 39.5 million in 1888 to 55 million by 1918.

Military expansion

By the 1890s, Japan was a colonial power. In 1894, Korea had asked both Japan and China to help curb an insurrection there. When both countries later sought to take over Korea, the Japanese swept the Chinese aside, and then demanded and received possession of Taiwan, as well as rights in Manchuria. Here they came into conflict with Russia. The Japanese victory in 1905 over a disorganized Russian fleet at the Battle of Tsushima Strait was the first time an industrial European power had been defeated by an Asian power. Japan had the world's attention. ▪

Emperor Meiji

Important not as a statesman or as the ruler of Japan in the sense of exercising actual power, Emperor Meiji (1852–1912)— whose personal name, never used, was Mutsuhito—was instead the symbol of the reborn Japan. Until the restoration of Meiji in January 1868, the emperors of Japan were little more than a symbol. Under the shogunate, they were obliged to remain invisibly at the royal palace in Kyoto more or less permanently. Strictly speaking, the "restoration" never happened: Meiji had already become emperor in February 1867, following the sudden death of his father, Emperor Komei.

For those ambitious *daimyo*, or feudal barons, who were determined to drive Japan into the modern world, elevating the emperor to a higher profile bestowed legitimacy on what was otherwise an act of usurpation. It is telling that one of their first acts was to force the emperor to move to Edo, which was renamed Tokyo in 1868, the former residence of the shogun. Meiji himself remained an impenetrable cypher to the end.

IN MY HAND I WIELD THE UNIVERSE AND THE POWER TO ATTACK AND KILL

THE SECOND OPIUM WAR (1856–1860)

Despite **China's great wealth**, Western powers are allowed **very restricted access** to Chinese ports.

Western merchants use **opium to pay for goods**, damaging China's economy.

The First Opium War is sparked by Chinese attempts to **stop the opium trade**.

The Second Opium War leads to further crippling territorial and trading concessions.

Unable to resist the West, **China sees its status diminished** internally and externally.

On October 6, 1860, after years of sporadic conflict known as the Second Opium War, an Anglo-French force seized the imperial capital of Peking (today Beijing), in China, to force the Chinese to submit to trading concessions. The point was underlined when the Europeans burned down the emperor's sumptuous Summer Palace. The Chinese agreed to talks, and the resulting Peking Convention not only increased the number of Treaty Ports open to Western trade, but British and French zones of influence were extended in south China and along the fertile Yangtze River.

See also: Stephenson's *Rocket* enters service 220–25 ▪ The construction of the Suez Canal 230–35 ▪ The Siege of Lucknow 242 ▪ The Meiji Restoration 252–53 ▪ The Taiping Rebellion 265 ▪ The Long March 304–05

The port of Canton, in southern China, was initially the only trading port open to Western merchants. After the two opium wars, Europe was given exclusive access to many more.

Less than 70 years before, Britain had sent an embassy to China to open trade talks, only to be rebuffed. Late 18th-century Qing China was the richest, most populous, and most powerful country in the world, and it could afford its complacency. By the mid-19th century, however, the nation was effectively bankrupt, racked by famines and revolts, and increasingly exploited and humiliated by the West.

Uprisings and revolts

China's problems were internal as much as external. A swelling population—100 million in 1650, 300 million in 1800, 450 million in 1850—provoked recurring famines. Between 1787 and 1813, there were three major uprisings. Its border provinces, conquered at huge expense in the 17th and 18th centuries, were in a near-permanent state of unrest.

In 1850, the Taiping Rebellion erupted across central China, resulting in the death of as many as 20 million people. When it was finally put down, in 1864, it was only after Western intervention. The Qing dynasty, its administration increasingly ineffective, had essentially lost control of China.

The West intrudes

It was this growing turmoil that the West exploited, weakening China further in the process. The first, modest trading concessions China had agreed to stipulated that all Chinese goods be paid for in silver. However, from the early 19th century, European traders, mostly by bribing officials, were increasingly able to use opium, cheaply grown in India, to pay for goods. By the 1820s, 5,000 chests of opium a year were entering China.

The Chinese attempt to end the opium trade and its debilitating effects led to a crushing defeat in the First Opium War of 1839–42, with the European powers, Britain above all, extracting substantial trading concessions. It was Western insistence in 1856 that these concessions be extended that led to the Second Opium War, concluded in 1860 by the Peking Convention. By 1900, a string of Western trading ports were scattered along the Chinese coast. Britain, France, Japan, and Russia now all controlled what had been Chinese tributary states on its borders. China, wracked by turmoil, was effectively disintegrating. ▪

The Boxer Rebellion

In the turmoil of late 19th-century China, it was inevitable that efforts would be mounted to end the growing dominance of the West. The imperial government in Beijing made a last-ditch attempt at reform on Western lines, but the chaos came to a head in 1899 with the Boxer Rebellion, mounted by the Militia United in Righteousness, a semi-secret society composed mostly of young men. Its goal, to be achieved in part thanks to their deluded belief that they were invulnerable to Western weapons, was the overthrow of all Western interests. The rebellion was variously supported and opposed by the imperial court, uncertain whether it represented a means of salvation or would merely provoke Western reprisals. The latter proved to be the case. An eight-nation military alliance, including Japan, was sent into China against the Boxers, and by September 1901 the rebellion had been crushed, amid scenes of indiscriminate violence.

I OUGHT TO BE JEALOUS OF THE EIFFEL TOWER. SHE IS MORE FAMOUS THAN I AM

THE OPENING OF THE EIFFEL TOWER (1889)

The opening of the Eiffel Tower on March 31, 1889 was a startling assertion of Parisian bombast in the years between the humiliations of the Franco-Prussian War in 1870–71 and the outbreak of World War I in 1914. This was the Belle Epoque, a time when Paris could—and did—confidently proclaim itself the City of Light, supremely cosmopolitan, the art capital of the world, and the epicenter of civilized living. Paris was the city reborn,

The Eiffel Tower was erected in time for the Universal Exposition of 1889. At the time, it was the world's tallest structure. It has since come to represent Paris to the entire world.

and over it now soared Gustave Eiffel's tower, at 984ft (300m) not just the tallest structure in the world, but a triumphant monument to technological progress.

The ideal city

Modern Paris was the creation of Napoleon III. Beginning in 1853, the French emperor had whole districts knocked down, replacing medieval buildings and tangles of tiny streets with imposingly vast boulevards. It was urban planning on a scale never seen before. Train stations were built, water supplies improved, sewers constructed, parks laid out, and stunning vistas created. The goal was a model city, one that would reflect not just the glory of France but also its mastery of the modern age.

It was a process mirrored in cities across the industrial West. In 1850, there were three European cities with populations larger than 500,000: Paris, London, and Constantinople. Fifty years later, there were nine with populations larger than 1 million, and by 1900, London was the largest city in the world, with 6.5 million inhabitants. The same dizzying growth occurred in the US,

See also: Stephenson's *Rocket* enters service 220–25 ▪ The construction of the Suez Canal 230–35 ▪ The opening of Ellis Island 250–51 ▪ France returns to a republican government 265

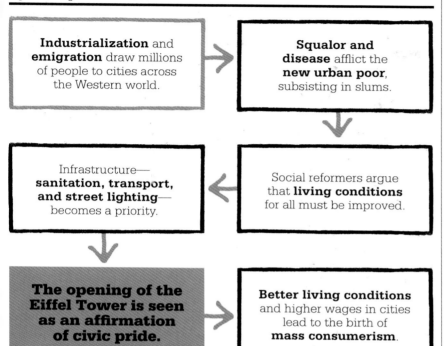

Industrialization and **emigration** draw millions of people to cities across the Western world.

→

Squalor and disease afflict the **new urban poor**, subsisting in slums.

↓

Infrastructure— **sanitation, transport, and street lighting**— becomes a priority.

←

Social reformers argue that **living conditions** for all must be improved.

↓

The opening of the Eiffel Tower is seen as an affirmation of civic pride.

→

Better living conditions and higher wages in cities lead to the birth of **mass consumerism**.

---

too—between 1850 and 1900, the population of Chicago, for example, tripled, from 560,000 to 1.7 million.

Difficulties and inventions
The initial consequence of this population explosion was quite staggering urban squalor. Diseases such as cholera and typhoid were commonplace. It became clear that the infrastructure demanded by any modern city had to include not just adequate public transport and well-lit streets, for example, but major improvements in public health—above all, sanitation.

The shift in the quality of life in these great metropolises was extraordinary. It was paralleled by the rapid development of mass consumerism, the direct consequence of improved living standards, shorter working hours,

and compulsory education, with basic literacy and numeracy now increasingly commonplace. It was an age of music halls and popular theater, and subsequently of the cinema; of the phonograph; of mass-circulation newspapers; and of a growing interest in sports.

Just as central to this age of growing affluence and increased leisure—at least for some—were the first department stores. These were a conspicuous part of a retail revolution that was coupled, from the 1870s, with an advertising explosion, with color posters mass-produced for the first time. And from the 1890s in the United States, cityscapes were further changed by a new type of building: skyscrapers. Just as the Eiffel Tower before them, they rapidly became symbols of the transformation in urban life. ▪

Underground railways

Between 1800 and 1900, the population density in New York rose from 39,183 per square mile to 90,366, and the congestion was worsening as public transportation took up valuable land. A solution favored in the United States was an elevated railway— a train line raised above the streets on steel girders. The first was opened in New York in 1868.

In the UK, the same space constraints led to the birth of the underground railway. The first, using conventional steam engines, was London's Metropolitan Railway, which opened in 1863 and linked Paddington and King's Cross stations with the City of London. Soon extended and combined with the District Line, by 1871 it encircled almost the whole of central London. The city's first electric underground service—faster, quieter, and much less dirty— was opened in 1890. Paris followed suit with the opening of the Métro, named after the London line, in 1900, and the first US underground service opened in Boston in 1897.

In 1890, London unveiled the world's first electric underground railway. It made transport within the city quick and reliable.

IF I COULD, I WOULD ANNEX OTHER PLANETS
THE BERLIN CONFERENCE (1884)

IN CONTEXT

FOCUS
The "scramble for Africa"

BEFORE
1830 France begins occupation of Algeria.

1853–56 David Livingstone crosses Central Africa.

1862 John Speke discovers the source of the Nile.

1879 H. M. Stanley is hired by Leopold II to survey the Congo.

1882 Britain takes over nominally Ottoman-ruled Egypt.

AFTER
1886–94 German territories in East Africa are established.

1890 The Anglo-French Convention grants France control of the Sahara.

1891–93 Cecil Rhodes brings Southern and Northern Rhodesia under British rule.

1899–1902 Boer War sees Britain wrest control of Orange Free State and Transvaal.

The **interior of Africa** is revealed by European exploration. Its **commercial possibilities** are alluring.

Unremitting competition between Europe's **colonial powers** sparks a sudden "scramble for Africa."

Europe takes full advantage of its **financial and military superiority** to impose itself on Africa.

At the Berlin Conference, new colonial possessions are created, supposedly in the interests of Christianity and "civilization."

By 1913, only **Liberia and Ethiopia** remain fully independent.

The Berlin Conference did not precipitate the sudden European takeover of Africa after 1880 but, rather, confirmed Europe's self-asserted right to impose itself on a continent deemed backward, ignorant, and savage. Called by Otto von Bismarck, Germany's chancellor, the conference was held over the winter of 1884–85 and attended by representatives from 14 countries. It was intended in part to legitimize a more or less enforced subjection of Africa and, by setting agreed rules of colonization, to avoid conflict between Europe's colonial powers, France and Britain most obviously.

See also: The construction of the Suez Canal 230–35 ▪ The Siege of Lucknow 242 ▪ The Second Opium War 254–55 ▪ The rise and fall of the Zulu kingdom 264 ▪ A Mahdist Islamic state is created in Sudan 265 ▪ The [Second] Boer War 265 ▪ Indian independence and partition 298–301 ▪ Nkrumah wins Ghanaian independence 306–07

It was also seen as a way to end the slave trade, not least by the actions of Christian missionaries. At the same time, it paved the way for Germany and Belgium, two nations with no history of colonial rule, to become major imperial powers. For Germany, this was little more than a logical next step in its challenge to Britain and France. If they could boast vast colonial possessions, Germany felt it should, too.

The European takeover
Before colonization, Africa possessed a variety of states and territories, some quite precisely defined, some amorphously tribal—there was an extreme contrast between the sophistication of Egypt, for example, and the Congo in tropical Africa. At the same time, much of the north was Muslim. The first European holdings in Africa were coastal trading forts, sustained by gold and the slave trade. The interior remained impenetrable, but as it was revealed from the early 1800s, European control of Africa gained momentum.

The subsequent heightening of tensions resulted in the near-complete reduction of Africa to European rule. African colonies were essentially artificial creations, lines drawn on maps to suit the colonizing powers. They took no notice of local histories and cultures, and any local resistance to colonization was invariably crushed by military means.

Belgian and German rule
In 1885, Leopold II, the King of the Belgians, proclaimed the establishment of the Congo Free State, an area 76 times larger than Belgium. Presented as a model colony, dedicated to humanitarian ends and free trade, in reality it was anything but.

Treated by Leopold II as his personal possession, the Congo witnessed brutalities on a near-genocidal scale. The exact numbers can never be known, but it is believed that between 2 million and 10 million Congolese died. Conditions in German southwest Africa, suddenly taken over after

Cecil Rhodes, portrayed in this Victorian cartoon as a giant straddling the entire African continent, was a great believer in colonization for the benefit of the British Empire.

1884 and today part of Namibia, were equally brutal. The true price of the riches produced by Africa for its European masters—ivory, rubber, gold, and diamonds—was extraordinary suffering. ▪

Cecil Rhodes

There was no more ardent exponent of British imperial rule in Africa than Cecil Rhodes (1853–1902), financier, statesman, and relentless imperialist. He envisaged a continuous body of British colonies that would run north and south across Africa, linking the two strategically vital extremities of Africa: Cape Town and Cairo. Having made his fortune mining and selling diamonds in South Africa, he dedicated the rest of his life to this audacious vision. He was able to carve out new British territories in Northern Rhodesia (now part of Zambia) and Southern Rhodesia (now Zimbabwe), which were both named after him. As the prime minister of Britain's Cape Colony from 1890, his relentless scheming to topple the Boer republics led to his eventual political demise in 1895. He remains perhaps the most striking example of the unashamed imperialist, not just permanently ready to extend British colonial control, but convinced that it was his duty to do so in the interests of what seemed to him a self-evident European fitness to rule.

MY PEOPLE ARE GOING TO LEARN THE PRINCIPLES OF DEMOCRACY, THE DICTATES OF TRUTH, AND THE TEACHINGS OF SCIENCE
THE YOUNG TURK REVOLUTION (1908)

IN CONTEXT

FOCUS
Modernizing Turkey

BEFORE
1798 French invasion of Egypt, which leads to the Ottoman loss of the country in 1805.

1830 Greece's independence signals the loss of the first Ottoman Balkan territories. France begins its conquest of Algeria.

AFTER
1912–13 Ottoman Turkey suffers humiliating defeats in the Balkan Wars.

1914 Ottoman Turkey enters World War I on Germany's side.

1920 Mustafa Kemal leads a revolt against the punitive Treaty of Sèvres imposed on the Ottoman government after defeat in World War I.

1923 The Treaty of Lausanne confirms the borders of modern Turkey; Kemal launches a modernization program.

Ottoman Turkey finds itself increasingly **unable to keep control** of its empire and match **Western powers**.

The Ottoman sultanate attempts **Westernizing reforms**, but these efforts are half-hearted.

The rule of **Abdul Hamid II** proves **repressive and corrupt**, ever more at the mercy of Western financial interests.

The Young Turk Revolution instigates modernizing reforms. It is unable to offer any lasting solution to Turkey's declining power.

Defeat in World War I shatters the Ottoman Empire but leads to the **formation of a secular republic**.

The Young Turk Revolution of July 1908 was instigated by nationalistic army officers dismayed by the territorial losses of the Ottoman Empire. It forced the sultan—the Ottoman ruler, the ineffective but repressive Abdul Hamid II—to reintroduce the constitutional monarchy that he had suspended in 1878 in favor of personal rule after only two years. In 1909, he was forced to abdicate in favor of his brother, Mehmed V, who was merely a figurehead.

The revolution did very little to halt the Ottoman decline, tending mostly only to highlight

See also: Stephenson's *Rocket* enters service 220–25 ▪ The 1848 revolutions 228–29 ▪ The construction of the Suez Canal 230–35 ▪ The Expedition of the Thousand 238–41

tensions between those who championed Turkish Islamic values and those liberals who believed only Western-style reforms could save Turkey.

Territorial decline

In 1800, despite repeated defeats at the hands of the Russians, Ottoman Turkey still ruled over a vast multinational empire that stretched across the Balkans, the Middle East, and North Africa. From 1805, it lost control of Egypt, which became effectively independent under one of the sultan's generals, Muhammad Ali.

In 1830, the year that France began its conquest of Algeria (completed in 1857), Greece won its independence; and by 1878, Serbia, Montenegro, Bulgaria, and Romania were similarly independent in all but name. In 1881, Tunisia, too, was taken over by France.

Following the Young Turk Revolution, the relentless decline of Ottoman Turkey continued. In 1911, Libya was lost to Italy, while

The Ottoman Empire became known as "the sick man of Europe" in the late 19th century, due to its inability to hold on to its lands. A defeat in World War I led to even more territorial losses.

the Balkan Wars of 1912–13 saw the surrender of almost all of Turkey's remaining European territories.

A fateful alliance

Following the Balkan crisis, the Ottomans' military government launched a drive to modernize the country along Western lines.

In October 1914, Turkey entered World War I as an ally of the Central Powers—Germany and Austria-Hungary—convinced that German military aid would allow it to reassert its potency. This was a calamitous mistake, and defeat in 1918 saw Turkey reduced to its Anatolian heartlands, its remaining Middle East territories lost, largely carved up between Britain and France.

The traumas of Turkey's defeat in World War I were underlined in 1920 by the Treaty of Sèvres, largely a Franco-British imposition. This confirmed the loss of Ottoman territory and also awarded much of western Turkey to Greece, provoking an immediate nationalist backlash led by Mustafa Kemal, as well as the overthrow of the last sultan, Mehmed VI.

The Turkey that emerged under Kemal, subsequently styled Atatürk ("Father of the Turks"), was exactly the centralized Western-style and, importantly, secular state that nationalist reformers such as the Young Turks had argued for. ▪

Kemal Atatürk

Mustafa Kemal, better known as Atatürk (1881–1938), the name he assumed in 1934, was the founder and first president of the republic of Turkey. Born in 1881, he took part in the Young Turk Revolution as an army officer. Later, he served with distinction in the Gallipoli campaign of 1915–16, which repulsed a joint Franco-British attempt to conquer western Turkey.

After the Turkish defeat in World War I, Atatürk established a provisional government. As leader of the Turkish Nationalists, he played a central role in driving the Greeks from mainland western Turkey. With the country's borders confirmed by the Treaty of Lausanne of 1923, the West effectively agreeing to the establishment of a new Turkish republic, Atatürk launched a radical program of social and political reforms intended to transform the nation into a modern, Westernized republic. However painful the process of dragging Turkey into the modern world, under Atatürk the country indeed emerged as a coherent, secular political entity.

DEEDS NOT WORDS

THE DEATH OF EMILY DAVISON (1913)

IN CONTEXT

FOCUS
Women's suffrage

BEFORE
1869 In the US, the National Woman Suffrage Association and American Women Suffrage Association are formed.

1893 New Zealand is the first country to grant women the vote.

1897 The National Union of Women's Suffrage Societies is formed in Britain. It campaigns peacefully for the right to vote.

1903 Emmeline Pankhurst forms the Women's Social and Political Union in Britain. Vote campaigns grow violent.

AFTER
1917 The National Women's Party begins a 30-month protest at the White House.

1918 All women 30 or over are granted the vote in Britain.

1920 The vote is granted to all American women 21 and over.

More women are educated and hold professional posts, **raising expectations** for them to have the right to vote.

Societies are established to **campaign for women's suffrage**, particularly in Britain and the United States.

Militant campaigners from Britain's Women's Social and Political Union are **arrested and imprisoned**.

Emily Davison's death raises the profile of women's suffrage across the world.

Women's war work **emphasizes their capabilities**. British **women win the vote** in 1918; American women, in 1920.

O n June 4, 1913, Emily Davison stepped onto the course at the Derby, England's premier horse race, and was knocked to the ground by a horse owned by King George V. She died four days later. It is unclear if this was a protest that went wrong or an active attempt at martyrdom.

However, the intended disruption was typical of the Women's Social and Political Union (WSPU), which Davison had joined in 1906.

Britain: the suffragettes
Women in the West had begun to feel that they, and by extension those elsewhere, should no longer

See also: The signing of the Declaration of Independence 204–07 ▪ The Battle of Passchendaele 270–75 ▪ The March on Washington 311 ▪ The 1968 protests 324 ▪ The release of Nelson Mandela 325

Emily Davison, George V's horse Anmer, and jockey Herbert Jones lie on Epsom race track after Davison's actions to draw attention to the cause. The suffragette was the only one to perish.

be regarded as second-class citizens. Extension of the right to vote to increasing numbers of men in countries such as Britain and the United States had left them asking why women should not be entitled to vote. In 1903, Emmeline Pankhurst founded the WSPU with the aim of using militant tactics to further this cause. Its slogan declared "Deeds not words," and the tactics of the suffragettes, as the vote-seeking women were by now mockingly known, became increasingly violent. Chaining themselves to public buildings and the disruption of meetings escalated into the smashing of shop windows, acts of arson, and bombings.

The more active members of the WSPU were repeatedly arrested and imprisoned: Pankhurst received seven prison sentences; Davison, nine. In 1909, the WSPU began to stage hunger strikes in prison; in response they were force-fed—a painful and degrading process.

The US: the suffragists

The experience of what in the US were known as the suffragists had clear parallels. The Women's Christian Temperance Union campaigned peacefully for women's rights, arguing that women could not influence political decisions— in this case, Prohibition—without having the right to vote.

However, the National Women's Party (NWP), established in 1916, imitated the militant tactics of Britain's WSPU. This was no surprise, given that its founder, Alice Paul, had been a member of the WSPU from 1907 to 1910 and had been sent to prison three times. The NWP's so-called Silent Sentinels, protesting outside the White House from January 1917, were also arrested and force-fed.

Success at last

At the outbreak of World War I, the WSPU stopped campaigning, mobilizing itself instead in support of the war effort. The contribution made by women during the war plainly demonstrated just how much wider their role could be than that traditionally expected of them as wives and mothers. In 1918, all British women aged 30 or over were granted the right to vote. In 1928, suffrage in Britain was extended to women aged 21 or over.

Meanwhile, in the United States, the NWP continued its protest into 1919, when Congress approved the Nineteenth Amendment, which was ratified the following year, granting women the same voting rights as men. ▪

Emmeline Pankhurst

The best known of all the suffragettes, Emmeline Pankhurst (1858–1928) epitomized a new breed of politically active women in the early 20th century. She was born into—and later remained in by marriage—an eminently respectable, somewhat left-leaning middle-class world in the north of England, which only served to cement her desire to further the cause of women's rights. This decision would prove explosive. She was single-minded, exceptionally active, and wholly unflinching in her refusal to compromise. Her leadership of the WSPU exhibited a determination to take the fight for women's suffrage into the heart of what she saw as the enemy camp. Her increasing readiness to use more violent methods to secure suffragette goals alienated many who may otherwise have been her natural supporters—women as much as men. Nonetheless, her absolute refusal to back down, coupled with the fervor she inspired in her followers, introduced a new mood of feminist militancy into a complacent masculine political world.

FURTHER EVENTS

THE PARTITIONS OF POLAND
(1772–95)

From 1569 to the 18th century, Poland and Lithuania were united in a large federated commonwealth that occupied an extensive area of northern Europe. In 1772, the powerful neighbors of the commonwealth—Austria, Prussia, and Russia—encroached on its territory in a series of annexations, diminishing it until they absorbed it fully in 1795. Russia took over the eastern half of the country; Prussia, the north; and Austria, the southern and central parts. The elimination of the Polish state bolstered these three great European powers, leaving Polish patriots to struggle for independence, which they achieved in 1918.

THE SIKH EMPIRE IS FOUNDED
(1799)

Maharaja Ranjit Singh brought together a group of Sikh states in and surrounding the Punjab region of northern India to create a powerful Sikh Empire in 1799. For its creation and defense, it drew on the Khalsa, the powerful united army that had been created in the 1730s by Sikh leader Nawab Kapur Singh. The empire lasted for 50 years before falling to the British. Although not long-lasting, it helped to strengthen Sikh unity and confirmed the close identification of the Sikhs with the Punjab region.

THE WAR OF 1812
(1812–15)

In 1812, the United States declared war on Britain over a number of issues, including trade restrictions, British commandeering of US merchant seamen, and the support lent by Britain to American Indian peoples who opposed US expansion into their lands to the west. The conflict took place on several fronts across North America and involved a failed US invasion of Canada, the burning of the city of Washington by the British in 1814, and a major US victory at New Orleans in 1815. After more than two years of fighting, the status quo was largely restored. However, the war gave the US a stronger sense of nationhood and confirmed that Canada remained part of the British Empire.

THE RISE AND FALL OF THE ZULU KINGDOM
(c.1816–87)

Shaka, ruler of a small Zulu chiefdom, was a dynamic leader who formed a Zulu state after 1816 by conquering and uniting a large number of groups of the Nguni people of southeast Africa. The Zulu kingdom had to contend with two belligerent groups of incomers: the Boers (descendants of Dutch settlers at the Cape) and the British. The British invaded Zulu territory in 1879 and, after suffering an initial defeat at Isandiwana, overwhelmed them with their firepower. The British divided up their kingdom, eventually adding Zululand to their empire.

THE TRAIL OF TEARS
(1830)

In 1830, the US Congress passed the Indian Removal Act, which granted Native Americans lands west of the Mississippi in return for them surrendering their lands within existing state boundaries in the east. Although the removal was in theory voluntary, in fact it resulted in the eviction of tens of thousands of people from their homelands on a forced westward march that became known as the Trail of Tears. The people forced to move were mainly Cherokee, Chickasaw, Choctaw, Creek, and Seminole. Among the Cherokee alone, some 4,000 died during the march.

THE IRISH FAMINE
(1845–49)

In the 1840s, the rapidly growing rural population of Ireland suffered a series of disastrously poor crops of their staple food, the potato. These bad harvests, due to potato blight that spread rapidly in the damp weather, led to about 1 million people dying of starvation, while another million emigrated either to Britain or North America. After the famine ended, emigration, especially to the US, continued as landlords evicted tenants as part of estate "rationalization." The famine was a terrible and pivotal event in

Irish history: Ireland's population has never risen back to its pre-famine level and bitterness about the British government's poor response to the famine remains.

THE TAIPING REBELLION
(1850–64)

By the mid-19th century, Qing rule in China had become corrupt, and there were many who wanted change. Among various anti-government groups was one led by a religious leader, Hong Xiuquan. Hong's followers attacked Nanjing and took the city in 1853. The rebellion grew and spread across virtually all of China until it became a war with hundreds of thousands of fighters involved. With European military aid, the Qings eventually managed to crush the rebels, and millions of soldiers and civilians were killed. Despite its failure, the Taiping Rebellion fatally weakened the Qing regime, which only survived for another half century, increasingly prey to foreign powers.

THE CRIMEAN WAR
(1853–56)

When war broke out between Russia and Turkey in 1853, France and Britain intervened in support of Turkey, sending a joint force to invade the Crimea and besiege the Russian port of Sebastapol. There were huge numbers of casualties, particularly on the Russian side, before Russia agreed to peace terms. Blunders such as the British cavalry's infamous, suicidal Charge of the Light Brigade made the war notorious for the needless waste of human life. The Crimean War also came to be associated with the

efforts of health reformers such as Florence Nightingale, who worked to improve the nursing service offered to the wounded and to improve training for nurses in both military and civilian hospitals.

FRANCE RETURNS TO A REPUBLICAN GOVERNMENT
(1870)

In 1870, Emperor Napoleon III of France surrendered at the Battle of Sedan, during the Franco-Prussian War, and he was taken prisoner. The French parliament declared a republic, expecting it to form an interim government until a new monarch was chosen. However, it proved impossible to decide on a constitutional framework for a new monarchy, or on who should take the throne. After elections in 1871, the Third Republic became permanent, with a president as head of state and a Chamber of Deputies, elected through universal male suffrage, to make the laws. The Third Republic lasted until 1940 and set the pattern for French government after World War II.

A MAHDIST ISLAMIC STATE IS CREATED IN SUDAN
(1885)

In 1881, the Sudanese leader Muhammad Ahmad declared himself the Mahdi (a messianic figure in some Muslim traditions) and launched a revolt against the government of Egypt, which ruled Sudan although Britain effectively controlled both countries. Ahmad laid siege to Khartoum, which fell in early 1885, in spite of a defense by Charles George Gordon, the

British Governor-General. The Mahdists were finally defeated by Lord Kitchener in 1898, after which Sudan was ruled jointly by Britain and Egypt.

THE [SECOND] BOER WAR
(1899–1902)

The war of 1899–1902 was the second conflict between the Boers (South Africans of Dutch descent) and the British. After initial Boer victories, the British defeated their enemies by applying a "scorched earth" policy in the areas of the country in which the Boers had fought successfully as guerrillas, and capturing women and children. Some 20,000 died in concentration camps, and the Boers lost their independence. The war reduced many surviving Boers to poverty, but it also spurred on their nationalism and indirectly led to Afrikaaner dominance of South Africa's government in the 20th century.

THE MEXICAN REVOLUTION
(1910)

Beginning in 1910 and initially led by Francisco Madero, the Mexican Revolution removed the dictator Porfirio Diaz, who had ruled for some 35 years. However, the new republic could not prevent armed factional struggles and civil war, which continued until the drawing up of a new constitution in 1917 and the election of a new government in 1920. The following two decades saw key reforms such as the redistribution of lands among peasants and Indian communities and, in 1938, the nationalization of oil.

THE MOD
WORLD
1914–PRESENT

ERN

Uprisings in Russia lead to the abdication of the Tsar. **Lenin** calls for a **revolution** and the **Bolsheviks** seize power in November.

Shares on the New York Stock Exchange **crash**. Billions of dollars are lost and the **financial disaster** plunges the world into the Great Depression.

Hitler **invades Poland,** and Britain and France declare **war on Germany**; the war lasts six years and is the deadliest in world history.

British India divides into two **independent nation-states**: Hindu-majority **India** and Muslim-majority **Pakistan**.

 1917

 1929

 1939

 1947

1919

1934–35

1942

1948

World War I ends (1918), and the Treaty of Versailles is signed in June 1919: the **Germans** are **stripped of land**, their **army is reduced**, and they have to pay **reparations**.

 Fleeing from Nationalists in southern China, 80,000 **communists** led by **Mao Zedong** head for the north on the perilous **Long March**.

 The Nazis meet at Wannsee to plan the annihilation of the **Jews**. More than 6 million are **killed in the holocaust**.

 The **Jewish state of Israel** is established in Palestine, which had previously been under British rule for three decades.

Historical perspectives on events close to the present day are inevitably shifting and uncertain. A historian writing in the mid-20th century might have characterized the modern era as a period of catastrophe, in which all the economic and political gains of liberal civilization had been squandered. However, by the early 21st century it was tempting to see continuity with the pre-1914 world, as a globalized capitalist economy and great technological innovation were combined with rapidly rising population and productivity.

The two world wars
The convulsions of the period from 1914 to 1950 were on an epic scale. Two world wars between them caused the deaths of between 70 and 100 million people, making them by far the most destructive conflicts in history. Both European civilization and science—the twin pillars of the traditional 19th-century idea of "progress"—were tarnished by association with this slaughter. Germany, often considered one of Europe's most "civilized" countries, descended into dictatorship and genocidal massacre. Science was used to create weapons of mass destruction, from poison gas to the atom bomb. Even in the interlude of relative peace between the world wars, global capitalism failed to function effectively, the economic misery of the Great Depression driving a retreat from democratic government and free markets.

To revolutionaries inspired by a Marxist vision, these upheavals seemed the death throes of the capitalist order. But the building of alternative "communist" societies, based on the model of a single-party state and a state-controlled economy, proved to be a costly experiment. In Russia, followed by China, communism succeeded in transforming relatively undeveloped countries into major industrial and military powers, but millions died as victims of the state, and citizens were denied fundamental freedoms.

A battle of ideology
World War II was followed by the Cold War confrontation between the "free world," led by the United States, and the communist bloc. Instead of disarmament, there was a potentially disastrous nuclear arms race. Meanwhile, the main European powers, economically weakened and demoralized, found themselves in no position to sustain

Egyptian leader Nasser declares the nationalization of the **Suez Canal**. Britain, France, and Israel **invade Egypt**, the US imposes a ceasefire, and the allies **withdraw**.

For 13 days the world is under the threat of **nuclear war** between Cuba and the US, during the Cuban Missile Crisis. The dispute is resolved by **diplomacy**.

The East German government lifts travel restrictions and thousands of people **tear down the Berlin Wall**; communism collapses.

On September 11, **Islamic extremists** launch a major **terrorist attack** on the US. Almost 3,000 people are killed.

1956   **1962**   **1989**   **2001**

1957   **1965**   **1991**   **2011**

Kwame Nkrumah wins **Ghanaian independence** from Britain through **peaceful** means. By the 1970s, most countries in Africa are independent.

The US sends troops to **South Vietnam** to prevent the spread of **communism** and is embroiled in the war for nine years.

The **first website** ("World Wide Web") goes live, built by British computer scientist Tim Berners-Lee to enable academics to **share information**.

The world's population exceeds **7 billion**; the global challenge is to improve **living standards** without destroying the **environment**.

---

their empires against colonized populations eager for freedom. The newly independent nations became an ideological and even, at times, military battleground between the capitalist and communist systems.

In the end, the issue was settled by economics. Capitalism showed its ability to generate economic growth on a vast scale, creating a booming consumer society in more advanced countries. In contrast, by the 1980s communist countries confronted economic stagnation and rising popular discontent. With great rapidity, communist regimes collapsed in the Soviet bloc, while communist China later became a powerhouse of capitalism.

In the wake of communism, the political scientist Francis Fukuyama coined the expression the "end of history" and argued that Western

liberal democracy was "the only game in town." Certainly, by the end of the 20th century liberalism was surfing a wave: in 1950, only a few nations in Europe were democracies; 50 years later, all of them were.

Progress and pessimism
From the 1960s, hotly contested campaigns for civil rights had progressed liberal ideals in areas such as racial equality and gender politics. Growing prosperity was also impressive. In Latin America and much of Asia, living standards had risen dramatically by the early 21st century. Despite the world's population increasing on a huge scale—from under 2 billion in 1914 to over 7 billion one century later—food supplies had not run out, as had once been predicted by many. Restricting environmental damage

was recognized as a major challenge for the future, a problem generated by humanity's growth and success.

Indeed, human progress in the 20th century was remarkable, from rising literacy and life expectancy to the development of air and space travel and computers. Yet there was no outbreak of general optimism. Environmental issues aside, it was all too evident that the future held potential dangers: the unsettled politics of the Middle East, sucking major powers into wars; brutal acts of terrorism; economic inequality generating mass migration; financial instability and market breakdown; epidemics spread by global travel— all provided plenty of material for pessimists. History offered no solid ground for predictions, suggesting only that the unexpected was to be expected. ■

YOU OFTEN WISH YOU WERE DEAD

THE BATTLE OF PASSCHENDAELE (1917)

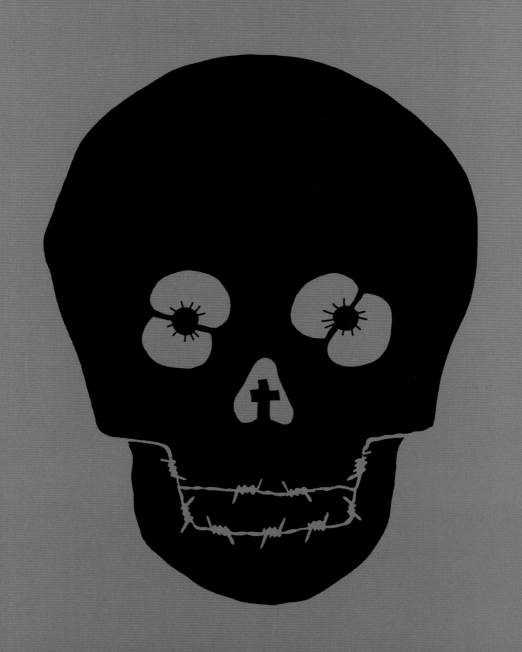

Passchendaele, officially known as the Third Battle of Ypres, was a large-scale attack against the German front line around Ypres, Belgium, during World War I. The Allies' aim was to advance into Belgium and free the German-held ports on the Belgian coast, which the Germans had been using to attack British shipping. The biggest challenge was to break through the defensive positions taken by the Germans on the West Flanders Ridge. Key to the breakthrough was seizing the village of Passchendaele.

Preparations for the battle began on June 7, 1917 with a heavy two-week bombardment of German positions. The infantry offensive began on July 31, 1917. Within days, the Allied forces were stuck in mud as torrential rain turned the area into a quagmire. By the time the Allies—made up of British, French, Canadian, and Australian troops—captured Passchendaele on November 6, the village was in ruins. The conflict cost 300,000 Allied lives, with a gain of 5 miles (8km), while the Germans lost 260,000 of their troops. It was

Soldiers at Passchendaele fought in appalling conditions. In the absence of anything better, these machine-gunners are using bomb indentations as makeshift shelter.

hailed as a victory by the British government but became a byword for the utter futility of war.

Secret diplomacy

Two main disputes led to World War I: one between Germany and France, and the other between Russia and Austria-Hungary. The long history of mutual antipathy between Germany and France came to a head in 1870 with France's humiliating defeat by Germany in the Franco-Prussian

Life in the trenches

At the outbreak of World War I, both sides anticipated fast-moving battles that would cover hundreds of miles. None expected a static fight with their forces deep in defensive trenches.

Early trenches were small furrows, but they grew more elaborate, fortified with wooden frames and sandbags. German trenches were more sophisticated and had electricity and toilets. Soldiers spent daylight hours avoiding enemy fire and endured periods of boredom and daily chores, broken up with spells in reserve and short rest periods.

The trenches sometimes filled up with rats and lice, as well as water, which turned to ice. Life in such conditions was exhausting, and soldiers had a repetitive diet of canned food and few comforts.

Snipers shot at any heads that appeared over the parapet, and raiding parties throwing grenades were a constant danger. The trenches were bombarded with shells, bullets, and poisonous gas. It was a relentless war of attrition—smelly, dirty, and riddled with disease.

See also: The Expedition of the Thousand 238–41 ▪ Russia emancipates the serfs 243 ▪ The October Revolution 276–79 ▪ The Treaty of Versailles 280 ▪ Nazi invasion of Poland 286–93

> The lamps are going out all over Europe. We shall not see them lit again in our lifetime.
> **Sir Edward Grey**
> **British Foreign Secretary (1914)**

war, which led to the annexation of most of the French provinces of Alsace and Lorraine.

In Eastern Europe, the Austro-Hungarian and Russian empires had a long-standing dispute over which of them had the strongest claim to power in the Balkans. Both depended on the area for access to the Mediterranean, and each eyed the movements of the other with intense suspicion.

Each state needed allies, and in 1882 Austria-Hungary, Germany, and Italy signed a Triple Alliance promising to give each other military support in case of war. Then, in the 1890s, Russia and France signed an agreement to protect one another in the event of a war against Germany. By the turn of the century, Kaiser Wilhelm II's provocative nationalistic speeches and naval expansion pushed Britain into closer ties with France. In 1904, Britain and France agreed an *entente cordiale*, or friendly alliance, which was broadened into a triple entente, embracing Russia, in 1907. The triple entente would become known as the Allied Powers.

The atmosphere generated by this international jostling led to an increase in military spending by European governments and the expansion of armies and navies.

War erupts

A spark was all that was needed to ignite the flame of enmity between these two alliances. It came on June 28, 1914, when a Bosnian Serb assassinated Archduke Franz Ferdinand, heir to the Habsburg throne, in Sarajevo. The Austrians suspected Serbia, their principal enemy in the Balkans, of the attack. After securing support from its ally Germany, Austria-Hungary presented Serbia with an ultimatum on July 23, demanding that the Serbs stop all anti-Austria-Hungary activities. Serbia accepted most of the demands, but Austria-Hungary declared war on Serbia on July 28. Britain called for international mediation, but the crisis quickly escalated into European war. When Russia mobilized against Austria-Hungary, Germany declared war on Russia on August 1, and on France two days later. Britain joined the war on August 4, after the Germans invaded neutral Belgium. The British Expeditionary Force (BEF), »

European powers are bound together in a **complex web of alliances**.

A European arms race leads to **larger armies** and more **destructive weapons**.

War breaks out, eventually drawing in all the major powers and causing **death on a scale previously unimaginable**.

The **relative equality of the armies** means that neither side can score a decisive victory.

The fighting on the Western Front becomes a bitter stalemate despite the enormous cost of battles like Passchendaele.

With both sides exhausted, **US entry into the war** on the side of the Allies facilitates a **breakthrough in the conflict**.

Huge artillery guns such as the Howitzer cannon were transported by horses and tractors. High-explosive shells fired in massive quantities were key to the war's high casualty rates.

a small professional troop led by Sir Douglas Haig, had arrived in France by August 22. It was deployed near the Franco-Belgian border, in line with pre-war military plans agreed with the French government.

Germany had to fight a war on two fronts. On the Western Front, in the first weeks of the conflict, the Germans invaded Belgium and France, but their advance was halted by the French and British at the Battle of the Marne. By the end of autumn, the two sides had reached a stalemate. Meanwhile, on the Eastern Front, the fighting remained fluid. Germany dominated, scoring a great victory against the Russians at Tannenberg, but its Austrian allies suffered several defeats. On the Western Front, however, a 400-mile (645km) trench line stretched from the Belgian coast in the north, down through eastern France to the Swiss border. The two sides faced each other across the open space between their front lines. This area—no-man's-land—had barbed wire fronting the trenches to slow the opposition. Continuous fighting from the

trenches, punctuated by appallingly bloody battles, failed to break the deadlock. More than 600,000 Allied troops were killed or wounded in the Battle of the Somme alone.

Total war

At the start of the conflict, both sides had been convinced it would be a short, decisive battle. No one

had anticipated a war of attrition. New mechanized weapons added to the high casualty rates. Tanks were used for the first time, and machine guns such as the German MG 08 Maxim could fire up to 600 bullets a minute. Aircrafts, first employed for reconnaissance, were later used for bombing. Both sides used poison gas. Horses were the backbone of logistical operations, but as the war progressed, railways and motor trucks were used to transport goods to the front.

Civilians were brought into the front line by the bombing of London and Paris by airships and bomber aircraft. By 1917, German submarines were sinking one in four merchant ships headed for Britain to try to starve the British into submission. Britain's naval blockade of Germany also led to

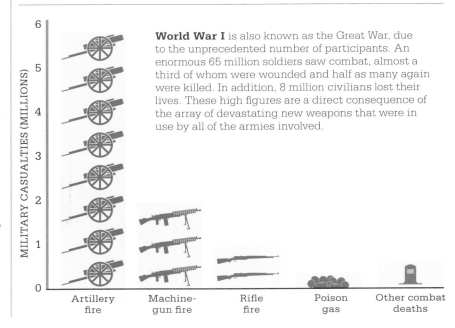

World War I is also known as the Great War, due to the unprecedented number of participants. An enormous 65 million soldiers saw combat, almost a third of whom were wounded and half as many again were killed. In addition, 8 million civilians lost their lives. These high figures are a direct consequence of the array of devastating new weapons that were in use by all of the armies involved.

MILITARY CASUALTIES (MILLIONS)

| Artillery fire | Machine-gun fire | Rifle fire | Poison gas | Other combat deaths |

> There was not a sign of life of any sort... Not a bird, not even a rat or a blade of grass.
> **Private R. A. Colwell**
> **Passchendaele (1919)**

acute food shortages. This was the first "total war," meaning not just soldiers but also civilian populations were involved.

Britain was forced to introduce conscription for the first time in its history. From January 1916, all single men aged 18 to 41 could be called up. Britain and France also assembled armies from their overseas colonies, such as India and Africa, and from the British dominions of Australia New Zealand and Canada. War brought many social changes, notably for women, who filled positions in factories and offices. Women were also increasingly employed in the munitions industry as governments switched to large-scale production.

Global conflict

The key belligerent states brought their vast empires into war with them, and the conflict soon became a world war. German colonies in China and the Pacific were invaded by Japan, which entered the war on the side of the Allies. Germany's

One of the social changes brought on by World War I concerned the role of women. The female population joined the war effort by working in places like munitions factories.

colonies in Africa were overrun by British, French, and South African troops. In May 1915, Italy joined the Allies, fighting Austria-Hungary and Germany in the Alps.

In early November 1914, the Ottoman Empire, an Islamic power, abandoned its neutrality and declared a military *jihad* (holy war) against France, Russia, and Britain. The US was drawn into the war by German submarine attacks on commercial ships at sea, such as the one on British liner *Lusitania* in 1915, with 128 Americans on board. After a German plot to persuade Mexico into an anti-US alliance was discovered, Congress declared war in April 1917.

When the Bolsheviks in Russia negotiated a peace treaty with Germany at Brest-Litovsk on December 22, 1917, it seemed Germany had won a significant victory. They also made gains on the Western Front in 1918, but then, in July and August, the Allies counterattacked, beginning an advance that would continue until November. Four million fresh

US troops helped defeat the Central Powers and bring the Germans to the peace table.

When the conflict ended, at 11am on November 11, 1918, the alliance led by France and Britain emerged victorious. More than 65 million troops had been involved in the war, of which at least half were killed or injured. The Russian, Austrian, and German empires had collapsed. After the war, the Treaty of Versailles redrew the map of Europe, leaving nations, particularly Germany, embittered. A public assembly of countries, the League of Nations, was founded to help maintain peace. However, the League proved toothless in the face of countries that chose to ignore it. When fascist Benito Mussolini came to power in Italy in 1922, he denounced the Treaty. In Germany, where the response to the Treaty was one of deep resentment, the Nazi Party began to gain momentum. Far from being "the war to end all wars," World War I had instead sown the seeds of future conflict. ■

HISTORY WILL NOT FORGIVE US IF WE DO NOT ASSUME POWER NOW

THE OCTOBER REVOLUTION (1917)

IN CONTEXT

FOCUS
Russian Revolution

BEFORE
1898 The Russian Socialist Democratic Labour Party is formed.

1905 Russia suffers a crushing defeat in a war against Japan, which leads to an uprising.

1914 Russia enters World War I and quickly suffers heavy losses in defeats to Germany on the Eastern Front.

AFTER
1918 Tsar Nicholas II and his family are executed.

1922 Lenin creates the Union of Soviet Socialist Republics (USSR) under the control of the Communist Party.

1929 Stalin becomes leader of the USSR and establishes a dictatorship.

In October 1917, Russia was in turmoil after suffering huge losses in World War I. There were food shortages, and workers in the cities faced low wages and appalling conditions. The February Revolution had ousted the Tsar, but the Provisional Government that replaced him faced imminent collapse.

Vladimir Lenin, a member of the revolutionary Bolshevik Party, took full advantage. He was committed to a workers' (proletarian) revolution and set out a series of proposals to overthrow the government in what became known as his April Theses. His simple slogan "Peace, land, and bread!" became a revolutionary

See also: The founding of St. Petersburg 196–97 ▪ Russia emancipates the serfs 243 ▪ Stalin assumes power 281 ▪ The Siege of Sarajevo 326 ▪ The Spanish Civil War 340

rallying cry. On October 24 (November 6, Gregorian calendar (GC)), there were attempts by the government to curb the activities of the Bolsheviks to prevent a coup. Orders were issued for the arrest of leading party members, and their newspaper, *Pravda* (The Truth), was closed down. Lenin, keeping a low profile in his apartment, urged action. "We must not wait! We may lose everything! The government is tottering. To delay action is the same as death," he wrote.

On October 25 (November 7, GC), the government tried without success to find armed support. The Petrograd Soviet of Workers and Soldiers party, of which the Bolsheviks were a faction, could rely on the support of troops in Petrograd (later St. Petersburg). The Bolshevik paramilitary Red Guards occupied the main telegraph office, post office, and power stations. Only the Winter Palace, the seat of the government, remained. The small unit of military cadets guarding it willingly surrendered to the revolutionary soldiers. The regime was overthrown, and power had passed to Lenin and his Bolsheviks.

Laying the groundwork

The October Revolution was the culmination of the civil unrest that had rumbled on for months. On February 23, 1917 (March 8, GC), in Petrograd, a riot had started, led by women frustrated at waiting hours for bread. They marched through the city, gathering support as they went. This grew into a general strike, and the demonstrations took on a more political nature. Red flags began to appear, and statues of »

Vladimir Ilyich Lenin

Born Vladimir Ilyich Ulyanov on April 10, 1870 (22 April, NS), the founder of the Bolsheviks and first leader of Soviet Russia was a bold theorist and tireless organizer. Lenin became an active Marxist revolutionary after his brother Alexander was executed in 1887 for conspiring to assassinate Tsar Alexander III, an event that caused Lenin to lose faith in God and religion. In 1895 he was arrested and exiled for three years to Siberia.

Lenin's chief aim was to organize the opposition to the Tsar into a single coherent movement. Following the Russian Revolution of March 1917, he returned to Russia believing his moment had come. In October, Lenin led the Bolsheviks against the government then, suppressing all opposition, became dictator of the world's first communist state.

Lenin's main challenge was civil war (1918–20). The Communists won, but Russia was brought to its knees. The strain of leadership also broke his health. After two strokes, one of which deprived him of speech, he died on January 21, 1924.

The Russian Revolution of 1905 forces a **range of reforms** from the autocratic Tsar Nicholas II.

↓

Dissatisfaction persists among the people.	**Russia suffers defeats** in World War I.	Economic hardship leads to **food riots**.

↓

In February 1917, the **monarchy is overthrown** and replaced by the Provisional Government. **The Tsar abdicates** in March.

↓

Lenin and the Bolsheviks demand total power for the proletariat, launching the October Revolution.

Tsar Nicholas II were toppled. Soldiers refused to obey orders to fire on the crowd, but police shot and killed 50 people.

Rise of revolutionary parties

With violence erupting on city streets, the Tsar abdicated in March, having relinquished power to the Provisional Government in February, with Prince Georgi Y. Lvov as its head. The government still represented only the middle classes and continued to back Russia's involvement in World War I. Groups such as the Petrograd Soviet of Workers and Soldiers, a council made up of workers and peasants agitating for change, grew stronger and gained power within the Provisional Government. Lenin, in exile for revolutionary activities, was anxious to return to Petrograd, convinced that the collapse of world capitalism was imminent. He received the help of the German government, which hoped that he could further destabilize the political situation in Russia for their war effort, and arrived secretly in a sealed train. Full of revolutionary zeal, he was determined to shape a new Russian government according to his ideas, and he accused his associates of not doing enough to overthrow the current regime.

Prime Minister Lvov resigned after the disastrous July Offensive on the Eastern Front. His successor, Alexander Kerensky, formed a new socialist government with the Petrograd Soviet, but he, too, insisted on keeping Russia in the war. After mass demonstrations in Petrograd encouraged by the Bolsheviks, Kerensky banned them and arrested many of their leaders. Lenin fled to Finland.

Revolution is nigh

In August, Kerensky faced a new threat. General Lavr Kornilov, Russia's army commander-in-chief, ordered troops into Petrograd. Kerensky believed that Kornilov was plotting to seize power. In desperation, he released the Bolsheviks, who armed those who wanted to prevent a counter-revolution. This was a massive boost for their cause. They were able to represent themselves to the people as defenders of Petrograd. By September the Bolsheviks had taken control of the Petrograd Soviet. Lenin seized the moment, returned to Russia, and renewed calls for revolution. He handed responsibility for military tactics to Leon Trotsky, a fellow Marxist. Peasants and farmers were revolting in rural areas, workers in the cities. Lenin decided the time was ripe for a Bolshevik seizure of power. The Bolsheviks took government buildings and the Winter Palace, where Kerensky's cabinet had sought refuge.

On the night of October 25 (November 7, GC), Lenin issued a brief address to the Russian people: "The Provisional Government has been overthrown. Long live the workers, soldiers, and peasant revolution!" After this initial triumph, Lenin was compelled to hold democratic elections, but the Bolsheviks received only a quarter of the vote. Lenin dissolved the elected government and sent armed guards to prevent it meeting again. In February 1918, he signed a peace treaty with Germany, but the terms were extremely harsh. Russia ceded the Baltic States to Germany, while Ukraine, Finland, and Estonia were transformed into independent states. Russia was also forced to pay six billion German marks in reparations. This move freed the Bolsheviks from the German threat, but the terms of the treaty were deeply unpopular. Many regarded it as a betrayal of their country.

Civil war

The Bolsheviks had gained power, but now they had to keep it. Lenin established a highly centralized government system, banned all

This painting of the storming of the Winter Palace portrays the dramatic moment in the October Revolution when the Bolsheviks seized the government building.

Vladimir Lenin addresses his troops in Moscow's Red Square in 1919, during the civil war that followed the October Revolution.

opposition, and started the Red Terror, a campaign of intimidation, executions, and arrests against anybody perceived to be a threat to the Bolsheviks.

The Bolsheviks were a minority in Russia, and their opponents marshalled their forces against them, primarily the Whites, made up of former tsarists, army officers, and democrats. The Bolsheviks were known as the Reds.

As various factions fought over the future of the country, a civil war characterized by extreme violence erupted in Russia and ran from 1918 to 1921. The Whites received help from Russia's former allies—Britain, France, the US, and Japan—which feared the spread of communism. At first, they made significant gains. However, they were badly coordinated, and Trotsky proved to be a brilliant military tactician.

In 1920, Lenin ordered a war against Poland to liberate the workers of eastern and central Europe, but at the Battle of Warsaw, after a magnificent counterattack, the Red Army was driven back.

A country in ruins

By 1921, the Whites had been defeated, and Lenin could finally turn his attention to rebuilding the Russian economy.

He faced a country on the verge of collapse. In the countryside, around 6 million peasants had died of starvation, and there was rioting in the cities. The Kronstadt naval rebellion in March 1921 further undermined the regime. Kronstadt was a naval town on an island off the coast of Petrograd. In 1921, 16,000 soldiers and workers signed a petition calling for "Soviets without Bolsheviks": freely elected Soviets and freedoms of speech and press. The Reds reacted ruthlessly, executing several hundred ringleaders and expelling over 15,000 sailors from the fleet.

In May 1922, Lenin suffered a stroke. In December, the Soviet government declared the establishment of the Union of Soviet Socialist Republics (USSR), a federal union consisting of Soviet Russia and neighboring areas that were ruled by branches

of the communist movement. From its inception, the USSR was based on a premise of one-party rule, prohibiting all other political organizations.

Lenin was disheartened by political infighting and worried about how the USSR would be run after his death. In late 1922 and early 1923, he dictated what became known as his "testament," in which he expressed regret at the direction the Soviet government had taken. He was especially critical of Joseph Stalin, then general secretary of the Communist Party. Stalin's aggressive behavior had brought him into conflict with Lenin.

Lenin died in 1924, but his legacy lives on. The Bolshevik Party's establishment of the world's first socialist state in the largest nation affected every country in the world. The victorious socialist revolution inspired workers with an alternative to capitalism and old imperialist regimes. ∎

The execution of the Tsar and his family was needed not only to… instil a sense of hopelessness in the enemy, but also to show that ahead lay total victory or total doom.
Leon Trotsky

THIS IS NOT PEACE. THIS IS AN ARMISTICE FOR 20 YEARS
THE TREATY OF VERSAILLES (1919)

IN CONTEXT

FOCUS
Peace after World War I

BEFORE
1914 The four empires of Austria-Hungary, Germany, Ottoman Turkey, and Tsarist Russia rule over vast lands.

1916 British and French diplomats meet secretly to determine the fate of the post-Ottoman Arab world.

1919 The Paris Peace Conference sets out the terms and conditions for post-war peace.

AFTER
1920 The Treaty of Sèvres carves up the Ottoman Empire to remake the Middle East.

September 3, 1939 World War II begins with the German attack on Poland.

October 24, 1945 The League of Nations, having disbanded, reforms after World War II as the United Nations.

After four years of global conflict, 16 million people had died and centuries-old empires and dynasties had collapsed. In January 1919, the victors of World War I met to discuss the terms of peace. US President Woodrow Wilson had devised a plan that he believed would bring a new order to Europe based on democracy. Wilson pushed for a League of Nations to act as arbiter and peacemaker in national disputes.

Britain and France wanted to ensure that Germany would never again be able to threaten European peace. The German Army was to be reduced and the Rhineland demilitarized. Germany was also asked to give up lands to France on its west and to Poland on its east and north. In addition, the Austro-Hungarian empires were to be split into new nations such as Czechoslovakia and Yugoslavia; and the Ottoman Empire was also to be carved up, to the advantage of the British and French.

You have asked for peace. We are ready to give you peace.
Georges Clemenceau
Prime Minister of France

War-guilt clause
Crucially, in a "war-guilt clause," the Germans had to admit to starting the war and also pay £6.6 billion in reparations. They signed the Treaty of Versailles on June 28, 1919 but stalled in paying compensation, so in 1923 France occupied Germany's industrial Ruhr Valley. However, in the interwar years, neither of those nations did anything to deter aggression by Nazi Germany. When Adolf Hitler took France in 1940, he ordered the master copy of the treaty to be burned. ∎

See also: The Young Turk Revolution 260–61 ▪ The Battle of Passchendaele 270–75 ▪ The Reichstag Fire 284–85 ▪ Nazi invasion of Poland 286–93 ▪ The founding of the United Nations 340

DEATH IS THE SOLUTION TO ALL PROBLEMS. NO MAN— NO PROBLEM
STALIN ASSUMES POWER (1929)

IN CONTEXT

FOCUS
Soviet Russia

BEFORE
1917 Lenin begins Russia's move toward communism.

1922 The Union Treaty joins Russia, Ukraine, Belarus, and the Transcaucasus into the Soviet Union.

1928 The first Five-Year Plan is adopted, with the state setting ambitious goals for the whole of the economy.

AFTER
1945 The Soviet Union defeats Nazi Germany and controls central Europe.

1989 East and central Europe reject communism, and the Berlin Wall comes down.

1991 The Congress of People's Deputies votes for the dissolution of the Soviet Union.

After the October Revolution in 1917, Russian leader Vladimir Lenin created a single-party state and appointed Joseph Stalin as general secretary. Stalin then used his position to launch his bid for supreme leadership, becoming dictator in 1929, five years after Lenin's death.

Stalin pushed the country into a period of rapid industrialization. He confiscated land belonging to rural farmers to turn it into large farms to be run collectively to make food for the new workforce. In 1931–32, he requisitioned grain from the peasants, which led to a severe famine in Ukraine, killing millions.

The People's Commissariat for Internal Affairs (secret police) was tasked with hunting out Stalin's political opponents. Thousands of Soviet citizens died in the 1930s' "blood purges," known as the Great Terror, and millions of non-Russians were deported to labor camps. Despite this, Stalin portrayed his country as a land of peace and progress, and himself as a man working for the benefit of the people.

I believe in one thing only, the power of the human will.
Joseph Stalin

The dictator looked for chances to expand communism beyond Soviet frontiers, and after World War II, it spread to Poland, Hungary, Czechoslovakia, East Germany, and others, becoming known as the Eastern Soviet Bloc. Communist parties came to rule in North Korea in 1948, China in 1949, Cuba in 1959, and Vietnam in 1975.

Stalin had risen to become one of the most powerful men in the world. Soon after his death in 1953, his nation was a superpower to challenge the United States. ∎

See also: The October Revolution 276–79 ▪ Nazi invasion of Poland 286–93 ▪ The Berlin Airlift 296–97 ▪ The fall of the Berlin Wall 322–23

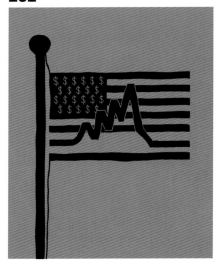

ANY LACK OF CONFIDENCE IN THE ECONOMIC FUTURE OF THE UNITED STATES IS FOOLISH
THE WALL STREET CRASH (1929)

IN CONTEXT

FOCUS
The Great Depression

BEFORE
1918 The global economy struggles to recover its stability after the disruption of World War I.

1922 The US economy starts to grow rapidly as factories mass produce goods.

1923 Prices in Germany spiral out of control in hyperinflation that destroys people's savings.

AFTER
1930 Mass unemployment hits the US, Britain, Germany, and other countries.

1939 The advent of World War II sees an increase in employment and government spending, speeding recovery.

1944 World leaders agree to set up the International Monetary Fund (IMF) and World Bank to finance economic development.

O ver six desperate days in October 1929, shares on the New York Stock Exchange crashed. The downturn began on October 23rd, when stocks from car manufacturer General Motors were sold at a loss and the market started to collapse. Panic set in, and the next day the market nose-dived.

On Tuesday October 29, which became known as Black Tuesday, stock prices plunged even lower. In total, $25 billion, approximately $319 billion in today's market, was lost. It was the biggest financial catastrophe ever, and it plunged the world into the Great Depression.

Roaring Twenties
The US had recovered quickly after World War I, and factories that had made supplies for the war effort switched to producing consumer goods such as cars and radios. The growth of new technologies and mass production saw economic output increase by 50 percent; the age of prosperity and consumerism that resulted became known as the Roaring Twenties.

Newspapers and magazines were filled with stories of people becoming rich overnight by dabbling in the stock market, and thousands of ordinary Americans bought shares, increasing the demand for them and inflating their value. Between 1920 and 1929, the number of shareowners rose from 4 million to 20 million.

By late 1929, there were signs of trouble within the American economy: unemployment was on the rise, steel production was declining, construction was slowing, and car sales had dipped. Still confident they could make a fortune, some people continued to invest on the stock market.

Speculators crowd around the front entrance of the New York Stock Exchange, deeply concerned about their financial investments in the days following the Wall Street Crash.

See also: The California Gold Rush 248–49 ▪ The Treaty of Versailles 280 ▪
The Reichstag Fire 284–85 ▪ The global financial crisis 330–33

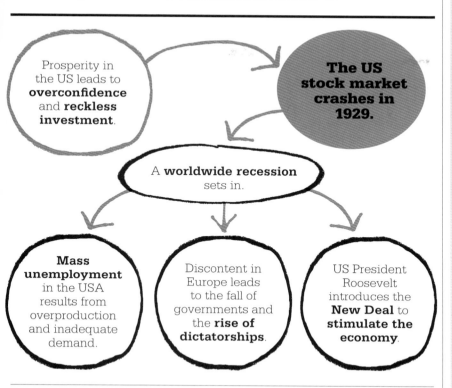

Prosperity in the US leads to **overconfidence** and **reckless investment**.

The US stock market crashes in 1929.

A **worldwide recession** sets in.

Mass unemployment in the USA results from overproduction and inadequate demand.

Discontent in Europe leads to the fall of governments and the **rise of dictatorships**.

US President Roosevelt introduces the **New Deal** to **stimulate the economy**.

Franklin D. Roosevelt

Franklin Delano Roosevelt (1882–1945) was the only president ever elected to serve four terms in office. He achieved this success despite suffering from polio in 1921, which crippled both legs and almost led him to give up on his political career.

Roosevelt won the election to be the governor of New York in 1929, and in 1932 was nominated as Democratic candidate for the presidency. Pledging a New Deal for the American people, he won a landslide victory. In his first 100 days, he introduced a program of social and economic reform to combat the Great Depression. These immensely popular measures won him a second landslide victory in 1936.

In 1941, the US was propelled into World War II, and Roosevelt took his place as one of the allied war leaders. He was one of the principal movers in the plans for a United Nations but died in March 1945, just before the UN's first meeting was convened in San Francisco.

However, when the stock prices began to drop in October 1929, panic set in. The ensuing crash triggered a worldwide recession known as the Great Depression.

The Great Depression

In the US, factories were closed and workers sacked. In the spring of 1933, the agricultural sector was on the verge of disaster: 25 percent of farmers were without work, and many even lost their farms. Unemployment went from 1.5 million in 1929 to 12.8 million, or 24.75 percent of the workforce, by 1933, a pattern seen around the world.

Unemployment in Britain rose to 2.5 million, 25 percent of the workforce, with heavy industry, such as shipbuilding, particularly badly hit. Germany suffered greatly, since its post-war economy was supported by huge American loans which it was unable to pay back.

A New Deal

The crash helped bring Democrat Franklin D. Roosevelt into the White House in 1932. His policy, the New Deal, introduced a program of social welfare for the poor and government expenditure on huge public projects that created new jobs.

The Great Depression marked the end of the United States' post-war boom. In Europe many turned to right-wing parties, such as Adolf Hitler's National Socialist Party in Germany, with its promise to restore the economy. In many countries, recovery came only with the increase in employment brought about by World War II. ▪

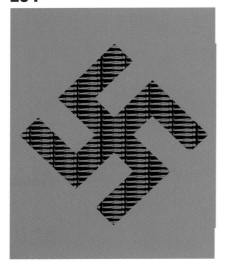

THE TRUTH IS THAT MEN ARE TIRED OF LIBERTY
THE REICHSTAG FIRE (1933)

IN CONTEXT

FOCUS
Rise of fascism

BEFORE
1918 World War I leaves Europe politically and economically unstable.

1920 The National Socialist (or Nazi) Party is founded in Germany, with racism a central tenet.

1922 Benito Mussolini is made Italian premier by King Victor Emmanuel III.

AFTER
1935 Mussolini invades Abyssinia (Ethiopia) as part of his ambitious foreign policy.

1936–39 The Spanish Civil War is fought.

1938 Adolf Hitler invades Austria. The Munich Pact grants Hitler control of the Sudetenland.

1939 Hitler orders the invasion of Poland, which triggers World War II.

Slowing European economies make everyday life harder.

German resentment festers over the terms of the **Treaty of Versailles**.

Extremist fascist and communist ideologies seem to offer easy solutions to **national problems**.

The Reichstag Fire is blamed on the communists and is used as a pretext to curb civil liberties and jail dissenters.

The **disintegration of formal structures** of government clears the way for **Adolf Hitler** to become dictator.

When fire broke out at the Reichstag, the German parliament building, just after 9pm on February 27, 1933, Chancellor Adolf Hitler claimed it was a communist plot to bring down the government—a cynical ploy that gave Hitler an excuse to decimate his communist rivals.

The timing was perfect: elections were due to take place in March 1933. While Hitler's National Socialist, or "Nazi," Party was the largest party in parliament, he lacked a working majority because the two next-largest parties (the Social Democrats and Communists) were both on the left, and he feared

See also: The Expedition of the Thousand 238–41 ▪ The Battle of Passchendaele 270–75 ▪ The Treaty of Versailles 280 ▪
The Wall Street Crash 282–83 ▪ Nazi invasion of Poland 286–93 ▪ The Wannsee Conference 294–95

> Ours is a fight to the
> finish until communism
> has been absolutely
> uprooted in Germany.
> **Hermann Göring**
> **Leading Nazi Party member**

his party would not fare well in the elections. Hitler rushed to blame the fire on a lone Dutch communist, which prompted suspicions that the Nazis were behind the arson, given that they had so much to gain from discrediting the communists.

The next day, the Reichstag Fire Decree banned the Communist Party. Hitler's response fed on fears of a communist takeover, and many Germans believed that Hitler's decisive action had saved the nation. By April, under pressure from the Nazis, the Enabling Act was passed by the Reichstag. This granted Hitler the right to make his own laws without involving the Reichstag, and it solidified his place as a fascist dictator with complete control over Germany.

Dictators seize power

Fascism emerged across Europe in the 1920s and 30s. As governments struggled with post-war economic hardship and the fear of communist revolutions, extreme right-wing movements—fascism in Italy and Nazism in Germany—were set up as defenders against communism.

They used paramilitary groups to intimidate opponents, and spread propaganda to gain popularity. In Italy, Benito Mussolini was seen as the only man able to restore order. Once appointed prime minister in 1922, Mussolini gradually assumed dictatorial powers, becoming *Il Duce*, the leader. By 1928, Italy was a totalitarian state.

In Germany, Hitler worked ceaselessly to transform the Nazis into a major political force. Relying on a mix of nationalist rhetoric, anti-communism, vicious anti-Semitism, and an unceasing call to reverse the peace terms made at Versailles in 1919, Hitler rode a wave of popularity. In 1933 he became Chancellor, then, shortly after, dictator, calling himself *Führer*.

Fascists united

In 1936 Hitler and Mussolini began to send military support to aid General Franco in the Spanish Civil War, which pitted right-wing nationalists against left-wing republicans. Franco's victory against the left-wing Popular

The Reichstag Fire is said to have burned so fiercely that flames could be seen for miles around. Hitler blamed the communists in an attempt to build support for his Nazi Party.

Front government emboldened the dictators and emphasized the weakness of Western democracies.

The Reichstag Fire was a key moment in Nazi history. It led to the absolute dictatorship of Adolf Hitler and the growth of fascism, setting Europe on the path to world war. ▪

Fascism across Europe

European fascism blossomed in a climate of economic disarray in the 1920s and 30s. Democracies lost legitimacy with their people, and fascist parties, offering a form of extreme right-wing nationalism, boasted that they would provide strength where weakness had prevailed.

In the 1930s, no European country, with the exception of the Soviet Union, was without a form of fascist party. Britain had Sir Oswald Mosley's British Union of Fascists (BUF). Ireland had the Blueshirts; France, Le Faisceau; and Denmark and Norway had many far-right parties. Engelbert Dollfuss's Fatherland Front was installed in Austria in 1934, while Greece was under the rule of General Ioannis Metaxas between 1936 and 1941. Portugal and Bulgaria also came under right-wing dictatorships, as did Romania.

By the end of the 1930s, authoritarian governments had assumed power in virtually every corner of central and Eastern Europe, and democracy was in decline.

IN STARTING AND WAGING A WAR, IT IS NOT RIGHT THAT MATTERS BUT VICTORY

NAZI INVASION OF POLAND (1939–1945)

IN CONTEXT

FOCUS
World War II

BEFORE
1919 The Treaty of Versailles at the end of World War I humiliates Germany and sows the seeds for future conflict.

1922 The Union of Soviet Socialist Republic (USSR) is founded.

1933 The Enabling Act gives Adolf Hitler dictatorial power in Germany.

AFTER
1942–43 The Soviets defeat the Germans at Stalingrad.

1944 The June 6 D-Day Landings, which were the largest amphibious military operation in history, begin the liberation of Western Europe.

1945 As Russian troops win the Battle of Berlin, Hitler commits suicide. The Germans surrender unconditionally.

In August 1939, Nazi Germany and the Soviet Union signed a non-aggression pact, also secretly agreeing to invade and then divide Poland between them. Russian leader Joseph Stalin had decided that in the event of war, Germany offered the best hope of Soviet security. One week later, on September 1, 1939, more than a million German troops invaded Poland from the west. Soon after, on September 17, Russian troops attacked Poland from the east. The context for this unprovoked assault, as declared by the German *Führer*, Adolf Hitler, was the pursuit of *Lebensraum*, "living space" deemed necessary for the expansion of the German people, whom Hitler saw as a superior "Aryan master race," with the right to displace inferior races.

The invasion lasted just over a month. Trapped between two huge, well-armed powers, the Polish air force and army fought valiantly, but they lacked modern aircrafts and tanks. The German Luftwaffe was quickly able to gain command of the skies. In the end, Poland's aviators and soldiers, fighting on two fronts, were overwhelmed.

The invasion ended in a resounding victory, and it added to Hitler's increasing belief that he was a military genius. Some areas in western Poland were absorbed into Germany, while territory east of the River Bug was annexed by the Soviet Union.

The Nazi regime in Poland

The Nazis imposed a brutal regime on the German part of Poland. Hitler was bent on the elimination of anyone who stood in the way of German domination.

As part of Hitler's plan for ethnic cleansing, around 5 million Polish Jews were rounded up and herded into ghettos. The invasion of Poland gave some forewarning of the violence that would soon be visited upon scores of countries and countless people around the globe.

The rise of the Nazi Party

Although World War II was triggered by Hitler's invasion of Poland, its origins can be traced back to Germany's defeat in World War I and the demand for reparation payments. The defeated nations lost land and prestige,

---

Britain and France want to ensure Germany is **incapable of starting another war**.

Hitler's government rebuilds **Germany's military** and promotes **extreme nationalism**.

 German forces invade Poland.

The **Treaty of Versailles** sets tight limits on German **armaments and armed forces**.

With **economic depression** crippling Germany, **Adolf Hitler** grows in popularity.

Britain and France declare war on Germany, leading to **World War II**, the **most destructive conflict** in history.

Adolf Hitler watches a victory parade in Warsaw following the invasion of Poland. He and Soviet leader Joseph Stalin agreed to the invasion and division of the country.

Fascism in Europe

Italy's Fascist dictator Benito Mussolini also had aspirations for foreign glory. In October 1935, he invaded Abyssinia (Ethiopia) in retaliation for the defeat the Italians had suffered there in 1896. By May 1936, Mussolini had conquered the country, facing no opposition from the Western powers.

Further evidence of Western democracies' weakness in facing up to the Fascist challenge was provided the same year, when both Mussolini and Hitler sent "volunteers" to fight in the Spanish Civil War, to aid nationalist General Franco in his campaign against left-wing supporters of the Spanish Republic. Britain and France took no action, and Franco's victory in 1939 bolstered the Fascist cause. »

causing deep resentment. Germany was forced to return Alsace and Lorraine to France, and all of its overseas colonies were annexed by the Allies.

Germany's Weimar Republic began its economic recovery in the 1920s, but it could not survive the blow that was inflicted by the US economic crash of 1929. This financial crisis aided the rise of the National Socialist (Nazi) Party, led by Hitler, who promised the German people he would make the nation great again.

Hitler had fought in World War I, and the experience of trench warfare, the shock of defeat, and the terms of the Versailles Treaty were to influence the rest of his life. He developed extremist views based on far-right nationalism; and by the time he became chancellor of Germany's coalition government in 1933, and dictator of the country the next year, he ruthlessly pursued his policies of nationalism, anti-Semitism, and anti-communism.

Hitler's *Lebensraum*

Under this creed, Hitler embarked on an ambitious foreign policy. In 1935, openly going against the terms of the Versailles Treaty, he began a massive program of re-armament. In 1936 he occupied the demilitarized Rhineland, but none of the major powers intervened. In March 1938, Hitler annexed Austria to Germany, before setting his sights on the German-speaking part of Czechoslovakia, the Sudetenland. British and French politicians wanted to avoid a repeat of the horrors of World War I and felt that the Sudetenland was not worth fighting for. In the Munich Agreement of September 29, 1938, the Sudetenland was handed to Hitler in exchange for his promise to end his land-grabs. British prime minister Neville Chamberlain declared that he had secured "peace for our time," only for the Nazis to invade the remainder of Czechoslovakia in March 1939.

German troops crossed the Polish frontier this morning at dawn and are since reported to be bombing open towns. In these circumstances there is only one course open to us.
Neville Chamberlain

The West intervenes

Hitler's invasion of Poland, which began on September 1, 1939, finally forced Britain and France into a war they had been trying desperately to avoid. Deciding that they needed to take a tougher stance against Hitler after his takeover of Czechoslovakia, the two nations had guaranteed Poland support in the event of German aggression. Honoring this promise, they declared war on Germany on September 3, which meant British and French colonies were also drawn into conflict: Britain's dominions Australia and New Zealand declared war immediately, the Union of South Africa followed on September 6, and Canada on September 10.

Germany quickly overran Poland with its tactic of *blitzkrieg* ("lightning war"), which utilized tank divisions supported by the Luftwaffe, the German air force. The British sent an Expeditionary Force (BEF) to France, but neither Britain nor France attempted an offensive against Germany. They were not ready for a large-scale attack, and some politicians still believed that peace terms could be negotiated.

This period became known as the "Phoney War." Expecting to be bombed, Britain began to evacuate its children from major cities. Air-raid shelters were built, and gas masks handed out. The Phoney War ended in April 1940, when Germany attacked and conquered Denmark and Norway. A month later, they turned on France, Belgium, and the Netherlands. The French army was poorly led and badly equipped. France had relied on the Maginot Line, a chain of fortresses along the frontier with Germany, to halt any attack. But the fortification did not extend along the Franco-Belgian border, and the Germans simply bypassed it at the north end. Within the space of six weeks, France had fallen to the German onslaught.

The Battle of Britain

Only a hesitation by Hitler, who may have wanted to rest his troops and spare them from a possible counterattack, prevented the destruction of British forces before they could be evacuated by sea from Dunkirk. Thousands of Allied soldiers were transferred across the Channel in all kinds of vessels in Operation Dynamo. Winston Churchill, the First Lord of the Admiralty and later Britain's wartime prime minister, told the British parliament, "The Battle of France is over. I expect the Battle of Britain is about to begin."

However, Hitler's attempts to invade Britain in Operation Sea Lion had to be abandoned after the Luftwaffe failed to win the battle of the skies. With the Luftwaffe triumphant in both Poland and France, the Germans had hoped that Britain could be beaten by air power alone. However, German crews were exhausted, intelligence was poor, and Britain's use of radar enabled the Royal Air Force (RAF) to track incoming planes and take off in time to meet an attack. The Battle of Britain in the summer of 1940

We shall defend our island whatever the cost may be. We shall fight on beaches, landing grounds, in fields, in streets, and on the hills. We shall never surrender.
Winston Churchill

Operation Dynamo, in June 1940, focused on the evacuation of Allied soldiers from the port of Dunkirk, in France, after they had become surrounded by German troops.

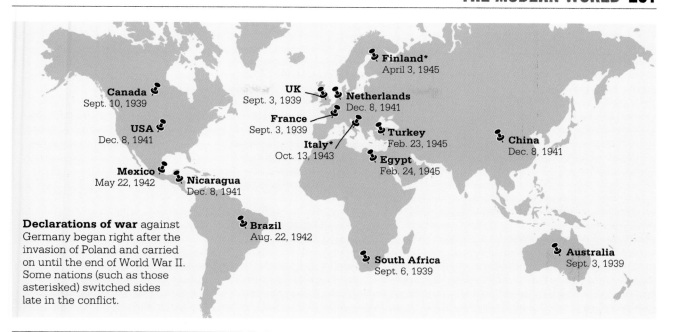

Declarations of war against Germany began right after the invasion of Poland and carried on until the end of World War II. Some nations (such as those asterisked) switched sides late in the conflict.

was the first real check on Hitler's progress, but Britain alone could not fight a power that now had control of almost the entire continent.

The world at war
What started as a European war gradually became a world war. In June 1940, Italy, emboldened by German successes, declared war on Britain and France, fulfilling the terms of the Axis agreement made between Hitler and Mussolini on May 22, 1939. But Italy's failures in Greece and North Africa forced Hitler to send German armies into these areas, as well as Yugoslavia.

On September 7, 1940 Germany began its first major air raid on London. The Blitz, as the bombing of the English capital became known, thrust civilians into the war and put relentless pressure on industry, ports, and British morale. With men joining the army, women were required to work in factories and on farms. Food rationing was introduced in Britain in January 1940, and

people were urged to grow their own food. Nazi-occupied Europe also experienced food shortages, which weighed most heavily on the conquered populations.

Collaboration or exile
In some locations, the Germans worked with existing governments and fully supported puppet administrations, such as the pro-Nazi Vidkun Quisling in Norway and the Vichy regime in southern France. Led by Marshal Philippe Pétain, Vichy was officially neutral, but it collaborated closely with Germany, fighting the French resistance, and implementing anti-Semitic legislation.

Germany had total control in Poland and eventual control of the Baltic states. Monarchs and politicians of more than a dozen occupied countries escaped to Britain. Polish ministers set up headquarters in London, and Belgium's government operations were transferred there. The Dutch royal family, under Queen

Wilhelmina, also sought refuge in London. When France fell to Germany, Charles de Gaulle, who opposed the newly installed Vichy government, became the voice of French opposition to the Nazi occupation.

In 1940 the biggest threat facing Britain was from German U-boats. As an island, Britain was dependent on its merchant ships to bring in vital supplies but also to export equipment to its fighting forces abroad, and German U-boats were sinking dozens of Allied ships each month. Merchant ships traveled in convoy to increase the chances of supplies getting through on each journey, but casualties were high.

Fighting the USSR
In June 1941, Britain gained a new ally when Germany invaded the USSR in Operation Barbarossa. Hitler had looked to the Soviet Union for new territory for the German people. It would also remove any future threat from the east but »

Operation Barbarossa, launched in June 1941, saw the invasion of the Soviet Union by Germany, in breach of the non-aggression pact the two countries had signed two years earlier.

fundamentally followed through on Hitler's plan to destroy communism. At first it looked as if Germany and its allies would be as successful against the Russians as it had been against the French. By winter, Germany had advanced to within 1 mile (1.5km) of Moscow, and Leningrad, the USSR's second city, was under siege.

Another powerful rationale for war in the east was one based on racist ideology and Hitler's hatred of Slavs and Jews. As German troops swept into Russia, they inflicted a terrible campaign of genocide against communists and Jews. Russian troops endured extreme hardships. German tanks plowed through the Red Army defenses. Prisoners of war were shot or left to starve. Fleeing civilians were butchered without a moment's hesitation. The harshness of the Russian winter slowed the Germans, and Russian counterattacks drove back their front line by several hundred miles. In the Battle of Moscow, from early October 1941 to January 1942, an estimated 650,000 soldiers from the Soviet Army lost their lives. In the spring of 1942, the Germans resumed their offensive in the USSR, driving the Red Army back and coming close to taking the Russian oilfields.

The Pacific and Africa

In December 1941, Japan entered the war by attacking the US fleet at Pearl Harbor, in the Hawaiian Islands, as part of its plan to drive American forces out of the Pacific. Germany—which had a tripartite agreement with Japan and Italy to provide mutual military assistance in the event any one of them was attacked by a nation not already involved in the war—immediately declared war on the United States. Britain now had two strong allies, the USSR under Joseph Stalin, and the US, led by President Franklin D. Roosevelt. Both were decisive in bringing about the defeat of the Axis powers. American industry became a triumph of wartime production, giving Americans in combat in Europe and Asia the tools they needed to fight the Axis.

Japan won quick victories in the Pacific. It successfully captured the Philippines, Malaya, Burma, Indonesia, and Singapore, Britain's main naval base in East Asia.

In North Africa, meanwhile, a renewed Axis offensive led by General Erwin Rommel brought the German and Italian armies within striking distance of Cairo and the Suez Canal. The first major Allied victory came in Egypt. In July 1942, Rommel was halted at El Alamein; in October, he was forced into retreat by the British 8th Army, led by Field Marshal Montgomery.

That same winter, the Red Army defeated the Nazis at Stalingrad. The Soviets encircled the Germans, forcing a surrender in February 1943.

History knows no greater display of courage than that shown by the people of the Soviet Union.
Henry L. Stimson
US Secretary of War

General Dwight D. Eisenhower led the Allied forces during the Normandy landings of June 1944. The invasion was a decisive step toward taking Europe back from the Nazis.

Hiroshima and Nagasaki

American planes dropped atomic bombs on the Japanese cities of Hiroshima and Nagasaki to force Japan to surrender and end World War II. On August 6, 1945, "Little Boy" was dropped on Hiroshima. The inhabitants below had no idea what was about to happen. People, animals, and buildings were incinerated in the searing heat. Some 70,000 died immediately. Despite this terrible event, Japan did not surrender.

Japan had cause to reconsider its position when the Soviets entered the war against them by crossing into Manchuria on August 9. When, that same afternoon, the US dropped "Fat Man" on Nagasaki, instantly killing 50,000, Japan was brought to its knees and agreed to the Allies' terms of surrender. These unprecedented attacks avoided a bloody ground assault by the Allies on the Japanese mainland, but many thousands lost their lives as a result of the long-term effects of radiation sickness.

The turn of the tide

In a conference at Tehran in November 1943, the Allied leaders agreed on a strategy to liberate Europe. While the Russians drove the Germans back in the east, and the British and Americans advanced slowly through Italy, a huge Allied invasion force arrived in Normandy in June 1944. Eleven months later, it had reached the river Elbe in northern Germany, while Russian troops were advancing block by block through Berlin. Germany was being hit repeatedly by British Lancaster bomber aircraft from Bomber Command and the US Eighth Air Force. Staring at defeat, Hitler committed suicide on April 30, and Germany surrendered unconditionally a week later.

The last act of the war came in August 1945, when the US, after fighting island by island through the Pacific, put an end to Japanese resistance by dropping

The Battle of Iwo Jima saw US troops fight against Japan's Imperial Army for possession of the tiny island in the Pacific Ocean, resulting in 100,000 Japanese casualties.

atomic bombs on Hiroshima and Nagasaki. The effects of the bombs were cataclysmic, inflicting unprecedented horror on the two Japanese cities.

Nations united

Hitler's invasion of Poland marked the start of World War II, the largest and most destructive war in history, by the end of which an estimated 60 million people had been killed. Like their predecessors in 1918, the Allies were determined that this should be the last war of its kind.

Representatives of 50 nations met in 1945 to set up the United Nations. There was hope that this would mark the start of a new era of international understanding. ∎

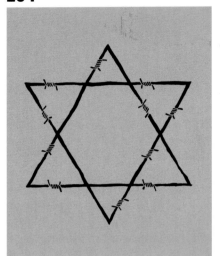

THE FINAL SOLUTION OF THE JEWISH QUESTION

THE WANNSEE CONFERENCE (1942)

IN CONTEXT

FOCUS
The Holocaust

BEFORE
1933 The first concentration camp is built in Dachau, near Munich. Its first inmates are communists, socialists, and trade unionists.

September 1935 As a result of the new Nuremberg Laws, Jews lose their civil rights.

1938 During Kristallnacht, the "Night of Broken Glass," the Nazis terrorize Jews across Germany and Austria.

June 1941 The German invasion of the Soviet Union is accompanied by the mass killing of Jews.

AFTER
May 1942 Gassings start at Auschwitz, in Poland.

1945–46 At the Nuremberg trials, 24 Nazi members are indicted and 12 sentenced to death.

Hitler becomes ruler of Germany and **introduces legislation** discriminating against Jews.

Hitler's takeover of Austria is followed by **widespread attacks on Jews**.

Germany conquers **Poland**, and Polish Jews are forced to move into **overcrowded ghettos**.

The Nazis look for **efficient ways** to kill millions after the **invasion of Russia**.

The Wannsee Conference organizes the Final Solution.

More than **6 million Jews** are killed in **the Holocaust**.

On January 20, 1942, 15 members of the Nazi Party and German officials met in the Berlin suburb of Wannsee to discuss the implementation of the "Final Solution of the Jewish Question"—the code name for the systematic annihilation of European Jews. During the conference, a tabulation of all the Jews in Europe was presented, country by country, as well as a target number for extermination: 11 million. The meeting lasted two hours and was matter-of-fact and dispassionate. After approving the "Final Solution" and the slaughter of the Jews, the men called for brandy and cigars.

See also: The Treaty of Versailles 280 ▪ The Wall Street Crash 282–83 ▪ The Reichstag Fire 284–85 ▪ Nazi invasion of Poland 286–93 ▪ The establishment of Israel 302–03 ▪ The Siege of Sarajevo 326

Auschwitz, in southern Poland, has become a byword for the Holocaust. Those prisoners subjected to forced labor were summarily executed when they became too weak to work.

The Wannsee Conference was far from the start of Nazi brutality against Jews. Adolf Hitler had come to power in 1933, spreading his belief that Germans were the Aryan master race, superior to all others, and that their blood should not be contaminated. He identified Jews as a race of people, not just a religious group. German Jews were banned from marrying non-Jewish Germans and subjected to increasing discrimination and segregation. From the time of the German takeover of Austria in 1938, Nazi brutality against Jews worsened. Jews wanting to flee German rule found other countries unwilling to accept them.

Gathering momentum
After Germany's invasion of Poland in 1939, the Nazi campaign against the Jews reached a terrifying new level. Herded into ghettoes, Polish Jews began to die in large numbers of starvation and ill-treatment.

When Germany invaded Russia in 1941, paramilitary death squads carried out mass killings of Jews in the conquered areas. To start with, victims were shot, up to 30,000 at a time, but the SS then began gassing Jews in the backs of vans. Poison gas was found to be a more efficient way to commit mass murder.

Until 1941, the Nazi leadership had envisaged solving the "Jewish problem" by deporting Jews to a distant location. By the time of the Wannsee Conference, however, they were committed to systematically killing Europe's Jewish population. Six dedicated death camps were built in Poland. Adolf Eichmann of the Nazi paramilitary corps, the SS, arranged the transport of Jews to the camps from right across Europe, including France, Greece, Hungary, and Italy. The Jews from the Polish ghettos were also taken there to be exterminated. Prisoners arrived at these huge killing factories by train and were gassed in shower rooms, their corpses burned in large crematoria. At the Belzec camp, about half a million Jews were

killed, and only seven prisoners are known to have survived. The death camp at Auschwitz, however, also had a labor camp attached, where those who were not killed on arrival were made to work. The Germans needed slave labor to support their war effort, and this offered Jews their best chance of survival. Along with other prisoners—including socialists, homosexuals, Roma, and prisoners of war—many Jews were sent to concentration camps. Their heads were shaved, and they were given a uniform to strip them of their identity. When the Allies liberated the camps in 1945, they found a vision of hell. The survivors were skeletal and traumatized.

State-sanctioned genocide
The Wannsee Protocol, the minutes of the conference, represents the unimaginable. For the first time, a modern state had committed itself to the murder of an entire people. As many as 6 million Jews lost their lives, and an estimated 5.5 million others—Slavs, homosexuals, communists—were also killed. ▪

The Nuremberg Trials

After the end of World War II, the Allies sought to bring the Nazis to justice. An international tribunal was held at Nuremberg, Germany, beginning in 1945. Newsreels captured from the Nazis revealed the gas chambers, the massacre of civilians, and the ill-treatment of prisoners. The trials were televised, showing to the world—and, in particular the German people—evidence of the horrors that had taken place in the concentration camps.

Adolf Hitler, Heinrich Himmler, head of the SS, and Joseph Goebbels, head of propaganda, had committed suicide, leaving 24 defendants facing four counts: crimes against peace, planning and waging wars of aggression, war crimes, and crimes against humanity. Most said they were "only obeying orders." Albert Speer, head of war production, was jailed for 20 years, while 12 of the other defendants were sentenced to death; the trials led to the setting up of a permanent international criminal court in The Hague, in the Netherlands.

ALL WE DID WAS FLY AND SLEEP

THE BERLIN AIRLIFT (1948)

IN CONTEXT

FOCUS
Cold War

BEFORE
1918–20 US troops fight against the Bolsheviks during the Russian Civil War.

1922 Russian revolutionary Vladimir Lenin creates the Communist International (Comintern) to promote international revolution.

1947 The Truman Doctrine pledges support for countries attempting to hold back communism.

AFTER
1961 The Soviets erect the Berlin Wall between East and West Berlin. It becomes an ugly symbol of the Cold War.

1985 Russian leader Mikhail Gorbachev campaigns for economic and political reforms: *glasnost* and *perestroika*.

1990 Germany is unified after the fall of the Berlin Wall.

At the Yalta and Potsdam conferences in 1945, the wartime Allies agreed to split defeated Germany into four zones, each separately administered by France, Britain, the USSR, and the US. The capital, Berlin, lay deep within Soviet-controlled East Germany. This, too, was split into four zones. On June 24, 1948, the Soviet Union imposed a blockade on West Berlin, cutting off all links by rail, road, and canal, to prevent vital supplies from reaching the population. In all, 2.5 million people faced a choice between starvation and accepting a communist regime.

Dozens of people in West Berlin stand waiting for the much-needed supplies that are about to be dropped from a low-flying US Air Force plane during the 1948 Berlin Airlift.

A clash between East and West had the potential to lead to another world war, but the Western nations devised a plan to use airplanes to drop supplies into Berlin. Over the next 14 months, 278,288 relief missions were flown to the city. At the height of the airlift, a plane landed every three minutes.

The Cold War

The era of cooperation between the victors of World War II was short-lived; the Western countries clashed with the Soviet Union (USSR), over the type of governments being set up in Europe. The USSR banned non-communist parties in every Eastern European country and created a block of satellite states subservient to Soviet leadership. The Western powers sought to create democracies that excluded communists from power. Germany remained divided into communist East and democratic West, an emblem of polarized Europe. In 1946, former British prime minister Winston Churchill summed up the situation when he stated that "an iron curtain has descended across the continent." This deep division between East and West became known as the Cold War, since it

See also: Russia emancipates the serfs 243 ▪ The October Revolution 276–79 ▪
Stalin assumes power 281 ▪ Nazi invasion of Poland 286–93

After World War II, the **communist East and democratic West**
disagree over the **future of Germany**.

⬇

The Western Allies plan to turn their **occupied zones**
into a **separate German state**.

⬇

The **Soviets cut road and rail links** into West Berlin
to force the capital into **surrendering**.

⬇

The West is **determined to have a presence**
in Berlin but cannot risk another **world war**.

⬇

The Berlin Airlift is a peaceful solution.

---

never escalated into direct military
conflict. The struggle over the
future of Berlin became the first
major crisis of the Cold War.

A plan to starve Berlin

In June 1948, the three Western
Allies announced plans to merge
their zones and introduce a new
currency. Stalin's response was
swift: his blockade sought to
starve Berlin into surrender
and wrest power away from the
West. The Western powers did
not want to give the Soviets
control of the Western sector
and were determined to stay.

The Berlin Airlift was a success,
and Stalin lifted the blockade in
May 1949. Spurred by the Berlin
crisis, Western European countries
formed a defensive alliance—the
North Atlantic Treaty Organization
(NATO). The communist states of
Eastern Europe organized a rival
alliance in the Warsaw Pact in 1955.

The crisis over Berlin
exacerbated the animosity
between the US and the USSR.
After World War II, Korea had also
been split—into a Soviet-occupied
northern zone and an American-
occupied southern one. The north,
backed by the USSR, invaded
the south in June 1950. The US
provided troops for a United
Nations army, which went to the
support of the South Koreans.
The Korean War ended in 1953,
but it, the conflict over Berlin, and
the Soviet testing of their first
atomic bomb in 1949, created a
climate of fear in the West over
communist expansion. ■

Joseph Stalin

The dictator of the USSR
from 1927 until his death,
Joseph Stalin (1878–1953)
was notorious for his ruthless
repression of dissent. His rise
to power began in 1903, when
he became a friend of Vladimir
Lenin, the first leader of Soviet
Russia. During and after the
Russian Revolution (1917), he
played a prominent part in
the Communist Party's rise
to power, and in 1922 he
advanced to become general
secretary of the Russian
Communist Party.

He became supreme
leader in 1927 and aimed to
transform the Soviet Union
into a major industrial force.
In 1928, he launched an
industrialization program
and introduced collective
farming. Millions died of
starvation, in labor camps,
or in a wave of purges directed
at his supposed opponents.

In the post-war years,
Stalin led the Communist Party
into a period of confrontation
with his former World War II
allies. Following his death,
Stalin was condemned by his
successors for his campaigns
of terror and murder.

AT THE STROKE OF THE MIDNIGHT HOUR, WHEN THE WORLD SLEEPS, INDIA WILL AWAKE TO LIFE AND FREEDOM

INDIAN INDEPENDENCE AND PARTITION (1947)

IN CONTEXT

FOCUS
End of empires

BEFORE
1885 The Indian National Congress (INC) is founded and campaigns for Indian rights.

1901 Australian colonies are united to form the Commonwealth of Australia.

1921 The Irish Free State (four-fifths of Ireland) gains independence from Britain.

1922 Egypt is given limited independence by Britain, but British troops remain to protect imperial interests.

AFTER
1947 The Commonwealth of Nations is formed—all former British colonies can take part.

1960 The Declaration of Decolonization asserts the rights of all peoples to self-determination.

For more than a century, India had been the crown jewel of the British Empire, but on the last stroke of midnight on August 14, 1947, it became an independent nation. In India's Constituent Assembly, Delhi, a special midnight gathering of parliament was convened. Jawaharlal Nehru, the first prime minister of India, rose to his feet to declare India's freedom. However, this independence also opened a social and geographic wound that has yet to heal.

The new Indian state was split into two independent nation states: Muslim-majority Pakistan and Hindu-majority India. Pakistan

See also: The formation of the Royal African Company 176–79 ▪ The Siege of Lucknow 242 ▪ Nkrumah wins Ghanaian independence 306–07

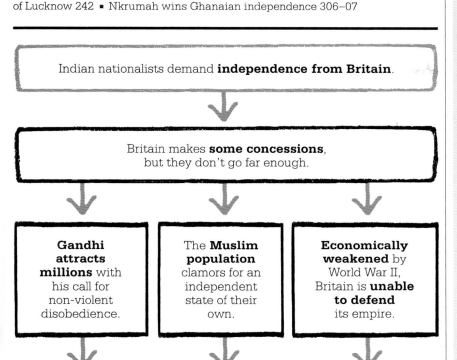

Indian nationalists demand **independence from Britain**.

⬇

Britain makes **some concessions**, but they don't go far enough.

⬇ ⬇ ⬇

Gandhi attracts millions with his call for non-violent disobedience.

The **Muslim population** clamors for an independent state of their own.

Economically weakened by World War II, Britain is **unable to defend** its empire.

⬇ ⬇ ⬇

Indian independence is achieved, and the country is split in two.

Mohandas Gandhi

The Indian national leader known as Mahatma, meaning "great soul," Mohandas Gandhi (1869–1948) led his country to independence from Great Britain. He came from a Hindu family and studied law in England before spending 20 years in South Africa trying to secure rights for the Indians living there.

Gandhi's involvement in Indian politics began in 1919, and he soon became the unquestioned leader of the independence movement. He preached the doctrine of Satyagraha (soul force, or passive resistance) which he applied against the British with great effect. He adopted a simple life believing in the virtue of small communities and campaigned against Indian industrialization.

Gandhi's life work was crowned in 1947, when India finally won independence, but the concessions he had made to the Muslims led to his assassination the following year by a Hindu fanatic, who blamed him for the partition of India, although Gandhi himself bitterly opposed the dismemberment of the subcontinent.

---

itself was split between northwest and northeast, because both wings had a Muslim majority. Immediately, millions of Muslims trekked to West and East Pakistan (the latter now known as Bangladesh), while millions of Hindus and Sikhs headed towards the newly independent India. Thousands never reached it, and many died from malnutrition and disease. Across India there were outbreaks of sectarian violence, with Hindus and Sikhs on one side and Muslims on the other.

By 1948, as the great migration drew to a close, more than 15 million people had been uprooted, and between 1 million and 2 million were dead. India was independent and India's Muslims had their own independent state, but freedom came at a great cost.

The road to independence

The spirit of nationalism in India gained ground in the mid-19th century and was strengthened in 1885 by the formation of the Indian National Congress (INC). During World War I, expectations for greater self-governance were raised when Britain promised to deliver self-rule in return for India's contribution to the war effort. But Britain envisaged a gradual progress toward self-government, beginning with the Government of India Act (1919), which created an Indian » parliament where power was shared between Indians and British officials. This did not satisfy Indian nationalists, and the British responded to their protests with sometimes brutal repression.

The push for independence from the 1920s to the 1940s was galvanized by the work of Mohandas Karamchand Gandhi. Gandhi not only launched the Satyagraha campaign, promoting non-violent protest, but also became an influential figure for millions of followers. In 1942 Gandhi led the "Quit India" campaign, calling for civil disobedience to disrupt Britain's efforts in World War II. The British immediately jailed Gandhi and other nationalist leaders.

By the end of World War II, it was clear that Britain lacked the means to defeat the nationalist campaign. Britain's officials in India were utterly exhausted, and Britain itself was almost bankrupt. Britain agreed to a fully independent India. While Gandhi and Nehru advocated Indian unity, the Muslim League, founded in 1906 to safeguard the rights of Muslims, demanded a completely separate Muslim state. Its leader, Mohammed Ali Jinnah, feared that Muslims could not protect their minority rights if left to live under Hindu rule. Congress rejected the proposal and violence on the streets between Hindus and Muslims began to escalate.

Pakistan is born

In 1947, Lord Louis Mountbatten flew into Delhi as Britain's final Viceroy of India. Faced with irreconcilable differences over the demand for a separate state for India's Muslims, he persuaded all parties to agree to partitioning the country into Hindu India and Muslim Pakistan.

From its birth, Pakistan faced many challenges. It had limited resources and a huge refugee problem. There were different traditions, cultures, and languages, and Jinnah, its first governor general, died the following year. In 1948, India and Pakistan fought over Kashmir, the only Muslim-majority area to remain within India.

> Ours is not a drive for power, but purely a non-violent fight for India's independence.
> **Mohandas Gandhi**

Colonies gain freedom

After World War II, the European colonial powers—mainly Britain, France, the Netherlands, and Portugal—recognized that change was inevitable. Some colonies won independence by peaceful means, such as in Burma and Ceylon (1948), but often, European powers tried to hold on to their colonies.

During World War II, Japan, itself a significant imperial power, drove the European powers out of Asia. After the Japanese surrender in 1945, nationalist movements in the former Asian colonies campaigned for independence rather than a return to European colonial rule. Dr. Ahmed Sukarno, leader of Indonesia's nationalist movement, declared the Independent Republic of Indonesia in 1945. The Dutch sent troops to restore their authority, and in two military campaigns that followed, an estimated 150,000 Indonesians and 5,000 Dutch soldiers died. International pressure eventually forced the Dutch to concede independence in 1949.

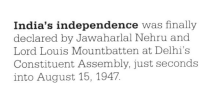

India's independence was finally declared by Jawaharlal Nehru and Lord Louis Mountbatten at Delhi's Constituent Assembly, just seconds into August 15, 1947.

The Japanese occupation of Malaya during the war had unified the Malayan people and greatly increased nationalistic feelings. Britain clamped down on protests, which led the militant wing of the Malaysian Communist Party to declare war on the British Empire in 1948. Britain responded by declaring a state of emergency and pursuing a bitter campaign against Chinese "communist terrorists." Independence was not granted to Malaya until 1957.

Unrest in Africa

In Kenya, the imposition of a state of emergency in 1952, in response to the Mau Mau (rebel) uprising, led to greater insurgency and the British rounding up of tens of thousands of Mau Mau suspects into detention camps. By 1956 the rebellion had been crushed, but the methods used by the British to regain control brought international condemnation. In central Africa, too, decolonization was born in violence. In Rhodesia, savage conflict erupted between the black majority and the fiercely racist white leadership, which had unilaterally declared independence in 1965.

> We are proud of this struggle, of tears, of fire, and of blood, to the depths of our being.
> **Patrice Lumumba**
> **First prime minister of the Congo (Zaire) (1960)**

The process of decolonization coincided with the new Cold War between the Soviet Union and the United States. The US became concerned that, as the European powers lost their colonies, Soviet-supported communist parties might achieve power in the new states. The US used substantial aid packages to encourage newly independent nations to adopt governments that aligned with the West. The Soviet Union deployed similar tactics in an effort to encourage new nations to join the communist bloc. Many resisted the pressure to be drawn into the Cold War and joined the "non-aligned movement." This movement began out of a 1955 meeting in Bandung, Indonesia, involving 29 African and Asian countries. Member countries decided they would not be involved in alliances or defense pacts with the main world powers, but focus on internal development instead.

Terrorism in France

France was determined to maintain its political status in Algeria. When independence was not realized after World War II, war broke out between Algerian nationalists and French settlers. In 1958 the National Liberation Front (FLN), the main nationalist group, led several terrorist attacks, first in Algeria, then in Paris. The crisis led to the return to power of Charles de Gaulle, the wartime leader of the Free French. In 1960, de Gaulle, to the horror of the French settlers, agreed to emancipate Algeria. After a long and bloody conflict in which an estimated 150,000 died, Algeria gained its independence in 1962.

Independence gained

During the 1960s and 70s, many of the countries that were once held as British colonies became

Mau Mau suspects captured in Nairobi's Great Rift Valley, Kenya, in 1952 are led away, with their hands on their heads, to be questioned by police and possibly held in detention camps.

independent states and joined the Commonwealth. The British Commonwealth, formed in 1931, became the successor to Britain's old empire, preserving Britain's global economic and political influence. In 1931 Britain extended dominion status to the already self-governing colonies of Canada (1867), Australia (1901), New Zealand (1907), and Newfoundland (1907). Britain and her dominions shared equal status, and they accepted the British monarch as head of the Commonwealth. In 1949 the British Commonwealth became "The Commonwealth," a free and equal association of independent states, but the end of the empire was drawing near. Britain fought a war to retain the Falkland Islands in 1982, and Hong Kong continued as a British dependency until 1997.

Gandhi had a profound influence on world politics. Other peaceful resisters—such as Martin Luther King Jr. and Tibet's Dalai Lama—emulated his methods. Around the world, the struggle for countries to secede from nations they belong to continues, as the likes of Scotland (United Kingdom), Quebec (Canada), and Palestine fight to be seen as nations in their own right. ■

THE NAME OF OUR STATE SHALL BE ISRAEL

THE ESTABLISHMENT OF ISRAEL (1948)

IN CONTEXT

FOCUS
Creation of Israel

BEFORE
1897 Zionism becomes an organized movement and calls for a Jewish state in Palestine.

1917 In the Balfour Declaration, Britain promises to help the Jews establish a homeland in Palestine.

1946 As part of its campaign of terrorism against Palestine and Britain, the Jewish underground army bombs the King David Hotel, killing 91.

AFTER
1967 During the Six Day War, Arabs unite against Israel, but Israel is victorious and captures swathes of Arab territory.

1993 Oslo Peace Accords try to initiate peace between Palestinians and Israel.

2014 Sweden becomes the 135th country to recognize the state of Palestine.

Zionist theorists envisage the possibility of a **Jewish homeland**.

Jews begin **settling in and developing** Palestine.

Jews escaping Nazi control flee to **Palestine**.

The United Nations grants the land of Israel to the Jewish people.

Many Palestinians are **forcibly displaced** and become refugees.

Wars break out periodically between Arab states and Israel.

As the sun rose on May 14, 1948, the British flag was lowered at Government House, on Jerusalem's Hill of Evil Counsel, ending the 26-year British mandate over Palestine. David Ben-Gurion, the longtime leader of the Jewish settlers, or Zionists, who had fled to Palestine from Europe, proclaimed the news of the establishment of the Jewish state in Palestine.

Israel's Muslim neighbors, united as the Arab League, rejected the state's creation and reacted with an attack. Troops moved in from Transjordan, Egypt, Lebanon, Syria, and Iraq. Hardened

See also: The Young Turk Revolution 260–61 ▪ The Treaty of Versailles 280 ▪ The Suez Crisis 318–21 ▪
The 9/11 attacks 327 ▪ The founding of the United Nations 340

> We shall live at
> last as free men on
> our own soil.
> **Theodor Herzl**
> **Zionist writer**

to fighting after years of protecting
their settlements in Palestine, the
Jews thwarted the Arabs.

A troubled land

Jews had immigrated to Palestine
to avoid persecution in Europe since
the 1880s, believing it to be the
land promised to them by God. With
the Balfour Declaration in 1917, the
British government supported a
Jewish homeland. The majority-
Arab population objected to the
settlers' claim on their country.

Facing increasing attacks, the Jews
formed local defense groups under
the umbrella term the Haganah.

Escalation of violence

In 1939, the rise of anti-Semitism
in Europe, particularly in Nazi
Germany, forced Jews to flee to
Jerusalem. Facing a much larger
influx of settlers than they had
anticipated, the British proposed
a restriction on the free settlement
of Jewish refugees in Palestine.

After World War II, violence in
Palestine escalated, and in 1947 the
British government said it would
terminate its rule and hand the
"Palestine problem" to the United
Nations. The Holocaust convinced
the UN that the Jewish people
needed a homeland, so they resolved
to partition Palestine into an area
for Arabs (about 44 percent) and the
rest for a Jewish state. The Jews
agreed with the plan, but the Arabs
refused it. Despite this, on May 14,
1948 the state of Israel was born.

Israel's immediate priority was to
build a credible defense force from
the Haganah. After the Six Day War

The flag of Israel was adopted in
1948, a few months after the birth of
the state. It was originally designed in
1891 for use by the Zionist movement
and has the Star of David at its center.

(1967), Israel controlled the Sinai,
Gaza, the West Bank, the Golan
Heights, and Jerusalem. It faced
many attacks from Arab neighbors,
in addition to threats from the
paramilitary Palestine Liberation
Army (PLO), formed in 1964.

Arab Palestinians repeatedly
called for an independent state in
the West Bank and Gaza. In the
occupied zones, they suffered from
poor living conditions, military
raids, and restricted movement. ▪

David Ben-Gurion

The founder and first prime
minister (1948–63) of the state
of Israel, David Ben-Gurion was
born in 1886 to Jewish parents
in Poland. In 1906, he immigrated
to Palestine, where he became an
active supporter of the struggle
for an independent Jewish state.
He led the Jewish campaign
against the British in Palestine,
authorizing acts of sabotage.

When he became the nation's
leader, he established the Israeli
Defense Force and guided the
modern development of Israel.
He promoted the use of Hebrew
as the language of the country.

His "Law of Return," announced
in 1950, granted permission for
Jews from around the world to
immigrate to Israel.

He briefly retired in 1953,
and in his later years in power
he initiated secret talks with
Arab leaders in an attempt to
gain peace for the Middle East.

In 1970, Ben-Gurion retired
fully from the Knesset (Israeli
parliament) and devoted himself
to writing his memoirs in Sde-
Boqer, a *kibbutz* (communal
settlement) in the Negev Desert
in southern Israel. He died in
1973 and is still a revered figure.

THE LONG MARCH IS A MANIFESTO, A PROPAGANDA FORCE, A SEEDING-MACHINE

THE LONG MARCH (1934–1935)

IN CONTEXT

FOCUS
Founding Communist China

BEFORE
1911–12 The Republic of China is born under Nationalist Sun Yat-sen; the last Qing emperor abdicates.

1919 The May Fourth Movement, a student-led protest, spreads ideas of nationalism and communism.

1921 The Communist Party founded in Shanghai promotes revolution based on Marxism.

AFTER
1958 Mao Zedong introduces the Great Leap Forward, a five-year economic plan.

1978 Premier Deng Xiaoping announces a new economic program to make China a major financial power.

1989 Troops kill hundreds of pro-democracy supporters in Tiananmen Square.

China is ruled by **regional warlords**, and there is **no central government**.

→ **Communist and Nationalist** parties unite against the **warlords**.

↓

The Nationalists have the upper hand, and the **Communists retreat**.

← Incompatible ideologies mean these two groups mostly **fight each other**.

↓

The exertion and triumph of the Long March cements Mao's leadership and becomes mythic.

→ The **Communists** regroup and survive to fight until the **People's Republic of China** is born.

In the autumn of 1933, the Chinese Communist Party (CCP) was on the brink of annihilation. Nationalists had taken control of the country and launched a major attack against their base in Jiangxi, a southeastern province. In October 1934, the Communists were forced to abandon their stronghold and break through the Nationalist blockade. Some 80,000 set out on an extraordinary journey of 3,700 miles (6,000km) that lasted 368 days. It became known as the Long March.

Guided by their future leader Mao Zedong, the Communists faced bombs and machine-gun fire

See also: The Second Opium War 254–55 ▪ The Treaty of Versailles 280 ▪ The Cultural Revolution 316–17 ▪ The global financial crisis 330–33

from the air and were constantly under attack by Nationalist troops on the ground. They traveled mostly at night, the unit splitting into different columns to make them harder to spot.

The Tibetan mountains, Gobi Desert, and miles of wilderness stood between them and their goal: to reach the safety of northern China and establish a new Communist base. Hundreds died of starvation: of the original 80,000 marchers, only about 8,000 survived. Far from being viewed as a failure, however, their feat was hailed as a triumph of endurance and ensured the survival of the CCP.

Unifying the nation

In 1895, China had suffered a heavy military defeat against Japan. Anti-Japanese feeling swelled following Japan's aggression against China during World War I. Huge protests erupted after the 1919 Treaty of Versailles handed former German colonies in China to Japan. In the wake of these protests, communist ideals gained support, and in

Mao Zedong rides his white horse alongside Communist Party members during the Long March of 1934–35. His role in the march ultimately saw him rise to leader of the nation.

1921 the CCP was founded. The Kuomintang, a Nationalist party, also grew and by the mid-1920s had begun unifying the country.

Massacre in Shanghai

Nationalists joined forces with Communists in 1926 under Chiang Kai-shek (Jiang Jieshi) in the Northern Expedition to regain territories controlled by regional warlords. During the expedition, as the CCP increased in strength, a bitter rivalry led to an attack by

Nationalists against the CCP in Shanghai, in April 1927. Hundreds of Communists were arrested and tortured. The massacre triggered years of anti-Communist violence, and the Communists retreated to the Jiangxi countryside.

The struggle for survival

After the Long March, the CCP regrouped in the north. Nationalists and Communists were forced into an uneasy alliance in 1937, when Japan invaded China. By 1939, large areas in the north and east had been conquered. After Japan's defeat in World War II, tension between Nationalists and Communists flared up again, leading to civil war in 1946. The Communists won after massive battles with more than half a million troops on either side. On October 1, 1949, Mao Zedong created the People's Republic of China.

The Long March was a feat of remarkable endurance. To the survivors, it provided a deep sense of mission and contributed to the perception of Mao as a leader of destiny and revolutionary struggle. ▪

Chiang Kai-shek

The foremost non-communist Chinese leader of the 20th century, Chiang Kai-shek (1887–1975) was a soldier who, in 1925, became leader of the Kuomintang (Nationalist Party), which had been founded by Sun Yat-sen.

During his various stints as China's premier, he ruled over a troubled country. He attempted modest reforms but was beset by intractable domestic strife, as well as by armed conflict with Japanese invaders.

Despite making attempts to crush his chief rivals, the Chinese Communists, when China was

attacked by Japan his followers forced him to make an alliance with the Communists against the invading Japanese. The alliance did not survive the end of the World War II, and in 1949 Chiang and his party were driven from the mainland to the island of Formosa, which by that time had become known to Westerners as Taiwan. While he was there, Chiang set up a government in exile, which he controlled until his death in 1975. His government was recognized by many states as China's legitimate government.

GHANA, YOUR BELOVED COUNTRY, IS FREE FOREVER

NKRUMAH WINS GHANAIAN INDEPENDENCE (1957)

African **nationalism gathers pace** during the early 1900s.

The ideology of **Pan-Africanism** gains adherents worldwide.

African experiences in **World War II** spur demands for **racial equality**.

Nkrumah wins independence for Ghana.

Nkrumah fails in his campaign for the **political unity** of Africa.

By the mid-1970s, most of Africa has gained **independence**, if not peace.

In February 1948, at a time when the Gold Coast, a British colony in West Africa, had been demanding independence for several years, a group of unarmed African ex-servicemen marched to the British governor with a petition of grievances. Ordered to stop, they refused, and the police opened fire.

In response to this, in 1949, nationalist Kwame Nkrumah formed the Convention People's Party (CPP), an organization fighting for self-governance. Nkrumah initiated a campaign of positive action inspired by Gandhi's philosophy of non-violent non-cooperation in India against

See also: The formation of the Royal African Company 176–79 ▪ The Slave Trade Abolition Act 226–27 ▪ The Berlin Conference 258–59 ▪ Indian independence and partition 298–301 ▪ The release of Nelson Mandela 325

the British. The strikes and protests they encouraged remained peaceful but paralyzed the country, and Britain agreed to elections in early 1951. The CPP won 35 out of 38 seats, and the Gold Coast moved rapidly toward independence, which was proclaimed on March 6, 1957 with Nkrumah becoming prime minister of the nation of Ghana. It was a moment of huge hope for a new kind of Africa.

The European powers that ruled Africa had been impoverished by World War II, and attitudes to colonialism were changing. Nations that had fought against fascism found it hard to justify imperialism.

A domino effect

Events in Ghana had a significant impact in West Africa. In 1958, Guinea voted to secede from France. Determined not to be left behind, Nigeria celebrated independence from Britain on October 1, 1960. By 1964, independence had also been granted to Kenya, Northern Rhodesia (Zambia), Nyasaland (Malawi), and Uganda.

Kwame Nkrumah, Kojo Botsio, Krobo Edusei, and other Ghanaian politicians celebrate the independence of their country, which was achieved peacefully and democratically.

The French fought an eight-year war to hold on to Algeria, finally conceding independence in 1962.

The Portuguese, the first European colonial power in Africa, fought a long war to hold on to their colonies of Angola, Mozambique, and Guinea from 1961 to 1974. The collapse of Belgian authority in the Congo in 1960 led to a wave of violence across the nation and the assassination of the first prime minister, Patrick Lumumba, in 1961.

Many African countries gained independence during the Cold War. Used as pawns between the capitalist and communist superpowers, they accepted loans and military aid: in the 1970s, Ethiopia was rewarded with billions of dollars' worth of Soviet military equipment. Civil wars were also numerous, such as the ethnic civil wars in Rwanda and Zaire, as well as the clashes between warlords over food supplies in Somalia.

Dictatorial rulers

Once independence was achieved, African nationalist leaders sought to consolidate power by banning political rivals. Coups and military governments predominated—such as that of Idi Amin in Uganda. By the early 1970s, only Zimbabwe and South Africa were still ruled by the white political elite. Corruption, however, existed in most African countries. Nkrumah wanted Ghana to be a beacon of success, but his Pan-Africanism failed, and Ghana's fortunes began to slide as he became increasingly dictatorial. ▪

Kwame Nkrumah

Ambitious and well-educated, Kwame Nkrumah had big plans for both Ghana and Africa as a whole. He went to college in the US and later traveled to England, where he became involved in the West African Students' Union. In 1948, he began traveling around the Gold Coast as leader of a youth movement calling for "self-government now."

Nkrumah's calls for positive action civil disobedience as head of the Convention People's Party led to his arrest, and he was

sentenced to three years in jail. While in prison, he won the general election, and five years later, in 1957, he became prime minister of the newly independent Ghana.

Nkrumah's popularity rose with the construction of new schools, roads, and health facilities, but by 1964 Ghana was a one-party state and Nkrumah its "life president." After two assassination attempts and increasing human-rights abuses, Nkrumah faced a coup in 1966 and went into exile in Guinea. He died of cancer in 1972.

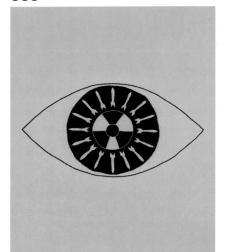

WE'RE EYEBALL TO EYEBALL, AND I THINK THE OTHER FELLOW JUST BLINKED
THE CUBAN MISSILE CRISIS (1962)

The USSR and US begin stockpiling **nuclear weapons**.

The **theory of MAD** acts as a deterrent to nuclear war.

A **struggle develops** for the control of satellite states, including Cuba.

Tension comes to a head in the Cuban Missile Crisis—nuclear war is only narrowly avoided.

The **scale of the threat** posed by nuclear war becomes more apparent.

World leaders **engage in diplomacy** and reduce stockpiles; **tensions cool**.

IN CONTEXT

FOCUS
Nuclear arms race

BEFORE
1942–45 The US sets up the Manhattan Project to develop the first nuclear weapon.

1945 The US drops atomic bombs on the Japanese cities of Hiroshima and Nagasaki, ending World War II.

1952–53 Both the US and USSR develop the H-bomb, 1,000 times stronger than the atomic bomb.

AFTER
1963 The USA and Russia agree to a nuclear test-ban treaty, and tensions lessen.

1969–72 Strategic Arms Limitation Talks (SALT I) yields a superpower agreement on missile deployment.

1991 Strategic Arms Reductions Treaty I (START I) reduces the numbers of US and Soviet long-range missiles.

For 13 days, from October 15 to October 28, 1962, the world teetered on the edge of nuclear destruction. Soviet leader Nikita Khrushchev had deployed nuclear weapons in Cuba, and US president John F. Kennedy demanded he remove them. Each threatened nuclear war.

This was no empty threat: from the 1950s, both superpowers had begun stockpiling vast nuclear arsenals. Strategists articulated the mutually assured destruction (MAD) doctrine, which held that if Russia attacked the West, the West would make sure that they retaliate. In short, there would be no winners.

See also: The October Revolution 276–79 ▪ Stalin assumes power 281 ▪ The Berlin Airlift 296–97 ▪
The launch of Sputnik 310 ▪ The Bay of Pigs invasion 314–15 ▪ The fall of the Berlin Wall 322–23 ▪ The 1968 protests 324

When Kennedy became president in 1961, he inherited a deteriorating relationship with Cuba. The US and Cuba had a history of mutual cooperation, but this had changed with the Cuban Revolution, when, on January 1, 1959, Fidel Castro overthrew the government of President General Fulgencio Batista.

Trade embargo

The US accepted Castro as ruler of Cuba, in spite of his communist leanings, and had a large economic presence in Cuba. However, Castro began to break the American hold on the economy, nationalizing all industry without compensation. In response, the US imposed a sweeping trade embargo, so Castro turned to the Soviet Union for support. Fearing communist expansion, the US tried to topple Cuba's government with the failed Bay of Pigs invasion in April 1961, involving CIA-backed Cuban exiles.

Also in 1961, the US deployed 15 nuclear-tipped Jupiter missiles in Turkey, in readiness to strike against the USSR should the need arise. Turkey shared a border with the Soviet Union, so this was viewed as a direct threat to Soviet territory.

An ultimatum

Khrushchev came under pressure from Soviet hard-liners to take a tough stance. This, and the desire to defend his Cuban ally from American aggression, led him to install missiles in Cuba that were capable of carrying nuclear warheads. On October 14, 1962,

Cuban president Fidel Castro and Soviet leader Nikita Khrushchev hold up their joined hands in a show of unity on an official state visit Castro made to Moscow in May 1963.

photographs taken by a U-2 spy plane showed nuclear weapon sites being built by the Soviets. Kennedy's military advisers sought an immediate attack on the missile sites, but Kennedy favored a naval blockade of Cuba to prevent the installation of more missiles. He issued an ultimatum to Khrushchev to withdraw and informed the world that nuclear war was an imminent possibility. Meanwhile, Khrushchev ordered the captains of Soviet ships to hold their course for Cuban ports.

Breaking the deadlock

Frantic diplomacy behind the scenes led to a deal that broke the deadlock: Kennedy agreed to remove missiles from Turkey in secret if Khrushchev dismantled all nuclear weapons in Cuba. The Soviet leader agreed—only if America would also abort its plan to invade Cuba.

On October 28, Khrushchev ordered his ships to turn around—a defining moment of the Cold War. The superpowers became more cautious, and the threat of nuclear war began to diminish. ▪

John Fitzgerald Kennedy

The 35th president of the US, John Fitzgerald Kennedy (1917–63) was the first Roman Catholic and, at 43, the youngest man ever to be elected to that office. As president, Kennedy brought a fresh and youthful style to politics, calling his program the "New Frontier." This included a challenge to venture into outer space and to eliminate poverty. His administration quickly won popular support.

Kennedy's years in power were marked in foreign affairs by Cold War tension. His greatest test was the Cuban Missile Crisis of 1962, where his firm stance against Russia brought him even greater popularity. His ambitious domestic reforms, however, on issues such as welfare and civil rights, were increasingly blocked by Congress.

While he was campaigning for the next presidential election, JFK was assassinated by Lee Harvey Oswald in Dallas, Texas, on November 22, 1963. Kennedy's death was a shock and a tragedy for Americans, at a time when tensions were just starting to ease between the United States and Russia.

PEOPLE OF THE WHOLE WORLD ARE POINTING TO THE SATELLITE
THE LAUNCH OF SPUTNIK (1957)

IN CONTEXT

FOCUS
Space race

BEFORE
1926 Robert Goddard launches the world's first liquid-fueled rocket.

1942 Germany successfully launches the first ballistic missile, the A4, or V-2.

AFTER
1961 Alan Shepard commands *Freedom 7* on the first Mercury mission, becoming the first American in space.

July 20, 1969 American Neil Armstrong becomes the first man to set foot on the moon.

1971 Russia's *Salyut 1*, the world's first space station, is launched.

1997 A US rover named Sojourner wheels on to the surface of Mars to explore the surface.

2015 *Mars Reconnaissance Orbiter* finds water on Mars.

On October 4, 1957, the USSR launched the world's first artificial satellite, *Sputnik 1*. Carrying a simple radio transmitter to relay information about conditions in space, the satellite remained in orbit until January 4, 1958, when it re-entered and burned up in Earth's atmosphere.

Sputnik symbolized far more than a scientific breakthrough. It was a sensational coup for the Soviets during the Cold War with the West. No shots were fired, but the military and political ramifications were immense. Americans felt more vulnerable to a nuclear attack. The USSR was now a Superpower, stunning the US and initiating the "space race," a frantic competition between nations for technological superiority.

The US catches up
Sputnik was a mass-media event that ushered in the "Space Age," capturing the world's collective imagination. There was a boom in science-fiction books, films, and TV dramas set in space. By 1958,

That's one small step for [a] man, one giant leap for mankind.
Neil Armstrong

the US had created NASA, the National Aeronautics and Space Administration, but they could only watch in envy as the Soviets sent Yuri Gagarin into orbit in 1961, the first human to go to outer space.

The US caught up by sending John Glenn into orbit in 1962, and by 1967 they had built a rocket, *Saturn V*, that was powerful enough to reach the moon. In 1969, 12 years after the launch of *Sputnik 1*, the American astronaut Neil Armstrong left *Apollo 11* and became the first man to walk on the moon. ∎

See also: The Berlin Airlift 296–97 ▪ The Cuban Missile Crisis 308–09 ▪ The fall of the Berlin Wall 322–23 ▪ The launch of the first website 328–29

I HAVE A DREAM

THE MARCH ON WASHINGTON (1963)

IN CONTEXT

FOCUS
Civil rights movement

BEFORE
1909 National Association for the Advancement of Colored People (NAACP) is founded.

1955 Rosa Parks refuses to give up her bus seat to a white man and, in doing so, ignites the civil rights movement.

1960 Four students at a whites-only restaurant counter are refused food, leading to sit-ins across the US.

AFTER
1965 Malcolm X, founder of the Organization of Afro-American Unity, is shot dead.

1966 Stokely Carmichael introduces the idea of "Black Power," turning away from non-violent protests.

1968 Martin Luther King is assassinated, leading to rioting across US cities.

The March on Washington on August 28, 1963 brought roughly 250,000 people— mostly African-Americans—to the nation's capital. They were calling for equality, an end to racial segregation, and for all Americans to have access to a good education, decent housing, and jobs that paid a living wage.

One of the speakers was the Reverend Dr. Martin Luther King, who had been arrested that April during anti-segregation protests in Alabama. "I have a dream," King cried, starting his famous speech.

Calls for equality

The abolishment of slavery after the American Civil War of 1861–65 led to emancipated slaves seeking American citizenship. However, while they were no longer slaves, they were not equal with whites, and they endured discrimination, segregation, and violent racist attacks. In the 1950s, a number of African-American groups fought back against discrimination with a policy of non-violence. In the 1960s,

There are those who say to you, we are rushing this issue of civil rights. I say we are 172 years too late!
Hubert Humphrey
Mayor of Minneapolis (1948)

civil rights marches in Birmingham, Alabama, led by King were central to the campaign. Some extremists, especially in the South, reacted with gruesome acts of violence.

After the March on Washington, US Congress passed the Civil Rights Acts of 1964, outlawing discrimination, and the Voting Rights Act of 1965. More than half a century later, however, many of the goals set on that day are still out of reach to black Americans. ∎

See also: The formation of Royal African Company 176–79 ▪ The Slave Trade Abolition Act 226–27 ▪ The Gettysburg Address 244–47 ▪ The release of Nelson Mandela 325

I AM NOT GOING TO LOSE VIETNAM

THE GULF OF TONKIN INCIDENT (1964)

IN CONTEXT

FOCUS
Intervention in Southeast Asia

BEFORE
1947 The Truman Doctrine, pledging American support for free peoples, guides US foreign policy in Southeast Asia.

1953 Cambodia wins its independence from France.

1963 President Ngo Dinh Diem of South Vietnam is killed in a US-backed military coup.

AFTER
1967 The Association of Southeast Asian Nations, ASEAN, is established to promote stability in the region.

1973 The Paris Peace Accords ends US combat in Vietnam, but does not end the conflict between North and South.

1976 The Socialist Republic of Vietnam is proclaimed, and Saigon is renamed Ho Chi Minh City. Many flee abroad.

Southeast Asian nations want **independence from colonial rule**.

→

The US fears **communism is spreading** across Southeast Asia.

↓

↓

After a war with France, **Vietnam splits** between a communist North and a US-backed South.

The US increases its **military presence** as a response to **communist successes** in the region.

↓

↓

Covert American activity culminates in a US warship being attacked in the Gulf of Tonkin.

↓

US president Johnson uses the incident to justify **military intervention in Vietnam**, widening the frontiers of the Cold War.

---

n the aftermath of World War II, the states of Southeast Asia struggled to create stable political systems, and the region became embroiled in the Cold War between the United States and the Soviet Union. In few places were the battle lines as sharply drawn as in Vietnam. After French colonial rule came to an end in 1954, Vietnam was divided at the Geneva Conference into North Vietnam, with a communist government under Vietnamese communist revolutionary leader Ho Chi Minh, and the US-backed South Vietnam. In 1960, Ho Chi Minh, with support from communist

See also: The construction of Angkor Wat 108–09 ▪ Stalin assumes power 281 ▪ Nazi invasion of Poland 286–93 ▪ The Berlin Airlift 296–97 ▪ The Long March 304–05

The US Navy destroyer *Maddox* was sailing off the coast of North Vietnam when it came under attack. This incident was the spark that led to the Vietnam War.

superpowers Russia and China, set up the National Liberation Front (NLF) in South Vietnam, and started a guerrilla war to unite the country under communist rule.

Tensions steadily rose until 1964. In August of that year, the US Navy destroyer *Maddox* was operating off the coast of North Vietnam in the Gulf of Tonkin, monitoring radar and radio from northern coastal installations, to support attacks made by the South Vietnamese navy. North Vietnam, believing the *Maddox* was linked to raids on its coastal targets, launched a torpedo attack. Two days later, the *Maddox* reported once again coming under fire. This second attack has since been disputed, but US president Lyndon B. Johnson, recognizing that South Vietnam could not prevail on its own against a communist-led guerrilla movement that already

controlled much of the country, used the skirmish to pass the Gulf of Tonkin Resolution in Congress. This allowed him to take any measures necessary to deal with threats to US forces in Southeast Asia.

US intervention

The US feared that if Vietnam became a communist regime, other countries in the region would soon follow. Using the Gulf of Tonkin Resolution, Johnson poured troops into the South and bombed North Vietnam by air. Huge numbers of

civilians were killed, but despite their technological superiority, the Americans failed to crush the Viet Cong guerrillas. American troops suffered high casualties and gradually became demoralized.

The specter of communism

The Vietnam War was the first televised war in US history. As the public watched horrific events unfolding, an increasing number opposed the conflict. Around the world, peace movements organized large anti-war demonstrations.

The communists' Tet Offensive of 1968, a series of fierce attacks on more than 100 cities and towns in South Vietnam, crushed US hopes of an imminent end to the conflict, and peace talks were initiated in 1969. In March 1973, the last American troops withdrew from Vietnam, and in April 1975 South Vietnam fell to the North.

US policy-makers consistently misinterpreted Asian nationalist movements for Soviet-inspired communism. Ultimately, however, what the US feared never came to pass, and with the exception of Laos and Cambodia, the region remained out of communist control. ▪

Pol Pot's brutal regime

During the Vietnam War, North Vietnam used Cambodia to channel soldiers and supplies to the South along the Ho Chi Minh Trail. In 1970, a joint US–South Vietnamese force invaded Cambodia to flush out the Viet Cong. The US also heavily bombed Cambodia. The military destabilization in Cambodia led to a surge of support for Pol Pot, the leader of the Kampuchean Communist Party, or Khmer Rouge, a guerrilla movement that seized power in 1975.

Pol Pot's brutal regime intended to style the country into a classless agrarian society inspired by Mao Zedong's Cultural Revolution in China. The entire population was marched to the countryside and forced to work as rice farmers. Over the next 44 months, around 2 million people—a quarter of Cambodia's population—died, either killed or starved. The fields where people died became known as the "Killing Fields". After three years of terror, Pol Pot was driven from power by a Vietnamese invasion.

A REVOLUTION IS NOT A BED OF ROSES

THE BAY OF PIGS INVASION (1961)

IN CONTEXT

FOCUS
Revolution and reaction in Latin America

BEFORE
1910 The Mexican Revolution is the first major social revolution of the 20th century.

1952 The National Revolutionary Movement (MNR) takes power in Bolivia.

1954 A military junta is installed in Guatemala in a coup organized by the CIA.

AFTER
September 11, 1973 Salvador Allende, president of Chile, dies during a coup led by army chief Augusto Pinochet.

1981 The US suspends aid to Nicaragua and supports fighters known as Contras, in an attempt to overthrow the left-wing Sandinistas.

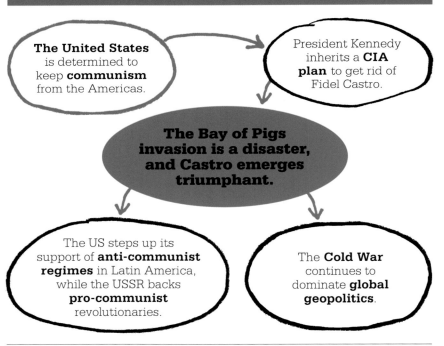

The United States is determined to keep **communism** from the Americas.

President Kennedy inherits a **CIA plan** to get rid of Fidel Castro.

The Bay of Pigs invasion is a disaster, and Castro emerges triumphant.

The US steps up its support of **anti-communist regimes** in Latin America, while the USSR backs **pro-communist** revolutionaries.

The **Cold War** continues to dominate **global geopolitics**.

On April 15, 1961, a force of Cuban exiles began an invasion of Cuba to try to topple Fidel Castro's left-wing regime and replace it with one more open to American interests. Eight American B-26 bombers flew from Nicaragua to destroy Castro's air force on the ground. The air raid seemed successful, but at least six of Castro's fighter planes survived.

The next day, Castro's air force sank two ships loaded with vital supplies. In the early hours of April 17, a group of around 1,400 Cuban exiles, codenamed Brigade 2506, launched an amphibious assault on the coast of southern Cuba, the Bay of Pigs. They were driven back by Castro's forces and ran out of ammunition. It only took three days to thwart the exiles' invasion.

See also: Bolívar establishes Gran Colombia 216–19 ▪ The October Revolution 276–79 ▪ Stalin assumes power 281 ▪ The Cuban Missile Crisis 308–09 ▪ The military coup in Brazil 341 ▪ Pinochet seizes power in Chile 341

The Bay of Pigs invasion was a disaster for the United States, and many anti-Castro forces were captured during the conflict.

Castro must go

After World War II, Latin America became a proxy battleground for two competing ideological systems: capitalism and communism. The US was determined to eradicate communism and supported right-wing dictators with anti-reformist regimes in countries such as Cuba, Honduras, and Guatemala.

During the 1950s, corruption and brutality within the Cuban Batista government forced a slow withdrawal of US support. When Castro defeated Batista in 1959, the US government had misgivings over Castro's communist leanings. By 1960, Castro had nationalized all US interests in Cuba without compensation and had broken diplomatic ties. To protect their economic assets and defeat communism, US policy-makers decided that Castro must go.

Within a year of Castro taking power, several counter-revolutionary groups were formed by Cuban exiles in Miami. The American Central Intelligence Agency (CIA) took an interest in these groups, providing them with training and equipment to topple the Cuban government.

The failure at the Bay of Pigs was largely down to poor planning and President Kennedy's reluctance to become too involved.

Pro-Cuba demonstrations

Castro forged a closer alliance with the Soviet Union, its ally against American aggression, enabling him to export his ideals across Latin America. The invasion incited pro-Cuba anti-US demonstrations from Chile to Mexico. Castro actively supported guerrilla warfare, and thousands of Latin American guerrillas went to Cuba for training. The revolution in Cuba inspired similar uprisings through the 1960s and 70s in Nicaragua, Brazil, Uruguay, and Venezuela, where there was disaffection with illiteracy, inequality, and poverty.

Latin America continued to preoccupy US foreign policy. The US intervened several times in an effort to contain communism. They supported military coups in Chile in 1973 and Argentina in 1976 and, fearing a communist takeover, funded the El Salvadoran military in the late 1970s to prop up their regime. In 1983, the US invaded Grenada; and in 1989, Panama. ▪

Cuba must not be abandoned to the communists.
John F. Kennedy

Fidel Castro

To his supporters, Fidel Castro (b.1926) was a revolutionary hero who stood up to the US. To his detractors, he was a dictator whose close ties with the Soviet Union brought the world close to nuclear war.

Jailed as a student in 1953 for his revolutionary activities, Castro was released two years later and went into exile in the US and Mexico. He returned to Cuba in 1956 with a small guerrilla band, among them the Argentine Marxist revolutionary Ernesto "Che" Guevara, and set to work undermining the regime of the dictator Batista. On January 1, 1959, he assumed absolute power. Castro was determined to improve literacy, offered free healthcare, and instituted land reforms.

Castro saw himself as a leader of the world's oppressed people and helped train anti-Apartheid forces in South Africa. In the 1970s, he sent troops to support communist forces in Angola, Ethiopia, and Yemen.

In 2008, wracked by ill health, Castro stood down as president of Cuba, leaving power in the hands of his brother Raúl.

SCATTER THE OLD WORLD, BUILD THE NEW

THE CULTURAL REVOLUTION (1966)

Mao Zedong fixes on ambitious plans to industrialize China.

In the **Great Leap Forward**, all Chinese society is **directed to this cause**.

Mao launches the Cultural Revolution.

Famine strikes, and mass starvation ensues. **Tens of millions die**.

Mao's death marks a key turning point in China's post-war history.

Deng Xiaoping's adoption of **capitalist ideas** allows China to move toward **superpower status**.

The Cultural Revolution was one of the darkest periods in Chinese history. Since taking power in 1949, Communist Party leader Mao Zedong had neither created his ideal China nor secured his power. To bolster his primacy and ignite revolutionary fervor, Mao decided to purge any opposition and transform capitalists and intellectuals into proletarians – ordinary workers. He ordered the Cultural Revolution, which would attack the "Four Olds": old ideas, old habits, old customs, and old culture. Squads of young communists, incited by Mao and known as the Red Guards, terrorized intellectuals,

In this propaganda poster dating from around the time of the Cultural Revolution, the Red Guards are shown with a copy of Mao's *Little Red Book*.

bureaucrats, and teachers. Some 36 million people were persecuted, and up to a million died in the turmoil, which lasted until 1976.

The Great Leap Forward
After creating the Chinese People's Republic in 1949, Mao launched reforms to transform China's semi-feudal, mostly agricultural society into an industrialized socialist state. In the late 1950s, in a bid to achieve rapid economic growth, Mao ordered the Great Leap Forward.

Industrial output climbed with steel and coal production, the rail network doubled, and more than half of all Chinese land was irrigated by 1961.

However, this development came at a terrible cost. Mao transformed rural China into a series of farming communes in which villagers pooled land, animals, tools, and crops. The authorities took vast amounts of grain from the communes to feed city workers, and this, along with a series of natural disasters, led to famine and starvation. The consequences were staggering: an estimated 45 million people died.

A new foreign policy
After the Cultural Revolution, Mao needed American expertise to restore China, and the US wanted an ally against the Soviet Union. In 1972, US President Richard Nixon traveled to Peking to meet with Mao. By the time Mao died in 1976, China had become a major oil producer with nuclear capabilities.

Deng Xiaoping, who led China from 1978 to 1997, was willing to use capitalist ideas to focus on economic growth. But while he initiated new and far-reaching measures, such as inviting foreign firms to invest in Chinese industry and supporting developing technologies, he also resisted pressure to make democratic reforms.

By the beginning of the new millennium, China's economic growth was spectacular. In 2001, the country was admitted to the World Trade Organization, and in 2008 it played host to the Olympic Games in Beijing. Some economists predict that by 2026 China will boast a gross domestic product (GDP) greater than Japan and Western Europe.

After Mao's death, the Chinese Communist Party condemned the Cultural Revolution as a disaster. However, as the country experienced a period of unparalleled economic growth, a sense of nostalgia for Mao's ideals, focused on the people and self-sufficiency, grew among farmers and members of the urban working class. Today, Mao's legacy continues to cast a long shadow over a modernizing China. ▪

Mao Zedong

Born in 1893 into a wealthy farming family from Hunan Province, Mao Zedong was the leader of Communist China from 1949 until his death in 1976. While working as a librarian at Peking University, he became a communist and helped found the Communist Party in 1921. Six years later, after leading an unsuccessful rebellion against nationalist leader Chiang Kai-shek, Mao was forced to retreat to the countryside, where he proclaimed the Chinese Soviet Republic in 1931. He took control of the Communist Party in 1935, after proving his leadership during the Long March, and defeated Chiang during the civil war of 1945–49.

A devoted Leninist, Mao became disenchanted with the Soviet policy of "peaceful coexistence" toward the West and developed Maoism, a stronger form of communism. However, his radical ideas and experiments with collectivization led to the death and suffering of millions. One of his last acts, in 1972, was to hold a meeting with Richard Nixon, the first American president ever to visit China.

WE SHALL DEFEND IT WITH OUR BLOOD AND STRENGTH, AND WE SHALL MEET AGGRESSION WITH AGGRESSION AND EVIL WITH EVIL

THE SUEZ CRISIS (1956)

IN CONTEXT

FOCUS
Modern Middle East

BEFORE
1945 Egypt, Iraq, Lebanon, Syria, Saudi Arabia, northern Yemen, and Transjordan form the Arab League.

1948 Israel is established in former Palestine, dividing Arabs and Jews.

1952 A military coup removes Egypt's King Farouk from power. Colonel Gamal Nasser seizes control two years later.

AFTER
1964 The Palestine Liberation Organization calls for an end to the Jewish state.

1993 The Oslo Accords provide for mutual recognition between the PLO and Israel.

2011 Protestors across Arab states demand reforms in a series of popular uprisings.

On July 26, 1956, Egyptian leader, Colonel Gamal Abdel Nasser, addressed a crowd in the city of Alexandria, declaring the nationalization of the Suez Canal, the waterway through which most oil bound for Western Europe had to pass. For Egyptians, the nationalization symbolized the liberation of their country from the British imperialist dominance it had been under since the 1880s. In response to Nasser's bold move, a secret plan was hatched by Britain, France, and Israel. France was eager for Nasser's downfall because of his support for Algerian insurgents against French colonial rule in Algeria. Israel had many reasons for

See also: The construction of the Suez Canal 230–35 ▪ The Young Turk Revolution 260–61 ▪ The Treaty of Versailles 280 ▪ The establishment of Israel 302–03 ▪ The 9/11 attacks 327 ▪ The Soviet invasion of Afghanistan 341 ▪ The Iranian Revolution 341 ▪ The USA and Britain invade Iraq 341

President Nasser of Egypt announces the nationalization of the Suez Canal to a quarter-million-strong gathering in Alexandria celebrating four years since the revolution.

toppling Nasser, including Egypt's denial of passage through the canal to any Israeli-flagged ships. The three conspired that Israel would attack Egypt, and Britain and France would intervene a few days later posing as peacemakers, taking control of the canal. On October 29, 1956, the Israelis began their assault. British and French troops invaded on October 31, but faced immediate diplomatic pressure to call a ceasefire. The United States, which was trying to cultivate good relations with Arab states, was appalled by the Anglo-French invasion, believing it threatened the stability of the whole region. President Dwight Eisenhower forced through a United Nations resolution imposing a ceasefire, and British and French troops had to conduct a humiliating withdrawal.

Splitting the land
The strong anti-Western sentiment in the Middle East dates back hundreds of years, fueled by the West's increased involvement in

the region. Colonialism in the 1800s and the division of the Ottoman Empire after World War I were bitter humiliations for peoples who felt their religion, Islam, was the highest form of divine revelation. In 1948, the partition of Palestine to form Israel spilt the land into two states, one Arab and one Jewish, and was rejected vehemently by Israeli Arabs and enraged the other Arab nations. The regular armies of the Arab states—including Iraq, Lebanon, Syria, Transjordan, and Egypt—attacked Israel in the first Arab–Israeli War in May and June 1948. The conflict ended in defeat for the Arabs and disaster for the Palestinians: more than half of the country's Arabs were uprooted as refugees, and they lost any possibility of a state of their own.

Ambitious plans
Egypt continued its stance of belligerence toward Israel by closing the Suez Canal to Israeli shipping. When Nasser ousted the regime of King Farouk in 1952 and sent him into exile, he imported arms from the Soviet Union to build his arsenal for future confrontations with Israel. Britain had agreed to withdraw its troops from the Suez area by June 1856, but as the last troops left Egypt, Nasser relied on funds from Britain and the US to fund ambitious plans to develop Egypt. This included the Aswan Dam project on the Nile. Nasser was angered when Britain and the United States withdrew its offer of loans to help him pay for the dam. The US and Britain backed out »

Israel is founded in Palestine.	There is a rise in **Arab nationalism**.

The Suez Crisis deals a blow to British and French imperialism and stirs anti-Western sentiments in the Muslim world.

The Israeli–Arab dispute widens to become an **Israeli–Arab conflict**.	The United States becomes the **main backer of Israel**.	There is an increase in **Palestinian liberation movements**.

Chaos and violence grip the Middle East.

because of Nasser's association with the Soviets and his unceasing diatribes against the West. Nasser felt insulted and immediately nationalized the Suez Canal. The move was popular in Egypt, as the canal was a source of Arab pride.

Nasser was a secular modernizer who advocated the separation of religion from political life, believing it the hallmark of Arab modernity, but this was not universally welcomed. The Muslim Brotherhood, founded in Egypt in 1928, argued for Islam to have a central role in government. After repeated calls for the application of Sharia law—a legal system based on Islam—and an assassination attempt against Nasser, the organization was finally banned in 1954.

In 1967, Arab countries suffered a crushing defeat at the hands of Israel in the Six Day War, in which Israel took the Sinai from Egypt, the Golan Heights from Syria, and the West Bank and East Jerusalem from Jordan, meaning Israel was now an occupier. In the 1970s and 80s, the Arab–Israeli conflict largely moved in the direction of peace: in 1979 the Israeli–Egypt peace deal ended 30 years of war. The rise of the

US President Jimmy Carter (center) looks on as President Anwar Sadat of Egypt shakes hands with Menachem Begin, Israel's premier, after signing a peace treaty at the White House in 1979.

Palestine Liberation Army (PLO) and of other Palestinian groups attacking Israel, however, as well as Israel's invasion of Lebanon in 1982, where many of the PLO were grouped, destabilized of the fragile peace continuously.

The Iran–Iraq War

Like many countries in the Middle East, modern Iraq was carved out of the ruins of the Ottoman Empire in the aftermath of World War I. Iraq was a land divided along ethnic lines between Arabs and Kurds, as well as sectarian lines between Sunni and Shia Muslims, the latter being the majority group. Saddam Hussein, a Sunni, became leader in 1979, and suppressed ethnic Kurds and Shias alike using immense brutality. He, like Nasser in Egypt, espoused Arab nationalism and ruled Iraq as a secular state.

In 1979, events in Iran inspired Islamists throughout the Middle East. The secular, Western way of life was swept away in an Islamic revolution in which the US-backed Shah was ousted. The new regime, under Ayatollah Khomeini, a Shia Muslim, based its laws and ideology on the strict teachings of the Koran. Saddam felt threatened by the Islamic revolution and a possible Shia uprising in his own country, so he invaded Iran on September 22, 1980 under the pretext of a territorial dispute over the Shatt al-Arab, a waterway that lies between the two countries.

The invasion triggered a bruising eight-year war that devastated both countries and increased tensions in the Middle East. Iran's principal ally was Syria, but Libya, China, and North Korea all also sent it weapons. Iraq's support came mostly from the Arab Gulf states, which viewed Iran as the greater danger to their security; Saudi Arabia and Kuwait

During the First Gulf War, Iraqi forces set fire to more than 600 Kuwaiti oil wells. Saddam Hussein's desire to control Kuwaiti oilfields had initially led to Iraq invading Kuwait in 1990.

provided billions of dollars in loans. Ultimately, there were no victors; and Iraq, now awash with armaments supplied by several Western nations, including Britain, France, and the United States, invaded the oil-rich state of Kuwait in 1990. The UN demanded their withdrawal, but Saddam announced that Kuwait had been annexed by Iraq. The United States, with support from coalition forces, sent in troops during the First Gulf War (1990–91) and freed Kuwait.

The 9/11 attacks

The continued US support of Israel led to profound grievances among Islamists. To them, the capitalist, secular US, with its greed for oil, symbolized all that was wrong with the West, and terrorist strikes on US targets grew. Al-Qaeda carried out the most shocking on September 11, 2001, against four targets in the United States, including the World Trade Center in New York City.

In response to the 9/11 attacks, a successful US-led international intervention brought down the Taliban regime in Afghanistan, which the US believed had given sanctuary to Osama bin Laden and

> We shall not be satisfied except by the final obliteration of Israel from the map of the Middle East.
>
> **Muhammad Salah al-Din**
> **Egyptian foreign minister (1954)**

al-Qaeda. After September 11, President Bush declared a "War on Terror" and, in 2002, with help from the British government, attacked Iraq on the premise of destroying "weapons of mass destruction" (WMDs) deemed a threat to national security. Western intervention in the Muslim world heightened the belief among Islamists that the West was the enemy of Islam.

The Arab Spring

The 9/11 attacks were inspired by a radical ideology and belief that the fundamental problems plaguing Arab and Muslim people could be resolved by attacking foreign powers that were seen to oppress Islam. In 2011, young Arabs—looking inward to promote change and blaming their own leaders for decades of political, economic, and cultural decline—were at the heart of uprisings across the Arab world. At its core, what became known as the Arab Spring was a new generation's attempt to change the state order. An extraordinary series of pro-democracy uprisings, the Arab Spring caused huge upheavals in the Middle East and North Africa. It started in Tunisia on December 17, 2010 when a street vendor set himself on fire in a protest against police brutality. Protestors throughout Tunisia demanded democracy, and President Zine el Abidine fled the country on 14 January. Disorder spread from Tunisia to Algeria, where there was unrest over lack of jobs.

On January 25, thousands of protestors took to the streets in Egypt, and after 18 days of protests there, President Hosni Mubarak resigned. By mid-February, civil unrest had swept through Bahrain, where it was brutally suppressed, and into Libya. Muammar Gaddafi's violent response to the dissidents led to civil war. An international coalition led by NATO launched a campaign of air strikes targeting Gaddafi's forces, and he was killed in October 2011.

Further uprisings occurred in Jordan, Yemen, and Saudi Arabia, but the worst violence against civilians was seen in Syria, where President Bashar Assad promised reforms but used force to crush the dissent—a move that merely hardened the protestors' resolve. In July 2011, hundreds of thousands of people took to the streets, and the country descended into civil war. By August 2015, the United Nations reported that more than 210,000 people had been killed in the conflict. Capitalizing on the chaos in the region, so-called Islamic State (also referred to as IS, ISIS, or ISIL), the extremist Muslim group born from a fringe of al-Qaeda, took control of huge swathes of territory across northern and eastern Syria, as well as neighboring Iraq.

Middle East instability

The Suez Crisis was the end of one era in the politics of the Middle East and the start of another. It marked the humiliating end of imperial influence for two European countries, Britain and France, whose role was soon taken over by the US. It stimulated Arab nationalism and opened an era of Arab-Israeli wars and Palestinian terrorism.

In modern times, the Middle East has never seemed so unstable. Wars are being fought over religion, ethnicity, territory, politics, and commerce, and these conflicts have led to the worst refugee crisis since World War II, with millions fleeing anarchy and fanaticism. ∎

Terrorism in the Middle East

Since the mid-20th century, terrorism has been synonymous with the Middle East. The Israel–Palestine conflict is one of the world's most challenging.

In 1964, Arab leaders formed the Palestine Liberation Organization (PLO), declaring Israel's establishment illegal. The PLO used terrorism to attack Israel and Western targets for their support of Israel. In 1970, Palestinian militants blew up three hijacked planes in the Jordanian desert, and in 1972 a group linked to the PLO hit the Israeli Olympic team during the games in Munich, Germany.

In 1983, Hezbollah, an Iran-backed fundamentalist Shiite Muslim group in Lebanon, blew up the Beirut barracks of both US Marine and French forces, killing 298 people. Hezbollah pioneered the use of suicide bombers in the Middle East.

Both Jews and Muslims have employed terrorism to derail the many attempts that have been made at peace in the region.

THE IRON CURTAIN IS SWEPT ASIDE

THE FALL OF THE BERLIN WALL (1989)

IN CONTEXT

FOCUS
Collapse of communism

BEFORE
August 1989 After 45 years, Poland sees the end of communist rule. Solidarity, a trade union, forms a new non-communist government.

August 23, 1989 Two million people form a human chain across Estonia, Lithuania, and Latvia in protest at Soviet rule.

September 11, 1989 Hungary opens its border with Austria to allow the departure of East German refugees.

AFTER
December 3, 1989 The US and USSR jointly declare that the Cold War has ended.

October 3, 1990 Germany is reunified.

December 1991 The Soviet Union disintegrates into 15 separate states.

Gorbachev is elected president of the USSR. He introduces **radical political and economic reforms**.

↓

This process of democratization **reduces Cold War tensions**.

Gorbachev has no intention of using **military force** to prop up satellite communist regimes.

↓

Uprisings take place across Eastern Europe, and **communist regimes** are ousted.

↓

The Berlin Wall is dismantled, followed soon after by the collapse of the Soviet Union.

For decades, the Berlin Wall, which separated East and West Berlin, stood as a reminder of the Cold War, the bitter division between Soviet communism and Western capitalism. On November 9, 1989, the East German government lifted travel restrictions, and thousands of people began converging at the wall. East German border guards yielded in the face of ecstatic crowds. On November 10, in extraordinary scenes, soldiers from both sides helped Berliners break through the wall. Over the next two days, more than 3 million people crossed the border.

The fall of the Berlin Wall meant liberation for many people. German reunification, the collapse of the Soviet Union, and the end of communism in Eastern Europe followed soon after.

Ruling the Eastern Bloc

At the end of World War II, the USSR had banned anti-communist parties in every Eastern European country, and created a bloc of satellite states under Soviet leadership, ruthlessly suppressing any opposition. In the fall of 1956, Hungary rose against its communist government, only to be crushed by Soviet tanks, and in 1968, the USSR invaded Czechoslovakia to remove a government it found too liberal.

In the 1960s, Germany was still divided between East and West, and its former capital Berlin split into the Allied-operated West and the Soviet-controlled East. Each had its own German administration: democratic in the West, communist in the East. Thousands of East Germans escaped to the West, and the country hemorrhaged its skilled workers. On August 13, 1961, the government sealed off East from West Berlin with a fence, which, over time, became a heavily fortified barrier dividing the city, the nation, and family and friends.

In 1985, Mikhail Gorbachev was appointed as General Secretary of the Soviet Communist Party.

Aiming for warmer relations with the West, he set out new reforms: *glasnost* (political "openness") and *perestroika* (liberal economic "restructuring"). Critically, he lifted the ban on Eastern Bloc countries reforming their political systems.

Collapse of communism

With the threat of Soviet military intervention removed, citizens in all Eastern Bloc countries protested to end communist rule. In June 1989, Poland's Solidarity, originally a banned trade union, was elected to lead a coalition government. As the push for reform gathered pace, the East German government declared that its citizens would be able to visit West Berlin through any border crossing, including the Berlin Wall.

The fall of the Berlin Wall was a momentous event. It marked an era that saw the end of the Cold War and the dissolution of the Soviet Union. It allowed millions to travel more freely, and previously stifled economies across Eastern Europe and the Former Soviet Union opened up to the world. Many former communist countries were welcomed into NATO and joined the European Union.

The world changed course in 1989. Communism was dead in the East, and a reunified Germany was about to take its place at the heart of Europe. ▪

The break-up of the Soviet Union

In 1985, Mikhail Gorbachev became leader of a stagnating Soviet Union. He laid out radical reforms—*glasnost* and *perestroika*—and in July 1989 he announced that countries within the Warsaw Pact could hold openly contested elections. Poles, Czechs, Hungarians, and others opted for democratic governments, destabilizing the Soviet Union itself.

In July 1991, the anti-communist Boris Yeltsin was elected president of Russia. A month later, with Gorbachev weakened by an attempted coup by hardline communists, Yeltsin took advantage. He banned the Communist Party in Russia and met secretly with the leaders of Ukraine and Belarus, who agreed to secede from the Soviet Union. On Christmas Day 1991, Gorbachev resigned, leaving Yeltsin as president of the new Russian state. The former empire split into 15 new independent states, and the USSR was no more.

ALL POWER TO THE PEOPLE
THE 1968 PROTESTS

IN CONTEXT

FOCUS
Radical post-war politics

BEFORE
1963 *The Feminine Mystique* by Betty Friedan reignites the women's rights movement.

1967 The killing in Berlin of student demonstrator Benno Ohnesorg sparks a revolt.

March 1968 Demonstrators in Italy protest against police brutality.

AFTER
1969 The Days of Rage demonstrations in Chicago use violence to protest against the Vietnam War and US racism.

1970s The radical group Japanese Red Army protests the presence of US military bases in Japan.

1978 The Italian Red Brigades take former prime minister Aldo Moro hostage as part of their left-wing terrorist campaign.

I
n 1968, a small demonstration over poor campus facilities at Nanterre University in a suburb of Paris, France, spread across the country. In March, riot police were called to deal with the unrest, and hundreds of students descended on Nanterre. By May, the uprising had moved to the center of Paris, and the number of protestors swelled to thousands. Tension erupted on the streets, as demonstrators called for revolutionary social change and the collapse of the government. Within a few days, 8 million workers went on a wildcat strike that brought France to a standstill.

A momentous year

France's journey to near revolution is the defining event of 1968, a year of global protest. Much was against the Vietnam War, but many people also marched against oppressive regimes. Politics became more radical: the "coming out" of sexual minorities, women's liberation, and sexual equality came to the fore. In the United States, groups such as the Black Panthers fought for racial equality; and the German Student Movement, led by Rudi Dutschke, opposed the older generation, who had been part of World War II.

The French protests lost steam as elections showed overwhelming support for the government. The revolutionary movements of 1968 ultimately failed, but they inspired a generation to question authority. In their wake came a rise in left-wing terrorist groups that used bombing and kidnapping while purporting to fight for social justice. ∎

What's important is that the action took place, when everybody judged it to be unthinkable.
Jean-Paul Sartre

See also: Nkrumah wins Ghanaian independence 306–07 ▪ The March on Washington 311 ▪ The Gulf of Tonkin Incident 312–13 ▪ De Gaulle founds the French Fifth Republic 340 ▪ The Red Army Faction's terrorist activity 341

NEVER, NEVER, AND NEVER AGAIN

THE RELEASE OF NELSON MANDELA (1990)

IN CONTEXT

FOCUS
End of apartheid

BEFORE
1948 The National Party (NP) takes power, adopting a policy of apartheid (separateness).

1960 Seventy black protesters are killed at Sharpeville; the African National Congress (ANC) is banned.

1961 South Africa is declared a republic and leaves the Commonwealth. Mandela heads the ANC's military wing.

AFTER
1991 F. W. De Klerk repeals apartheid laws; international sanctions are lifted.

1994 With the first democratic elections, South Africa joins the UN General Assembly.

1996 The Truth and Reconciliation Commission begins hearings on human rights crimes committed in the apartheid era.

N elson Mandela received a life prison sentence in 1964 for his role in anti-apartheid protests held in Sharpeville, South Africa. Mandela was a militant member of the African National Congress (ANC), set up to campaign against apartheid, a system of racial segregation enforced by the white ruling government. While in prison, Mandela had become a symbol of the struggle for racial equality. On his release in 1990, he was greeted with euphoria.

Friends, comrades, and fellow South Africans, I greet you all in the name of peace, democracy, and freedom for all.
Nelson Mandela

When the Nationalist Party was elected to power in 1948, white Afrikaners implemented a brutal apartheid policy—black people were segregated and could not vote. Many in the anti-apartheid movement advocated non-violent protest, which helped rally white South Africans to their cause. Apartheid was globally condemned, and tough international sanctions were imposed.

A new dawn

In 1990, President F. W. De Klerk astounded the world by lifting bans on the ANC. Seeing the need for fundamental change, he had been in secret negotiations for two years to end the apartheid system.

Multiracial elections were held in 1994, and Mandela won by a huge margin. His release was one of the defining moments of the late 20th century, ending 300 years of white rule in South Africa. It transformed the country into a multiracial democracy without the bloody civil war that so many had feared. ∎

See also: The Slave Trade Abolition Act 226–27 ▪ The Berlin Conference 258–59 ▪ Nkrumah wins Ghanaian independence 306–07 ▪ The March on Washington 311

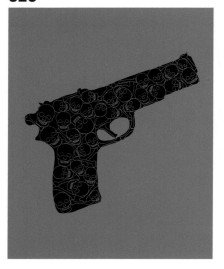

CREATE AN UNBEARABLE SITUATION OF TOTAL INSECURITY WITH NO HOPE OF FURTHER SURVIVAL OR LIFE
THE SIEGE OF SARAJEVO (1992–1996)

IN CONTEXT

FOCUS
Conflicts since the collapse of the USSR

BEFORE
November 9, 1989 The Berlin Wall collapses, leading to the reunification of Germany.

1989 Romania overthrows the ruthless regime of Nicolae Ceauşescu.

1990 In Poland, Hungary, and Czechoslovakia, newly formed center-right parties take power.

1992–95 The war in Bosnia-Herzegovina results in the death of some 100,000 people.

AFTER
1998–99 War breaks out in Kosovo between ethnic Albanians and ethnic Serbs. NATO troops intervene.

2014 Fighting erupts between Russians and Ukrainians in eastern Ukraine.

The Siege of Sarajevo, Bosnia, was one of the most appalling tragedies in Yugoslavia's civil war (1991–2002). During the 44-month siege, the city's food and electricity supplies were cut off, and the civilian population was bombarded by nationalist Bosnian Serbs. Thousands of Bosnian Muslims were targeted and murdered.

A new wave of nationalism
Yugoslavia was comprised of six socialist republics: Croatia, Montenegro, Slovenia, Bosnia and Herzegovina, Macedonia, and Serbia, each with its own prime minister and constitution. Overall power in Yugoslavia was held by a president, notably communist leader Josip Broz Tito from 1953 to 1980.

After the 1991 break-up of the Soviet Union, a nationalist revival swept Eastern Europe. Croatia and Slovenia's call for independence was opposed by Serbia, and Vukovar, in eastern Croatia, was destroyed by the Yugoslav army under Serbian leader Slobodan Milosevic. When

Bosnia also declared independence in 1992, the violence intensified. Bosnian Serbs aimed to create a separate ethnically pure Serbian state, the Republika Srpska, carving it from the new Republic of Bosnia-Herzegovina. Nationalist Bosnian Serbs, supported by neighboring Serbia, launched a campaign to expel non-Serbs, and during the Siege of Sarajevo, they targeted the majority Bosnian Muslim population.

The Bosnian War ended in 1995, but fighting continued in Kosovo, where ethnic Albanians began a separatist movement against the Serbs. Ethnically based nationalism also led to bloody anti-Armenian pogroms in the Nagorno-Karabakh region and in Baku, the capital of Azerbaijan. In Georgia, violence erupted between the Georgian and Abkhazian population.

The wars in Yugoslavia forced the issue of the global community's responsibility to resolve disputes that threaten wider instability or cause unacceptable human suffering or rights violations. ∎

See also: The October Revolution 276–79 ▪ Nazi invasion of Poland 286–93 ▪ The fall of the Berlin Wall 322–23

TODAY, OUR FELLOW CITIZENS, OUR WAY OF LIFE, OUR VERY FREEDOM CAME UNDER ATTACK
THE 9/11 ATTACKS (2001)

O n September 11, 2001, a group of Islamic extremists launched a devastating attack against the US. Two hijacked airliners crashed into the World Trade Center in New York; another hit the Pentagon, in Washington, DC; and a fourth plane crashed in Pennsylvania. Almost 3,000 people were killed.

The seeds of extremism
September 11 was not the first terror attack on American soil by Islamic extremists. On February 26, 1993, a bomb was detonated at the World Trade Center by men thought to have links to al-Qaeda, a militant Islamist organization. Some Muslims had been radicalized and adopted international terrorism during the struggles over Israel. In 1979, the Soviet invasion of Afghanistan led to the worldwide mobilization of Muslim militants to fight the invaders. Around that time, Osama Bin Laden formed al-Qaeda. Intelligence reports suggested that he was the mastermind behind September 11. He was killed in 2011.

We love death more than you love life.
al-Qaeda motto

The civil war in Syria since 2011 and the power vacuum left by the departure of US forces in Iraq has led to the emergence of ISIS, the so-called Islamic State of Iraq and Syria, which has taken control of several towns in the region.

The events of September 11 mark the largest terrorist attack ever on US soil. Subsequent attacks in London, Madrid, and Paris, carried out by a diffuse network of regional terrorist groups, have added a chilling dimension to the threat of Islamic terrorism. ∎

See also: The Young Turk Revolution 260–61 ∎ The establishment of Israel 302–03 ∎ The Suez Crisis 318–21

YOU AFFECT THE WORLD BY WHAT YOU BROWSE

THE LAUNCH OF THE FIRST WEBSITE (1991)

IN CONTEXT

FOCUS
Communication and computing

BEFORE
1943–44 John Mauchly and J. Presper Eckert build the Electronic Numerical Integrator and Calculator (ENIAC), the forebear of digital computers.

1947 The transistor allows for small, powerful electronics, enabling later developments such as the home computer.

1962 The Telstar 1 satellite is launched, sending TV signals, telephone calls, and fax images through space.

1980s The first mobile phones come onto the market.

AFTER
2000s The boom in wireless communication connects nearly all of humankind.

2003 The invention of Skype allows for free communication over the Internet.

The US military sets up the **Advanced Research Projects Agency Network** (ARPANET). The **ARPANET** grows and develops to become **the Internet**.

The first website is launched to help users navigate the Internet.

 The Web becomes a **global telecommunications tool** used by millions.

 The Internet radically changes how the world **shares information and conducts business**.

The first website was titled "World Wide Web" and gave basic information about the World Wide Web project and how to create Web pages. It was built by Tim Berners-Lee, a British computer scientist at the European Organization for Nuclear Research (CERN) in Geneva, Switzerland.

Berners-Lee was interested in facilitating the exchange of ideas between scientists in universities and research institutes, and he first proposed his idea for a worldwide network of computers sharing information in 1989. His site went live in 1991 and was accessed by a small group of fellow CERN

See also: The opening of the Amsterdam Stock Exchange 180–83 ▪ Darwin publishes *On the Origin of Species* 236–37 ▪ The Berlin Airlift 296–97 ▪ The launch of Sputnik 310

Sir Tim Berners-Lee, creator of the World Wide Web, was fascinated by computers from a young age. Today, he is an advocate for an open and free Internet.

researchers. Crucially, Berners-Lee persuaded CERN that the World Wide Web should be given to the world as a free resource.

Although it revolutionized the computer and communications world like nothing before, the World Wide Web was only possible by bringing together several existing technologies: the telephone, television, radio, and Internet.

The Internet

The Soviet Union's launch of the Sputnik 1 satellite in 1957 spurred the US Defense Department to consider means of communication after a nuclear attack. This led to the formation of the ARPANET (Advanced Research Projects Agency Network) in 1969, a system initially of four computers. In the mid-1980s, this growing network of interconnected computers became known as the Internet. Both the Internet and the World Wide Web were limited to academic and research organizations.

It wasn't until the 1993 launch of a user-friendly Web browser called Mosaic that the Web took off for more general use. Mosaic could show pictures as well as text, and users could follow Web links simply by clicking on them with a mouse. The Web became synonymous with the Internet, but they are distinct from one another. The World Wide Web facilitated navigation of the Internet and helped make the Internet such an effective mode of communication.

The computing revolution

The introduction in 1981 of IBM's 5150 personal computer drove a revolution in home and office computing. Smaller and cheaper than the large office computers, it and its successors had access to the Internet and email. With personal computers, the Internet saw huge growth. The first search engines began to appear in the early 1990s; Google, which is now almost synonymous with Web searches, arrived a little later, in 1997. The launch of online marketplace Amazon in 1994 revolutionized the way people shopped, allowing the purchase of everything from books and CDs to hotel rooms and airline tickets from the comfort of home.

The Internet brought about significant changes to the way businesses operated; globalization escalated, and the world seemed to become a much smaller place, with communication improved by the speed and efficiency of the Internet. Jobs were outsourced, and companies effectively became "nationless," since it was easier to operate from anywhere in the world.

The next wave of technological advances saw devices become smaller and more mobile due to electronic components on tiny integrated circuits, or "chips."

The future is now

Nowhere has the introduction of microchip technology had more impact than the introduction of the Apple iPhone in 2007. So-called smartphones have made the Internet a mobile resource, with wireless connectivity offering on-the-go access to news and satellite navigation, for example. Information and ideas can be shared from anywhere at the touch of a button via social-networking sites such as Facebook and Twitter. Smartphones have also had an impact on education, healthcare, and culture, and have changed the political landscape through use by protestors organizing rallies via social media to undermine regimes. Uprisings such as the Arab Spring, which began in 2010, were partly powered by activists who communicated across the Internet. Internet activism, or "clicktivism," has since become a powerful way to share ideas, raise awareness, or support a cause. With more than 3 billion users, the World Wide Web has transformed every aspect of modern daily life. ▪

The information highway will transform our culture as dramatically as Gutenberg's press did the Middle Ages.
Bill Gates

A CRISIS WHICH BEGAN IN THE MORTGAGE MARKETS OF AMERICA HAS BROUGHT THE WORLD'S FINANCIAL SYSTEM CLOSE TO COLLAPSE

THE GLOBAL FINANCIAL CRISIS (2008)

IN CONTEXT

FOCUS
Globalization and inequality

BEFORE
1929 The Wall Street Crash leads to the Great Depression, the worst economic crisis of the 20th century.

1944 Delegates of 44 countries meet at Bretton Woods, New Hampshire, to reshape the global financial system.

1975 France, Italy, Germany, Japan, Britain, and the US form the Group of Six (G6) to foster international trade.

1997–98 The Asian financial crisis, starting in Indonesia and spreading around the world, is a precursor to events in 2008.

AFTER
2015 World leaders pledge to eradicate world poverty by 2030.

The turn of the 21st century brought troubling signs of a worldwide recession. Low interest rates and unregulated credit had induced more and more people to get into unsustainable debts. Bankers, particularly in the US, offered mortgages to customers with a poor credit history. These mortgages were called "subprime mortgages". It was hoped that if people could not keep up with their mortgage payments, their houses could be repossessed and sold at a profit, but this depended on house prices rising. In 2007, interest rates crept up, and house prices fell. People began defaulting on their monthly repayments. Across the US,

See also: The Wall Street Crash 282–83 ▪ The 1968 protests 324 ▪ The launch of the first website 328–29 ▪ Global population exceeds 7 billion 334–339

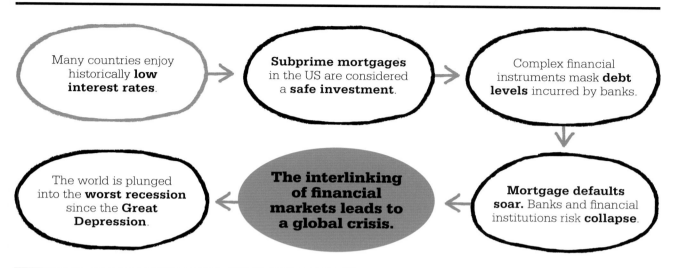

Many countries enjoy historically **low interest rates**.

Subprime mortgages in the US are considered a **safe investment**.

Complex financial instruments mask **debt levels** incurred by banks.

Mortgage defaults **soar**. Banks and financial institutions risk **collapse**.

The interlinking of financial markets leads to a global crisis.

The world is plunged into the **worst recession** since the **Great Depression**.

houses were repossessed at a great loss, with bankers fearing they would not get their money back.

The crisis spreads to Europe

In August 2007, the French bank Paribas revealed that it was at risk from the subprime mortgage market. Bankers had gambled with trillions of dollars of investment on risky mortgages that might now be worthless. Panic set in, and banks stopped lending to one another. British bank Northern Rock faced a shortage of readily available cash, and was forced to ask the British government for an emergency loan.

Around the world, shares began to plummet. In September 2008, US mortgage lenders Fannie Mae and Freddie Mac were rescued by the American government, while Lehman Brothers, a powerful investment bank heavily involved

in the subprime mortgages market, was forced to file for bankruptcy. The US government considered Lehman Brothers too insolvent and did not bail it out.

The turmoil in financial markets led to a severe economic downturn in most Western economies. Share prices plummeted, and world trade decreased because governments

spent less. Ireland became the first European country to fall into recession, a period of economic decline. Iceland's government resigned in October 2008 after the country became almost bankrupt. Some governments—such as those in the US, China, Brazil, and Argentina—planned stimulus »

Lehman Brothers, an investment bank with a long history of trading, filed for bankruptcy on September 15, 2008, after getting involved in the failing subprime mortgage market.

packages to boost their economies. They increased government spending and decreased taxes. Others, especially in Europe, opted for austerity, freezing public spending, and increasing taxes. Protests and strikes swept through Europe in response to these measures. Portugal, Spain, and Greece came under pressure from the European Union (EU) to lower their debts. The EU spent billions propping up weak economies in an attempt to keep the Eurozone, and the euro, viable. But the effect of the economic crisis was devastating, and many people lost their homes and jobs. It was the worst economic downturn since World War II.

Post-war economy
After World War II, most of Europe, Japan, China, and the Soviet Union, all devastated by war, needed time to recover. The US, which had experienced a huge rise in manufacturing for the war effort and was spared destruction, continued manufacturing at higher levels than ever before and dominated the world economy. The post-war economic planners sought a new economic order based on industrial strength and a stable dollar. In 1944, the International Monetary Fund (IMF) was formed to foster the revival of global trade. The US's strong post-war economy and the Marshall Plan of 1947, a US-led initiative to aid Western countries, invigorated world trade through encouraging capitalism and the free exchange of goods between nations. Signed in 1947, the General Agreement on Tariffs and Trade (GATT) dictated that tariffs be removed to open up markets around the world.

The Asian tiger
Japan, meanwhile, saw massive economic growth. The Japanese government implemented reforms based on efficiency and restricted foreign imports. They did not sign up to the GATT agreement until 1955. Japan invested in its coal and steel industries, as well as shipbuilding and car manufacturing. In the 1960s, Japan specialized in high-tech products such as cameras and computer chips. Countries such as South Korea, Taiwan, Singapore, and Malaysia experienced similar growth with an emphasis on electronics and technology. These successes became collectively known as "Asian tiger economics."

The role of oil
By the 1970s, the world was divided between rich industrial countries and poor developing nations, and oil had become increasingly important. In 1960, the Organization of Arab Petroleum Exporting Countries (OAPEC), including Saudi Arabia, Egypt, Iraq, and Iran, was founded. As oil reserves in other countries dwindled, the states around the Persian Gulf, where this resource had remained plentiful, became dominant. In October 1973, when Egypt and Syria invaded Israel during the Yom Kippur War, OAPEC embargoed oil to any country helping Israel, and prices tripled. Without oil, industrial output dropped. The United States introduced strict fuel rationing, which ended in March 1974, when the oil embargo was lifted.

A new economic model
The oil crisis in the mid-1970s led to a deep global recession, soaring inflation, and high unemployment. In response, a new "neo-liberal" economic policy was adopted, transferring control of economic factors from the public to the private sector. Welfare programs were perceived to be one cause of economic failures, and there were drastic cutbacks. Deregulation became the driving force behind world economics, sweeping away many governmental controls and freeing up organizations to trade across a wider range of territories. The need for this was particularly felt in the United States, which faced stiff competition from a world now fully rebuilt from the

> September and October of 2008 was the worst financial crisis in global history, including the Great Depression.
> **Ben Bernanke**
> Former head of the Federal Reserve

destruction of World War II. Some of the rigid laws and regulations that had been put in place to protect consumers were now considered to be interfering with free enterprise.

The global push for deregulation resulted in the adoption of new markets, greater competition, and openness, especially as the world adapted to the end of the Cold War and the collapse of the Soviet Union. The example set in east Asia influenced policy makers in other Asian countries, such as India and China. Mexico and Brazil lowered their barriers to trade and embarked on economic reform, leading to a dramatic improvement in living standards. As East and West Germany reunited in 1989, after the fall of the Berlin Wall, the European Union (EU), an economic union of 28 European countries, emerged as a major force in the world economy. Also in the 1980s, China opened up to foreign trade, and huge sums of foreign investment poured in, leading to extraordinary growth.

Global economy

The world economy is now far more open. Internet use allows people to order goods in one part of the world and have them delivered elsewhere within a matter of days. World trade is made up of global partnerships, with multinational companies that boast huge turnovers. Across the globe, people tend to migrate to cities to find work, resulting in an increase in urbanization.

One complaint that is often aimed at globalization is that some companies exploit cheap labor and behave unethically in their bid for profit. Another is that globalization has contributed to the extraordinary accumulation of wealth by a few individuals and, thus, increased inequality. Some countries have also remained extremely poor—areas of sub-Saharan Africa, for example have fared badly and been left behind, in debt to wealthier nations.

Economic recessions have occurred throughout history, but the financial crisis of 2008–11 was the worst—at least since the Great Depression of 1929—and maybe the worst ever. Many felt it was an avoidable disaster caused by widespread failures in government regulation and heedless risk-taking by investment bankers. Only massive monetary and fiscal stimuli prevented catastrophe. Household and business debts remained high, and there was widespread fury directed at bankers, whom many felt had survived relatively unscathed. Austerity measures provoked civil unrest. Demonstrations were held against capitalism; the Occupy Movement spread, with tens of thousands marching in New York, London, Frankfurt, Madrid, Rome, Sydney, and Hong Kong. While financiers argued over the causes of the Global Recession, the impact on the lives of ordinary people had profound, lasting consequences. ∎

People took to the streets to protest against the actions of banks and multinationals, which were seen as the trigger of the financial crisis.

An era of protest

The global economic crisis that began in 2008 generated much anger at institutional symbols of power and greed, and there was an upsurge of popular protest. Demonstrations united those venting at bankers and capitalists, anti-globalization protestors, and environmentalists. There was growing anger at the level of inequality, corporate greed, and the lack of jobs.

When the G20, an international forum for finance ministers, met in the financial heart of London in 2009, they were faced with thousands of angry protestors. Social media became critical in the organization of large gatherings and the occupation of physical spaces. As protests spread throughout Europe, they used the banner of "Occupy," a movement set up in New York to protest against social and economic inequality. There were riots in Rome, strikes in Greece, demonstrations in Portugal, and occupations in the public squares of Barcelona, Moscow, Madrid, New York, Chicago, and Istanbul.

THIS IS A DAY ABOUT OUR ENTIRE HUMAN FAMILY

GLOBAL POPULATION EXCEEDS 7 BILLION (2011)